NEW GENERATION OF ELECTRIC VEHICLES

Edited by **Zoran Stević**

New Generation of Electric Vehicles
http://dx.doi.org/10.5772/45641
Edited by Zoran Stević

Contributors

Omar Ellabban, Joeri Van Mierlo, Cristina Inês Camus, Davide Tarsitano, Ferdinando Mapelli, Adolfo Perujo, Geert Van Grootveld, Harald Scholz, Zlatomir Zivanovic, Zoran Nikolic, Monzer Al Sakka, Hamid Gualous, Noshin Omar, João Pedro Trovão, Paulo G. Pereirinha, Zoran M. Stevic, Liang Zheng, Richard Anthony Guinee

Published by InTech

Janeza Trdine 9, 51000 Rijeka, Croatia

Publishing Process Manager Iva Simcic
Technical Editor InTech DTP team
Cover InTech Design team

Additional hard copies can be obtained from orders@intechopen.com

New Generation of Electric Vehicles, Edited by Zoran Stević
p. cm.
ISBN 978-953-51-0893-1

For EU product safety concerns:
IN TECH d.o.o., Prolaz Marije Krucifikse Kozulić 3, 51000 Rijeka, Croatia,
info@intechopen.com or visit our website at intechopen.com.

Meet the editor

Zoran M. Stević is an associate professor at University of Belgrade, Serbia and he is chief of Department at the Technical Faculty in Bor. He received his PhD diplom from Electrotechnical faculty in Belgrade, University of Belgrade. His research areas include application of information technology in measuring and process control, power electronics, electrochemistry, IR thermography and e-health. Prof. Stević published over a 200 papers in national and international journals and presentations at international and national scientific conferences and symposiums. Also, he was a project coordinator and a member of working group at many scientific and technical projects.

Contents

Preface

The rapid development of science and technology leads to improvement of human life, but also creating of new crisis situation. Mankind is confronted with risks that have not been seen before in human history. Global worming is one of the typical examples. Although majority of the experts that are studying climate changes claim that global worming is a fact and that it is caused by human, there are also scientists that doubt those statements. One of the main problems related to critical situations is – mater of responsibility. World Governments have to consult with experts and to estimate when to announce risk situation. Strong political initiative is needed to start dealing with serious ecological problems such as global worming.

Electric drive vehicles present one of the most important technological advances having in mind spread of this kind of nature pollution. Lately there is increased world interest for so called hybrid vehicles that have reduced fuel consumption and much less pollutants emission than regular vehicles. Hybrid vehicles can in broadest sense be described as vehicles utilizing combination of production and storage of energy. Good properties of conventional vehicles (long range and acceleration, very good supply network) are combined with electrical vehicles (zero emission, quiet operation, regenerative use of braking energy).

After 1970, environmental problems and oil crises increased the interest in electric vehicles. Especially in the United States people develop interest in EV and made it a habit to use widely electric vehicles for golf courses, for airports, for parks and fairs. According to some sources, one third of vehicles intended for driving on gravel roads were with electric traction. So there was a need to develop a new industry. Late 20th century contributed to an even greater exacerbation of conditions around the EV application. Scientists have become aware that environmental pollution is becoming larger, the emission of exhaust gases and particles affect climate change and that non-renewable energy sources under the influence of high demand and exploitation are becoming more expensive and slowly deplete. On the development of modern EV have been working both large and small manufacturers of motor vehicles. EV still has significant problems arising from low-volume production so that these vehicles are still expensive and thus less attractive. In the first place it is air-conditioning for passengers and a relatively small possibility of storing electricity in batteries.

Barriers associated with implementation of the electrical technologies may be significant. These barriers may include technological feasibility, cost, legal and regulatory issues, public concerns, infrastructure, energy efficiency and other factors. Reduction or removal of these barriers and providing appropriate incentives will have a strong influence on the desirability and effectiveness of these programs.

Goal of this book is to bring closer to the readers new drive technologies that are intended to environment and nature protection.

The book is divided into two sections. The first section deals with the current status and trends of development of electric vehicles, while the second section deals with the modeling and computer design of EV.

Prof. Zoran Stević
University of Belgrade,
Technical faculty in Bor,
Serbia

Section 1

Present and Future of Electric Vehicles

Chapter 1

Present and Future Role of Battery Electrical Vehicles in Private and Public Urban Transport

Adolfo Perujo, Geert Van Grootveld and
Harald Scholz

Additional information is available at the end of the chapter

http://dx.doi.org/10.5772/54507

1. Introduction

"Electricity is the thing. There are no whirring and grinding gears with their numerous levers to confuse. There is not that almost terrifying uncertain throb and whirr of the powerful combustion engine. There is no water circulating system to get out of order – no dangerous and evil-smelling gasoline and no noise." [1]

The OECD estimates that more than 70% of the developed world population lives in urban environments[2], which explains a larger concentration of vehicles there. In the EU-27, there were about 230 million passenger vehicles in 2007 and the new vehicle sales were nearly 16 million vehicles in that year. Notwithstanding the improvements in regulated air pollutants from road transport, the urban population remains at higher risk levels by directly suffering the impact of conventional vehicles because of their closeness to the pollutant source. On one hand urbanization means that people when travelling in their urban environment will typically travel less than 100 km a day. And on the other, that a large percentage of all transport and delivery of goods will take place in urban areas. Acceleration and deceleration frequency, traffic jams, thus energy efficiency and pollution per km are worst within urban traffic. Many business cases exist for urban electrified road transport because these offer a lower Total Cost of Ownership (TCO) than conventional means already today. The above

1 Thomas Alva Edison (February 11, 1847 – October 18, 1931)

2 See e.g., p. 17 in " Trends in Urbanisation and Urban Policies in OECD Countries: What Lessons for China?", OECD and CDRF, http://www.oecd.org/urban/roundtable/45159707.pdf

reasons make the urban area the cradle where the electrification of road transport can deploy its full potential of positive impact, both environmentally and energetically.

There are several bottlenecks on the take-up by economic operators and the public at large of this technology, mainly: price of purchasing of an electric vehicle (EV), its limited range (range anxiety) and long charging time. Most of them are related to the present available battery technology. Improved batteries, maybe together with super-capacitors (so called hybrid power-packs) will most likely represent the core of the developments. The integration of the electrically recharging vehicle into the smart electric grid of the future, which calls for automatic communication technologies, is another frontline of research. Advances in these areas will probably reduce the obstacles for battery powered EVs in near future.

In the last 30 years the batteries' energy density (Wh/kg) has increased by a factor of four in three very well distinctive development waves: i.e. the development in 1995 of Ni-Cd batteries (with about 70 Wh/kg), that of Ni-MH in 2000 (~100 Wh/kg) and the third wave with the development of Li-ion batteries in 2005 leading to currently about 200 Wh/kg. With the present battery's energy density a pure battery electric vehicle (BEV) can drive ca. 150 km with one charge, already opening the door for a substantial portion of series-produced EV models notably in urban environments. This already achievable all-electric range is larger than most of the daily average distance of city dwellers (in the USA about 90% of automobiles travel about 110 km daily and in Europe this distance is even smaller, as GPS-coupled monitoring analyses of ten-thousands of urban based cars have meanwhile proven also experimentally).

In any introduction of a new technology the role of stakeholders (public, commercial and private) is very important and their needs have to be understood and addressed. Because of the role of EVs in reducing the level of ambient pollution in urban conglomerations, this chapter will also look to different efforts and programs that some stakeholders as for instance different municipalities and regional and national governments, are setting up in order to actively support and stimulate the introduction of EVs.

Finally, the chapter will address how the above developments will support the introduction of EVs in the urban environment; it will also describe how reduced TCO will translate into more business cases and how this will impinge in a more general electrification of public transport with the consequent improvement of urban ambient air quality, noise levels, etc.

2. City vehicles

There is a very noticeable development effort on small city vehicles indicating that for the automobile industry (OEMs) the urban area represents the main niche in order to roll out the electrification of road transport. This effort is a globalised one with examples not only in Europe, but also in the US, China, Japan and India. In many cases demonstration is implemented by consortia of OEMs, or OEMs together with a university, or in public-private partnership.

Table 1 gives some examples of these cars besides the already launched ones in the market like, e.g. in Europe, the Smart for two Electric Drive, i-Miev, Peugeot-ION, Citroen C-Zero, Think City, etc..

We can conclude that OEM's are focusing on specific market segments within cities:

- The Smart for two for instance is part of a Car sharing project in Amsterdam;
- The HIRIKO will be used in Bilbao (Spain) to study the interest of the public for 'mobility on demand';
- The Renault Twizy is focused on very low purchase price and young customers;
- The VW Nils and the Audi concept are focused on individual transport.

It is noteworthy that for the city cars the OEMs are in particular concentrating in pure electric vehicles (BEV). Also hybridization of small cars is in development, and some technologies involved in hybridizing down-sized conventional engines, like capacitor banks of a few hundred Farad of capacity, might be cross-fertilizing the advent of advanced technologies also for pure electric solutions.

In the appendix further information on market share, number of BEVs per country and other data is presented.

3. Rechargeable Energy Storage Systems (RESS) for vehicles

Rechargeable Energy Storage Systems (RESS) in vehicles include a variety of technologies, each one providing different sizes and different levels of maturity/development. Among these technologies we can name: Electrochemical Storage (Batteries, capacitors and notably super capacitors), Fuel-cell (often containing also a buffer battery) electricity provision with e.g., a hydrogen or *on-board reformer* fuel storage system, and (more in a niche situation) Compressed Air Energy Storage (CAES), and Flywheels. It is noteworthy to indicate that whatever is the chosen RESS for *electrified* vehicles it will be a key enabling technology for the penetration of this class of vehicles, because it influences in a decisive way their weight, energy efficiency, maintenance complexity and thus longevity and usability – and thus generally their acceptance-level achievable in the market.

In figure 1 some RESS are presented. From this figure the benefits of a hybrid power pack can be seen. These packs combine a high power density of fuel cells and batteries with a high energy density of (super) capacitors. Also the flywheels can be located in this figure.

This section intends to give an overview on battery, super-capacitors and hybrid power-pack (batteries plus super capacitors) developments that in a near future will probably reduce the obstacles, questions and doubts that potential users might have, and thus helping to bridge the gap between early adopters of the technology and the public at large.

Model	Characteristics	View
Peugeot BB	Concept car, 4 seats, range of 120 km	
VW E-Up	4 seats and a range of 130 km (announced to come on the market in 2013)	
Toyota iQ FT-EV	Range 150 km (it will be in 2012 on the market)	
Gordon Murray T-27	Range up to 160 km, weight under 680 kg, now entering the investment phase	
Kia Pop	Range of 160 km, still a concept car	
HIRIKO	New concept of urban mobility, developed by MIT, it will be introduced in Bilbao in 2013	
VW Nils	One seater, light weight city car. It is a concept, for 2020	
Audi City Car	It is still a concept car	
Mahindra REVAi	Range of 80 km and a lead battery. It is a cheap car, coming soon to the EU market.	
Visio.M city EV (BMW & Daimler)	The aim is to develop a car with low price and low weight	
Renault Twizy	A low priced and low weight (500kg) city car. The battery is leased. The range is 100 km. The car is on the market since 2012.	

Table 1. Some examples of small city vehicles either in the process of being launched into the market or at concept stage

http://dx.doi.org/10.5772/54507

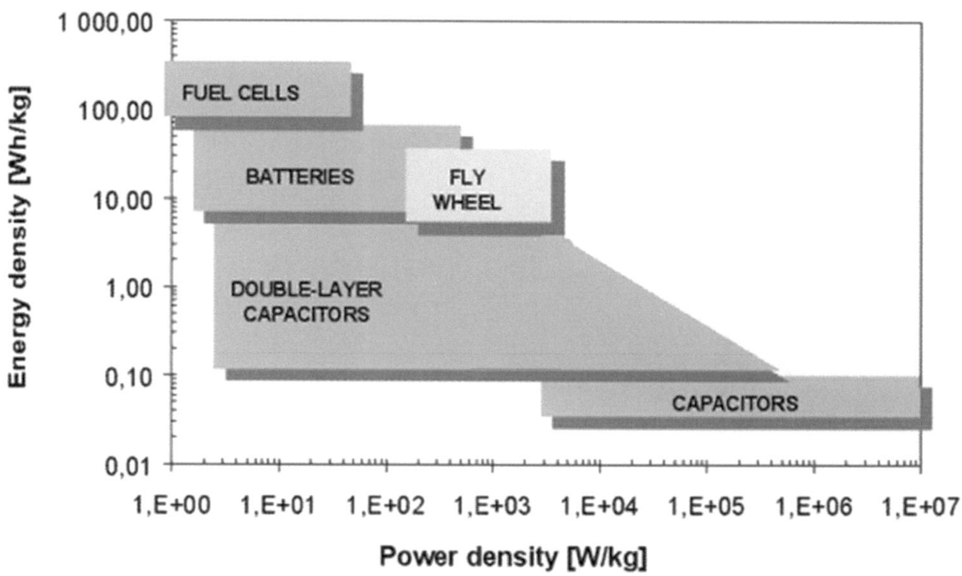

Figure 1. Energy density versus power density of different systems[3]

3.1. Batteries

There are many possible chemistries (battery technologies) that are considered as possible viable options to be used in an electrified vehicle (either BEV or HEV). They range from the very well-established, but comparatively heavy lead-acid batteries to others still in its research stage as Li-air, Al-air or Fe-air batteries passing through Li-ion batteries that represent currently the most used battery-type in commercial BEV.

It is not the intention of this chapter to give an exhaustive insight[4] on the chemistries of each of these batteries but rather to indicate the advantages and drawbacks as well as the possible gains in the future of new battery types still at the laboratory stage in terms of cost and specific energy/power, as these will strongly influence the viability of electrified vehicles.

Figure 2 shows a possible battery technology development roadmap indicating some characteristics of the here discussed batteries technologies.

---

3 http://www.mpoweruk.com/alternatives.htm

4 For a more exhaustive review of storage technologies see e.g.,: "Outlook of Energy Storage Technologies" (IP/A/ITRE/FWC/2006-087/Lot 4/C1/SC2) and "White Paper. Battery Energy Storage Solutions for Electro-mobility: An Analysis of Battery Systems and their Applications in Micro, Mild, Full, Plug-in HEVs and EVs" EUROBAT Automotive Battery Committee Report.

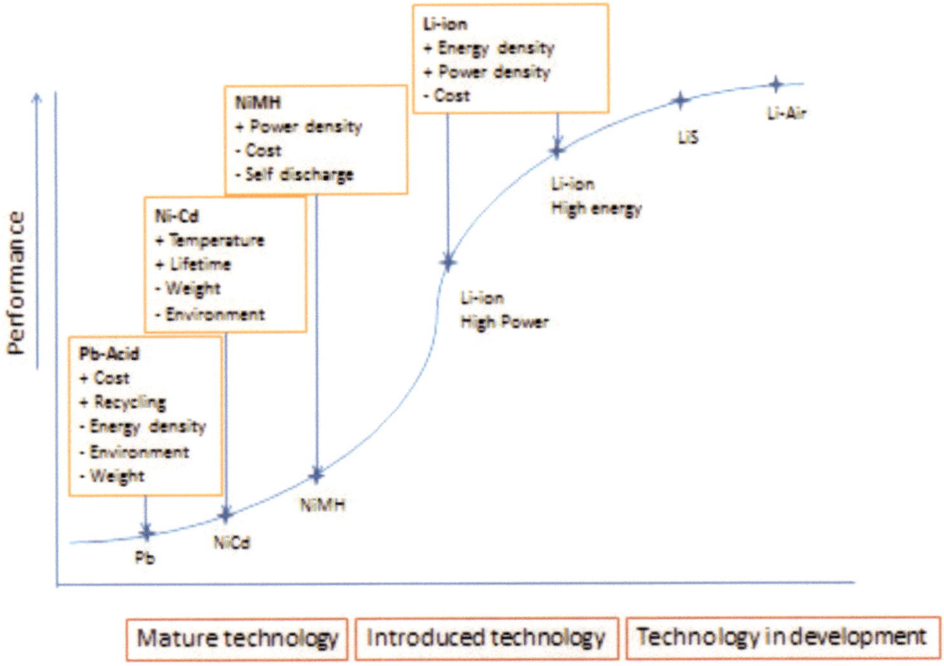

Figure 2. Battery technologies roadmap and characteristics

McKinsey argued in a paper that there are three important factors that could accelerate the development of electric vehicles. These are the manufacturing at (large industrial) scale, lower component prices, and boosting of battery capacity [1].

Table 2 shows some target performance parameters stated for batteries in electrified vehicles for the years 2015, 2020 and 2030 in a Technology Roadmap published by the IEA in 2009 [2].

	Energy density (Wh/kg)	Power density (W/kg)	Costs (Euro/kWh)
2010	100	1000 - 1500	1000 - 2000
2015	150	1000 - 1500	250 - 300
2020	200 - 250	1000 – 1500	150 - 200
2030	500	1000 - 1500	100

Table 2. Expectations on battery performances [2]

Table 2 indicates that the energy density is expected to improve by a factor 5 and that the costs are expected to be reduced by a factor 10 within the next 20 years. These two parameters (energy density and costs) are seen to be the limiting factors of today's BEV. By increasing the energy density the range an electric vehicle can drive will be extended substantially leading to fewer stops for recharging. This should boost EV usability especially in typical urban use. Decreasing the costs of the battery will lead to substantially cheaper electric vehicles, enabling more purchases by the public and fleet investors, due to more sound business cases for commercial use of BEVs.

The cycle-stability is an equally important parameter in applied battery chemistry. The attractiveness for automotive applications is not only dependent on the costs, the power density and the energy density of a battery, but also on the number of battery cycles that can be guaranteed.

3.1.1. Lead-acid batteries

The use of lead-acid batteries in electrified vehicles is mainly in industrial vehicles (e.g. forklifts, which must be heavy) because although at very affordable cost levels (100 – 150 $/kWh), the weight of lead representing about 60% of the weight of the battery translates into a low specific energy (30-50 Wh/kg), making this technology not competitive for most of electric road transport vehicles (even HEVs). It also suffers from a limited lifetime (3 – 5 years). It remains to be seen if lead-acid battery companies can substantially enter the market of micro-hybrid cars in view of small intermediate storage batteries as compared to the concurring battery technologies or modern, compact and lighter capacitor banks / supercapacitor units. At stake is a potential for growth of micro-hybridisation for small cars in the medium term (5-10 years).

3.1.2. Nickel-metal hydride batteries

The use of Nickel-metal hydride batteries (NiMH) had been considered a sufficiently good intermediate stage for application in electrified vehicles (see e.g., the more than one million Toyota Prius sold with NiMH technology, and ca. two million hybrid cars running on NiMH world-wide.) Clearly outperforming NiCd batteries, they were the choice as long as there were still concerns on the maturity, safety and cost of Li-ion batteries. As NiMHs' specific energy (< 100 Wh/kg) cannot meet the requirements for full electric vehicles, it has been mainly used in hybrid vehicles (both HEVs and PHEVs) of limited storage capacity requirements. For PHEVs, NiMH on-board storage capacity arrived at electrical ranges of typically 30 km. There exist concerns on the supply of rare earths (typically mischmetal) and nickel in their anode respectively, cathode. The relatively high content of Ni and possibly rising Ni prices limit further the prospects of reducing their cost and thus use in future EVs.

3.1.3. Lithium-ion batteries

These batteries represent the most actual, wide-spread application in new BEVs world-wide. Nowadays BEVs with ranges above 150 km have all in common that the on-board storage is provided by Li-ion battery packs, often containing some sort of thermal control devices. The name of Li-ion batteries covers a large number of chemistries; indeed, if only a small number of them are actually in use, the list of potential electrode materials is quite large. On the other hand, possible electrolytes range from the mostly used solutions of lithium salts in organic liquids to ionic conducting polymers or ceramics additions to polymers. The current advantage position for this technology is based on its relatively high specific energy (it has reached 160 Wh/kg respectively, 450 Wh/l) however, at present, cost is still a drawback (700 – 1000 $/kWh). The main efforts are thus directed to decrease its cost and to increase its performance level keeping the system safe. There seems to exist a trade-off between performance of the cathode material and its safety. While cathodes made of $LiFePO_4$ depict good safety records its performance in terms of specific energy is poorer than, for example, $LiCoO_2$. However, the latter has a worse safety performance. $LiFePO_4$ also have a comparatively high amount of useful charging cycles during their life-time.

Present research concentrates on the development of an advanced Li-ion batteries exploring the capacity limits of the system through the development of new cathode and anode materials in combination with higher voltage (up to 5V) which will require new electrolytes and binders. Breakthroughs are expected from the combination of so called 5V or high capacity (and then lower voltage) new positive electrode materials and intermetallic new anodes [3].

3.1.4. High temperature Na - β alumina batteries (Na-S and Na-NiCl$_2$)

The first prototypes of this battery type were introduced at the end of the 60s and contained sulphur as the positive electrode and the sodium β"-alumina as solid electrolyte. This material is an electronic insulator and exhibits sodium ion conductivity comparable to that of many aqueous electrolytes. However, to achieve enough electrochemical activity the Na-S battery operates between 300 and 350°C. Because of safety concerns, a derivative of this technology, based on the use of $NiCl_2$ instead of sulphur and termed ZEBRA battery [4], was later developed and evaluated for use in automotive applications. It has the advantage of being assembled in the discharged state and hence without the need of handling liquid sodium. As far as performance is concerned, its specific energy is relatively close to that of Li-ion batteries (115 Wh/kg), it has strongly improved its specific power (400 W/kg) and it has a relative low cost (600 $/kWh) although still between 4 to 6 time higher than the target set in many EV developing programs [5].

3.1.5. Other battery technologies

There are other battery technologies in the research stage that might in future meet the targets needed in electrification of road transport. We can mention among others Li-S [6], [7] and Li-air [8], [9] batteries (see figure 3). In particular they have demonstrated a specific energy[5] about 300 Wh/kg. However, other aspects as life-time, achievable cycles over lifetime and specific power still need further research to meet the challenge.

In the line of using ambient air (oxygen) as the cathode, other materials such as Zn, Al and Fe can be used instead of Li. However, those systems are still in their infancy and at different stages of development. Developments on their recharge ability, air electrodes (porous design) cycle stability and safety are among the areas to be addressed.

---

5 This value is at cell level (research object) as for a battery pack is expected to be lower due to the extra weight of materials used for packing and interconnection of the cells.

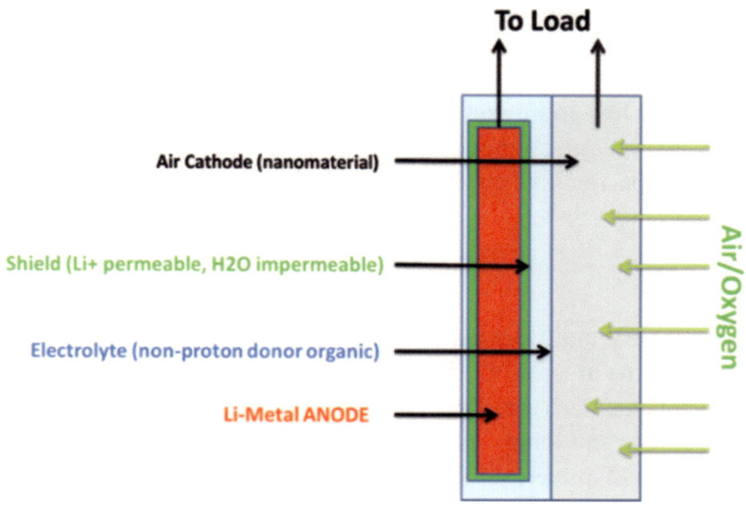

Figure 3. Scheme of a Li-air battery

3.2. Electrochemical capacitors

These devices are sometimes referred to as 'ultra-capacitors' or 'supercapacitors' but these latter are rather commercial names.

Electrochemical Double Layer Capacitors refer to devices that store electrical energy in the electric double layer (EDL), which is formed at the interface between an electron conducting surface and an electrolyte. The EDL may be considered as a capacitor with two electrodes; the capacitance is proportional to the area of the plates and is inversely proportional to the distance between them. Their capacitance is very large because the distance between the plates is very small (several angstroms). The energy stored by such capacitors may reach 5 Wh/kg but they are power systems which can deliver their storage energy in a few seconds (up to 5s). Therefore, they are intermediates between batteries (high energy, low power density) and conventional capacitors (high power low energy density) and thus, they are complementary to batteries and are not in competition with them.

Supercapacitors are already used in transportation applications. They have been announced to be used in the starter/alternator of micro-hybrid cars and are under study by many car manufacturers (Toyota, BMW, Renault, PSA). Recently Ford and Ricardo UK announced[6] the results of the HyBoost project, powering a small additional electric turbo-charging turbine for a down-sized thee-cylinder engine via such a fast ultracapacitor device of ca. 200 F capacity. Together with their outstanding cycle life, another key feature of EDLC systems is that, unlike Li-ion batteries, they can be recharged as fast as discharged. This is why they are used today in large-size applications for energy recovery in trams in Madrid, Paris, Mannheim and Cologne. There is hope, that a certain cross-fertilization in this area will happen

6 http://www.theengineer.co.uk/in-depth/analysis/hyboost-programme-promises-engine-efficiency/1010742.article.

between different improved road transport technologies, which may enable mass-production of EDLC systems sooner than later.

Supercapacitors (ultra-capacitors) have the ability to charge in a very short time however, its energy density is quite low and therefore by using only supercapacitors the electric range of an EV would not be sufficient. Consequently, the ideal situation would be combining both batteries and supercapacitors, which however requires a much more complicated voltage management.

3.3. Challenges

The performance of BEV and its competitiveness are closely linked to the performance of available battery systems in term of their specific power, efficiency and battery cost. In a recent paper Gerseen-Gondelach and Faaij [10] explored the performance of batteries for electric vehicles in the short and longer term. They review the different battery systems in term of performance and cost projection including sustainability aspects and learning curves. They concluded that well-to-wheel (WtW) energy consumption and emissions of BEVs are lowest for those with lithium-ion batteries, and that in the medium term only Li-ion batteries will have a specific power level of 400 W/kg or higher. Other battery systems like Li-S, Li-Air need efficiency improvements towards 90% to reach Well-to-Wheel (WtW) energy consumption of the BEV as low as found with Li-ion batteries. The author argued that already today, despite improvable efficiency levels, all batteries-types can enable similar or lower WtW energy consumption of BEVs compared to traditional internal combustion engine (ICE) vehicles: The WtW emissions are 20 – 55% lower using the EU electricity generation mix. Battery prices turned out to be of course the main parameter for improving the economics of BEVs e.g., if ZEBRA batteries attain a very low cost of 100 $ /kWh, such BEVs become cost competitive to diesel cars for driving ranges below 200 km. Such cost assumptions however were judged "unlikely" for the next and medium term.

With years of market introduction passing, an issue becoming provable will become battery ageing. With their use in extended time, batteries' performance can significantly reduce in terms of peak power capability, energy density and safety. Different auto manufacturers have set goals or targets for calendar life, deep cycle life, shallow cycle life and operating temperature range. However, it is still an issue of technological research to what extend current battery technologies can meet them.

Some examples of these targets are: for calendar life, the goals are typically for 15 years at a temperature of 35 °C, but current targets are for 10 years at which point a battery retains at least 80 per cent of its power and energy density. For deep cycle life, where the charge cycles go from 90 to 10 per cent of SOC[7], the goal is typically 5000 cycles, while the shallow cycle life expectation is 200,000 to 300,000 cycles. Goals for the temperature range as extreme as -40 to +66 °C can be found, such the question arises, whether batteries shall be specified for ambient conditions harsher than it has been done for any normal conventional ICE-vehicle. One extra difficulty that some of the results obtained on batteries performance are valid only for some specific

7 SOC means State of Charge

charging and discharging rate and some specific range of ambient temperature exposure. It is still not clear if the test rates are more or less severe than the actual cycles a battery will be subjected to in an EV, and the interaction of ambient temperature with deep SOC cycling is also an unknown factor. A lot of pre-normative research is in front of us.

4. Cities are the natural environment to develop and to implement e-mobility

Cities are very important for the development and implementation of e-mobility, because the energetic and environmental benefits of BEVs replacing conventional vehicles are largest in city traffic. Moreover,

- About 70% of Europeans live in urban areas [11]. Most of the people live in cities with more than 50,000 inhabitants, and there are about 1,000 of such cities in Europe.
- Cities contribute substantially to the economics of Europe, 85% of European GDP is generated in cities [12].
- They contribute substantially to new knowledge (for instance from research being done on universities) and innovations by (high-tech) small and medium enterprises. Therefore cities have the potential to contribute to a better international competiveness of Europe.

The service sector is the most important source of employment in European urban economies. For example, in London, Paris, Berlin, Madrid and Rome the service sector accounts for between 80% and 90% of total employment. Examples of services are: government, telecommunication, healthcare/hospitals, waste disposal, education, insurance, financial services, legal services, consulting, information technology, news medias, tourism, and retail sales.

Providing and using these services lead to large transportation needs and activities of people and goods, and this, in turn, leads to a high use of energy and to the generation of anthropogenic emissions, like CO_2, NO_x, ozone, fine particles, noise, etc.

Let's focus in some of these aspects.

The energy consumption in European cities is high. About 80% of Europe's energy is used in cities [13]. It is expected [14] that this number will increase in future, because the urban population will grow and also the economic activities and the prosperity will grow.

We have about the same figures for CO_2. Cities are the largest emitters of CO_2. About 75% of the European's CO_2 is emitted in cities. On average the CO_2-emissions for European cities are in the neighbourhood of 1 ton CO_2 per capita per year [15]. Of course these emissions are dependent on the modalities of transportation which are used in the different cities. The higher the share in public transport, walking, cycling the lower the CO_2-emissions will be per capita.

Some examples: In Berlin [16], in 2008 32% of the people choose a car for transportation, 29% walked, 26% public transport and 13% took the bike. In London [17], in 2007 41% choose the car, 25% public transport, and 30% walked.

In some situations the concentration of NO_x and fine particles exceed the air quality limits. These situations are also called: hot spots. NO_x contributes to the formation of smog. Also acid rain can be formed out of NO_x.

Figure 4, depicts a street canyon in Copenhagen [18]. In many European cities the dispersion of air pollution is restricted by the geometry of buildings. This creates so-called street canyons. These canyons lead to elevated concentrations of local pollution, and therefore people living in (or in the neighbourhood of) these hot spots have a higher risk for getting ill.

Figure 5 depicts the concentration of NO_x in ambient air in a city (London) [19]. As can be seen from this picture the NO_x-concentrations exceed the maximum regulated value, which is 40 μg/m^3.

Figure 4. An example of a street canyon in Copenhagen [18]

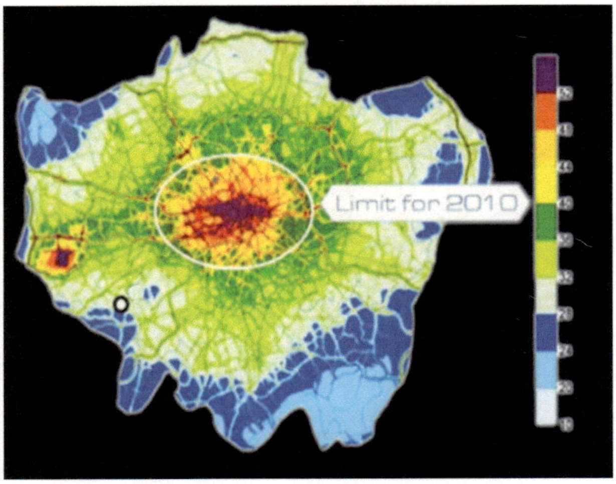

Figure 5. NO_x-emissions in the city of London [19]

Figure 6 shows the NO_x-concentrations in European regions [19]. The intensively populated zones can be recognized easily. These are mainly cities and intensively used highways between the cities.

4.1. E-mobility can tackle these problems; many stakeholders are willing to contribute

The big advantage of e-mobility is that it gives direct results for improving ambient air quality. An electric vehicle does neither emit NO_x and PM, nor VOC (volatile organic compounds). So, when electric vehicles are introduced to replace conventional vehicles, these emissions decrease directly and ambient air quality will improve. Because ozone is formed by a photo-catalytic reaction between VOC and NO_x, also the ozone concentration will be reduced.

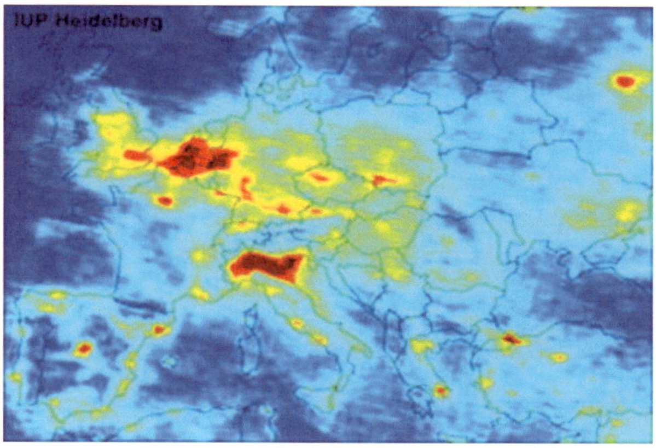

Figure 6. NO_x-emissions in Europe

A large number of stakeholders have parallel interests in the development and implementation of e-mobility in cities. The citizens want a clean city to live in. So, the ambient air quality needs to improve in several situations.

These ambient air problems are also a main driver for the politicians and administrations of cities to stimulate electro-mobility. The Covenant of Mayors which is signed on February 2009 is a good example of this. The main goal of this covenant, which now has about 4,000 signatories, is to increase energy efficiency and to use renewable energy sources. Within the framework of this covenant the Sustainable Energy Action Plans (SEAP) play a central role. A Sustainable Energy Action Plan (SEAP) is the key document in which the Covenant signatory outlines how it intends to reach its CO_2 reduction target by 2020. Already more than 1400 of these plans are submitted. A lot of these plans contain actions on the stimulation of electro-mobility in cities.

Also business leaders are major stakeholders. A first reason for that is that e-mobility can lead to sound Total Costs of Ownership (TCO). This means that economic activities can be done more cost effectively with e-mobility than with the petrol based vehicles. A second reason is that the spin-off of this technological development can be enormous. It is already stated that there are about 1,000 middle large cities in Europe. This is a big market for small and medium sized enterprises that develop new technologies for implementing e-mobility-systems.

4.2. Cities as living labs: some European experiences

Cities can be regarded as a living lab. This means that they have the possibility to test new concepts under real life circumstances. The behaviour of consumers working with new concepts can be studied, and the feedback of the consumer can be used by the supplier to modify and improve the concept. So, a cyclic process can be organized leading to the rapid development of new concepts. The administrations can take the lead in organizing these processes. They have all the ingredients to do so: the consumers, the suppliers, the infrastructure, and also the challenges and the solutions.

There are a lot of interesting projects going on in European cities on the development and implementation of electric vehicles. Some examples are the projects started within the European Green Cars Initiative [22].

Most of them concern electric mobility, for instance the Green eMotion project [23]. This project is supported by 43 partners from industry, the energy sector, electric vehicles manufacturers, municipalities as well as universities and research institutions. The goals of Green eMotion are:

- Connecting ongoing regional and national electro mobility initiatives;
- Comparing the different technology approaches to ensure the best solutions prevail for the European market;
- Creating a virtual marketplace to enable the different actors to interact;
- To demonstrate the integration of electro mobility into electrical networks (smart grids);
- Contribute to the improvement and development of new and existing standards for electro mobility interfaces.

In several projects ICT is introduced to facilitate the implementation of electromobility. One of these is the project MOBI.Europe [24]. In this project the users of electric vehicles are getting access to an interoperable charging infrastructure, independently from their energy utility and region. It is built on the e-mobility initiatives of Portugal, Ireland, the Spanish region of Galicia and the Dutch city of Amsterdam.

Another project is the smartCEM [24] project in which four European cities/regions are participating: Barcelona (ES), Gipuzkoa-San Sebastian (ES), Newcastle (UK) and Turin (IT). The goal of this project is to demonstrate the role of ICT[8] solutions in addressing shortcomings of e-mobility, by applying advanced mobility services, like EV-navigation, and EV-efficient driving.

---

8 ICT = Information and Communication Technologies

One part of the VIBRATe (VIenna BRATislava E-mobility) [25] project is to identify the possibilities of connecting two neighboring metropolitan areas—Bratislava (Slovakia) and Vienna (Austria) with a "green" highway. This highway will interconnect the two cities with a network of public charging stations for electric vehicles. In this project IBM is working together with Západoslovenská energetika, a.s. (ZSE) and the concerned municipalities.

Autolib [26] is an electric car-sharing program which is launched in Paris at the end of 2011. This program will start with 250 vehicles. The amount of vehicles will grow to 2,000 in the summer of 2012. This number will grow to 3,000 in the summer of 2013. In this car-sharing program the compact Blue car is introduced. This four-seat car is the result of a collaboration of the Italian car designer Pininfarina and the French conglomerate Groupe Bollore.

Car2Go [27] is a subsidiary of Daimler AG that provides car sharing services in several cities in Europe and North America. In November 2011 a fleet of 300 smart for two electric vehicles was deployed in Amsterdam.

In London the "Electric 10" is formed. This is an initiative of 10 companies that use electric commercial vehicles for their activities. The Electric 10 partnership was formed in autumn 2009, bringing together 10 major companies who are already using electric fleet vehicles on daily basis: Sainsbury's, Tesco's, Marks and Spencer, UPS, TNT Express, DHL, Amey, Go Ahead, Speedy, Royal Mail. The Municipality of London is working with these companies to learn from their experiences and encourage others to take their lead [28]. The use of electric vehicles for goods delivery not only benefits the environment, it also has a positive total cost of ownership (TCO).

4.3. Cities have the power to implement; and they are already doing so

City administrations have the possibility to develop new concepts under real life circumstances, as we have seen in paragraph 4.2. and to set projects to bring e-mobility to a reality. Of equal importance is that they also have the ability to implement using their legal instruments. Many cities are already doing this.

The instruments they use can be divided in three categories [29]:

Financial incentives

Examples of financial incentives are exemptions from vehicles registration taxes or license fees. Or exemptions from congestion charge. Another financial incentives are of operational nature e.g. the electric vehicle gets a discount on parking costs.

Non-financial incentives

There are cities which give non-financial incentives. For instance, free or discount cost for a parking place in the city centre. Or that the owner will get access to restricted highway lanes. An important incentive is also to get easy access to public charging facilities.

Their purchasing power

Municipalities are not only regulators. They also have a vehicle fleet and they give licenses to public transport systems. With these possibilities they also can stimulate the e-mobility.

They can buy electric vehicles for their municipal fleet and they can add hybrid buses to public transport systems. Municipalities can install charging stations on the public area, like: libraries, parking garages, city halls, or other public buildings.

4.4. Some remarks to this section

In paragraph 4.2 a total of 7 projects which are presently going on in Europe are described shortly. It should be stated that these are just illustrations. There are many more interesting projects on e-mobility. What we see is a steep increase in the amount of battery electric vehicles (BEV) in Europe [30]. In 2010 in total 765 BEVs were introduced on the EU-27-market, and in 2011 already ca. 9,000 BEVs. This took place predominantly in France, Germany, UK, the Netherlands, and Austria. The main BEVs types were Peugeot-ION, Mitsubishi-i-MIEV, Smart for two, Nissan-Leaf, and Citroen-C-Zero.

We expect that this steep increase will continue, because of the battery developments we described in the beginning of this chapter and also because of the strong efforts of stakeholders, like member states, municipalities, car manufacturers, and the EU. Indeed, in the 2011 Transport White Paper 'Roadmap to a Single European Transport Area – Towards a competitive and resource efficient transport system' (COM (2011) 144 final), the European Commission proposes 10 goals for a competitive and resource efficient transport system which serve as benchmarks for achieving the 2050 60% GHG emission reduction target. One of these goals is to halve the use of 'conventionally-fuelled' cars in the urban transport sector by 2030 and to phase them out by 2050, thereby also reducing the transport system's dependence on oil.

5. Enabling technologies for the introduction of electricity in road transport

Another reason why electric vehicles are promising is because of the fact that it can contribute to the development and introduction of smart grids. With smart grids the share of green electricity by means of wind and solar can be better managed to increase. Electric vehicles can serve as storage for electricity (spinning reserves) in those times when the households don't need the amount of electricity produced at a certain moment, and the vehicles can deliver electricity to the grid in times when the households need more electricity than produced at that moment. The benefits are that with these smart grids the CO2 emissions will decrease as well as the use of fossil energy. The CO2 emissions will go down even more, because from well-to-wheel-analyses it can be seen that in most cases the CO2-performance of electric vehicles is better that petrol based vehicles [21].

It is generally considered that smart grid and V2X where X represents another vehicle (V2V), the grid (V2G) or sometime the user's home (V2H) are essential technologies for the early introduction of electrified vehicles as these provide an added value to the vehicle respectively, reduce its TCO.

http://dx.doi.org/10.5772/54507

5.1. What is a smart grid?

The concept of Smart Grid[9] was developed in 2006 by the European Technology Platform for Smart Grids, and concerns an electricity network that can intelligently integrate the actions of all actors connected to it - generators, consumers and those that do both - in order to efficiently deliver sustainable, economic and secure electricity supplies. Interoperability of EVs to Smart grids promises an increase of the EVs' overall energy efficiency and cost benefits.

Decentralized supply of electricity is growing. There are at least three types of decentralized supply options:

- More and more wind turbines are in operation;
- (micro) Combined Heat and Power (CHP) is up-coming.
- The generation by means of solar PV is increasing;

This development means that the fluctuations over time in the supply of electricity would be increasing in a near future with the consequent challenge to harmonize it with the demand of electricity.

There are some options to deal with this challenge. The first one is to influence the regulate the supply. When at a certain moment more wind and solar electricity is produced then the supply of electricity from fossil sources should be limited. To realize this real-time communication between consumers and producers should take place. This can be done by means of smart meters. To reduce the supply of electricity from fossil sources is, however, not always an easy task.

A second option is to realize a situation in which the fluctuations which might appear on the supply side will be match on the demand side. This can be realized by introducing fluctuating prices, which again can be realized by means of smart meters. So when the supply is high then the price will be low and then the smart meter can for instance start charging an electric vehicle or it can start other appliances e.g. the washing machine. And when the demand is high then the price will be high and then the electric vehicle will supply electricity to the house. Thus, by means of the price mechanism and the smart metering the supply and the demand can stay in balance, despite of the fluctuations occurring in the supply side.

A third option is by introducing a storage facility. This can be done by means of fly-wheels, ultra-capacitors, compressed air, and batteries. In this third option the batteries of the electric vehicles can play a role (see section 3).

A fourth option is that when there is an oversupply of electricity, it is used for the electrolysis of water and the formed hydrogen is either used directly or it is coupled with CO2 to produce methane. When there is a shortage of electricity the hydrogen and/or methane can be used to produce electricity by means of a Combined Heat and Power Plant (CHP). Of course the hydrogen can also be used to fuel a Fuel Cell Electric Vehicles (FCEV's).

Hence, with a smart grid it is possible to:

9 http://www.smartgrids.eu/documents/TRIPTICO%20SG.pdf

- Better facilitate the connection and operation of electrical generators of all sizes and technologies;
- Allow consumers to play a part in optimizing the operation of the system;
- Provide consumers with more information and options for choosing an energy supply;
- Significantly reduce the environmental impact of the whole electricity supply chain;
- Organize a symbiotic relation between the grid and the electric vehicle. The vehicle can be charged when the price is low, and the vehicle can contribute to the grid when electricity is needed there;
- Charge the electric vehicle with low-CO2-containing electricity, which contributes to low CO2-emissions when driving the vehicle;
- Maintain or improve the existing high levels of system reliability, quality and security of supply;
- Maintain or improve the efficiency of existing services;
- Foster the development of an integrated European market.

5.2. Some European efforts on a practical scale on smart grids including electric vehicles

There are considerable efforts in Europe (the same thing can be said on other developed markets; i.e. USA, Japan…) on smart grids[10] by supporting and carrying out many projects. In several of these projects electric vehicles are included and studied. Some examples are:

InovCity concept in Évora (Portugal)

The goal of this project is that the entire municipality of Évora will be connected to an intelligent electricity system which includes 30,000 customers.

Some characteristics of this project are:

- The project is initiated by EDP[11] Distribuição, with support from national partners in industry, technology and research (EDP Inovação; Lógica; Inesc Porto; Efacec; Janz and Contar);
- The electricity grid is provided with ICT, so that the grid can be controlled automatically. This is done by monitoring the grid in real time;
- Increase of renewable energies (PV solar cells, micro wind turbines) is facilitated by the intelligent electricity grid;
- The Energy Box plays a central role in this system. All consumers will have such a box, and this box connects the consumers to the intelligent grid. In the box the amount of electricity used and/or produced is recorded. And by means of the box the consumers can

10 A survey can be found on http://www.smartgridsprojects.eu/map.html

11 http://www.inovcity.pt/en/Pages/media-center.aspx

program devices, like washing machines, when the price of electricity is low. This process of programming devices can be automated fully;

- The electricity grid is also facilitating the charging and discharging of electric vehicles. The batteries of these vehicles will serve as a buffer when there is an oversupply and the batteries will serve as a producer of electricity when more electricity is needed in the homes.

Endesa's Smartcity Málaga Project (Spain)

The goals of SmartCity Málaga[12] are to implement and integrate distributed energy resources, energy storage, electric vehicle charging and discharging facilities, and intelligent public lighting devices.

The characteristics of the project are:

- Endesa in cooperation with 11 partners is rolling out state-of-the-art technologies in smart metering, communications and systems, network automation, generation and storage, and smart recharging infrastructure for e-vehicles;
- More than 17,000 smart meters are installed;
- 11 MW of renewable generation capacity which consist of solar PV, wind energy and co-generation;
- A storage facility consisting of batteries;
- A network of recharging points for vehicle-to-grid-technology;
- By means of ICT all these devices are connected to the Network Control Center, where these are monitored and controlled.

Harz.EE-Mobility (Germany) [13]

This project has been initiated by Siemens CT in cooperation with 14 partners – including research institutes, the Deutsche Bahn (German Railroad Company), and wireless provider Vodafone. The goal is to make Germany's Harz district a model region for electric mobility. Wind, solar, and other alternative energy sources already contribute more than half of the power generated in the Harz district. Sometimes in windy periods some wind turbines have to switch off. This problem could be solved using electric vehicles as small energy storage units allowing for useful demand shift.

The project focuses on Vehicle-to-grid-technology (V2G). Electric cars would recharge their batteries whenever winds are strong, especially at night. Conversely, during calm periods they could feed electricity back into the grid at higher prices. Ultimately, V2G aims at bidirectionality of both, car / grid communication and their energy flow.

In this project an energy management system is developed. All the 2,000 energy generation devices are connected and automatically controlled (PV, wind turbines, biogas, and

12 http://www.endesa.com/en/aboutEndesa/businessLines/principalesproyectos/Paginas/Malaga_SmartCity.aspx

13 *http://www.siemens.com/corporate-technology/en/research-cooperations/mobility.htm*

electric vehicles). The project also monitors and studies the movement profiles of electric vehicles. With this information is can be predicted how many electricity in what period is needed to recharge the vehicles. This will also be important control data for the electricity generation devices.

The above examples indicate the important role that information and communication technology play in an early uptake of electrical vehicles by their seamless integration in the electrical distribution and control network ("smart grid" of the future).

6. Discussion

This chapter has focussed on the technological requirements that electrical vehicles need in order to break into the (primarily urban) main stream as a valid personal or commercial transport means. However, their cost/price and environmental impact have not been addressed. This section intends to indicate some of the recent efforts that can be found in the open literature, both to forecast when this vehicle technology will become possibly a preference of the user and what policies could be put in place to better address the environmental benefit of increasingly electrifying road transport.

In a recent paper [31] Weiss *et al.* have forecasted the price for hybrid-electric and battery-electric vehicles using *ex-post* learning rates for HEVs and *ex-ante* price forecasts for HEVs and BEVs. They forecasted that price breakeven with these vehicles may only be achieved by 2026 and 2032, when 50 and 80 million BEVs, respectively, are expected to have been produced worldwide. They estimated that BEVs may require until then global learning investments of 100–150 billion € which is less than the global subsidies for fossil fuel consumption paid in 2009. Their findings suggested that HEVs, including plug-in HEVs, could become the dominant vehicle technology in the next two decades, while BEVs may require long-term policy support. In line with what it has been pointed out in this chapter, the authors indicated that the performance/cost ratio of batteries is critical for the production costs of both HEVs and BEVs. If current developments persist, vehicles with smaller, and thus less costly, batteries such as plug-in HEVs and short-range BEVs for city driving could present the economically most viable options for the electrification of passenger road transport until 2020.

More studies on specifically urban electrification of road transport might move the quantitative arguments to some extent, and show that there are several niches of earlier cost-effectiveness even for BEVs.

There is a debate on how to consider the environmental impact of this class of vehicles. Unlike their counterpart fossil-fuelled vehicles, the emissions generated by electrified vehicles are produced “upstream”, that is where the electricity is generated. Should they be considered to have GHG emissions of "0 g/km"? Lutsey and Sperling [32] argue that considering electric vehicles as 0 g/km and assuming 10% of cars sold by 2020 to be electric, this could result in a loss of 20% of the conventionally calculated benefit from USA regulations aimed at reducing vehicle GHG emissions – so one has to pay attention of what is summed up. They also found that if upstream emissions were included, an electric vehicle powered from the America electricity grid would on average emit about 56% less CO2 than their petrol counterpart (104 g/mile compared to 238 g/mile). It is clear that the exact amount will depend upon the particular electricity generation fuel mix and thus generation efficiency in the given State where the BEV was charged. These authors support the idea of using a full life-cycle analysis as regulatory option rather than the "0 g/km". This approach, although more complicated, would ensure that GHG regulations were scientifically rigorous and could accommodate future energy technology development.

7. Conclusions

In the 2011 Transport White Paper 'Roadmap to a Single European Transport Area – Towards a competitive and resource efficient transport system' (COM (2011) 144 final), the European Commission proposes 10 goals for a competitive and resource efficient transport system which serve as benchmarks for achieving the 2050 60% GHG emission reduction target. One of these goals is to halve the use of 'conventionally-fuelled' cars in the urban transport sector by 2030 and to phase them out by 2050, thereby also reducing the transport system's dependence on oil. Among the possible options to support this target, the electrification of road transport seems to be a winning one - as we have indicated in this chapter. We have addressed the technological challenges that electrified vehicles have to face in order to overcome the present status quo. These are mainly due to the storage system on board of a BEV. There are promising technologies that can positively support the introduction of electric vehicles in our streets and roads (e.g. V2G and interoperability with smart grids through standardised communication). Finally, the areas of cost and environmental impact has been also addressed by commenting recent efforts in both forecasting the price reduction in the future and addressing full life-cycle analysis as possible policy options to include the full picture of the impact of vehicles in GHG emissions.

Author details

Adolfo Perujo, Geert Van Grootveld and Harald Scholz

European Commission, Joint Research Centre, Institute for Energy and Transport, Sustainable Transport Unit, Ispra (Va), Italy

References

[1] Russell Hensley, et al. Battery technology charges ahead. McKinsey Quarterly. July 2012

[2] Technology Roadmap, Electric and plug-in hybrid electric vehicles, International Energy Agency, 2009 (page 43).

[3] Developments in battery chemistries: 2009 European Energy Fair Zürich: www.investorrelations.umicore.com

[4] P. Tixador; IEEE/CSC & ESAS EUROPEAN SUPERCONDUCTIVITY NEWS FORUM, No. 3, (2008), www.esas.org

[5] Matsushita, EP 0569 037, 7 July 1993

[6] Mikhaylik Y. V. et al. High Energy Rechargeable Li-S Cells for EV Application. Status, Challenges and Solutions. ECS Trans. 2009; 25 (35) 23-34

[7] Kolosnitsyn V. S. and Karaseva E. V. Lithium-sulfur batteries: Problems and solutions. Russian Journal of Electrochemistry, 2008; 44 (5) 506-509

[8] Abraham K. M. and. Jiang Z. A Polymer Electrolyte-Based Rechargeable Lithium/Oxygen Battery. J. Electrochem. Soc. 1996; 143 (1) 1-5

[9] Advanced Batteries Technology, March 2010, including announcement of IBM project launch

[10] Gerssen-Gondelach, Sarah J., Faaij, André P. C. Performance of batteries for electric vehicles on short and longer term. Journal of Power Sources 2012; 212 (0) 111-129

[11] Promoting sustainable urban development in Europe ACHIEVEMENTS AND OPPORTUNITIES, http://ec.europa.eu/regional_policy/sources/docgener/presenta/urban2009/urban2009_en.pdf

[12] EUROCITIES: http://www.eurocities.eu/eurocities/issues/climate-adaptation-issue (accessed 3 July 2010).

[13] EUROCITIES: http://www.eurocities.eu/eurocities/issues/energy-efficiency-issue (accessed 3 July 2010).

[14] International Energy Agency: World Energy Outlook 2008: Chapter 8. Energy use in cities http://www.iea.org/weo/docs/weo2008/WEO_2008_Chapter_8.pdf (accessed 3 July 2010).

[15] International Association of Public Transport: PUBLIC TRANSPORT AND CO2 EMISSIONS http://www.uitp.org/news/pics/pdf/MB_CO23.pdf (accessed 3 July 2010).

[16] City of Berlin: Mobility in the City: Berlin Traffic in Figures: Edition 2010 http://www.stadtentwicklung.berlin.de/verkehr/politik_planung/zahlen_fakten/download/Mobility_en_komplett.pdf (accessed 3 July 2010).

[17] Transport for London. Travel in London Key trends and developments. Report number 1. http://www.tfl.gov.uk/assets/downloads/corporate/Travel-in-London-report-1.pdf (accessed 3 July 2010).

[18] ASSET (Assessing Sensitiveness to Transport): Sensitive urban areas http://www.asset-eu.org/doc/Copenhagen.htm (accessed 3 July 2010).

[19] Photo courtesy of Wikipedia. The picture is derived from www.dft.gov.uk

[20] REBOURS, Yann and KIRSCHEN, Daniel. What is spinning reserve? The University of Manchester report (Release 1, 19/09/2005). Available at: http://www.eee.manchester.ac.uk/research/groups/eeps/publications/reportstheses/aoe/rebours%20et%20al_tech%20rep_2005A.pdf.

[21] Edwards, R., Larive, J.F., Beziat, C. Well-to-wheels Analysis of Future Automotive Fuels and Powertrains in the European Context. WELL-to-WHEELS Report Version 3c, July 2011. European Commission Joint Research Centre, Institute for Energy.

[22] European Green Cars Initiative. http://www.green-cars-initiative.eu/about (accessed 3 July 2010).

[23] Green eMotion. http://www.greenemotion-project.eu/about-us/index.php (accessed 3 July 2010).

[24] ElectroMobility pilot projects launched at the European Parliament (February 2012)http://www.icarsupport.eu/media/news-2/electromobility-pilot-projects-launched-at-the-european-parliament/(accessed 3 July 2010).

[25] Green Car Congress. Energy Technology Issues and Policies for Sustainable Mobility. http://www.greencarcongress.com/2012/04/vibrate-20120409.html (accessed 3 July 2010).

[26] European Travel. http://www.msnbc.msn.com/id/45553670/ns/travel-destination_travel/t/paris-launch-electric-car-sharing-program/ (accessed 3 July 2010).

[27] Wikipedia, the free encyclopaedia. http://en.wikipedia.org/wiki/Car2Go (accessed 3 July 2010).

[28] NYC Global Partners' Innovation Exchange website. Best Practice: Electrical Vehicle Development http://www.nyc.gov/html/unccp/gprb/downloads/pdf/London_ElectricVehicles.pdf (accessed 3 July 2010).

[29] Forbes. The Global Electric Vehicle Movement: Best Practices From 16 Cities. http://www.forbes.com/sites/justingerdes/2012/05/11/the-global-electric-vehicle-movement-best-practices-from-16-cities/ (accessed 3 July 2010).

[30] E-Mobility in the EU, Facts and Figures, European Commission, DG Joint Research Center, to be published

[31] M. Weiss, et al. On the electrification of road transport - Learning rates and price forecasts for hybrid-electric and battery-electric vehicles. Energy Policy (article in press)

[32] Lutsey N. and Sperling D. Regulatory adaptation: Accommodating electric vehicles in a petroleum world. Energy Policy 2012; 42 308-316

Chapter 2

The Contribution and Prospects of the Technical Development on Implementation of Electric and Hybrid Vehicles

Zoran Nikolić and Zlatomir Živanović

Additional information is available at the end of the chapter

http://dx.doi.org/10.5772/51771

1. Introduction

Population growth in the world had a constant value since the beginning of a new era to the 19th century when the population was 1 billion. The technological revolution is largely influenced by that in this century, the population increase by 68 %. The population in the world increased by about 270 %, or over 6 billion people just in the 20 century. Although the UN in [1], estimates three possible scenarios of population growth in this century, the picture 1, is the most possible one that predicts that the world population will increase by 2050. to about 8,9 billion, and afterwards it will be a slowdown so that by the end of the 21st century, and in the next few, does not expect the growth of population in the country. In any case, in the near future over the next four decades strong growth of the population is expected. With the growth of population in the world there is a need to increase transportation of people, goods and raw materials as a prerequisite for the growth of production and consumption and the standards of living.

The 19th century was the age of industrial revolution. Thee more factors enabled the industrial revolution. The first was the new steam and textile technology and then the new agriculture and population growth crating both the labor force for the new industrial factories and the markets to buy their manufactured goods. Development of a superior transportation system for getting raw materials was basis that colonials provided raw materials for the factories as well as more markets for their goods.

The result of all this was an industrial revolution of vast importance in a number of ways. For one thing, it would spawn the steam powered locomotive and railroads which would revolutionize land transportation and tie the interiors of continents together to a degree never before imagined. It would trigger massive changes in people's living and working conditions as well as the structures of family and society. No invention of the 1800's played a more vital role in the Industrial Revolution than the steam locomotive and railroad, triggering the biggest leap in transportation technology in history. Railroads cut travel time by 90 % and dramatically reduced freight costs, see [2].

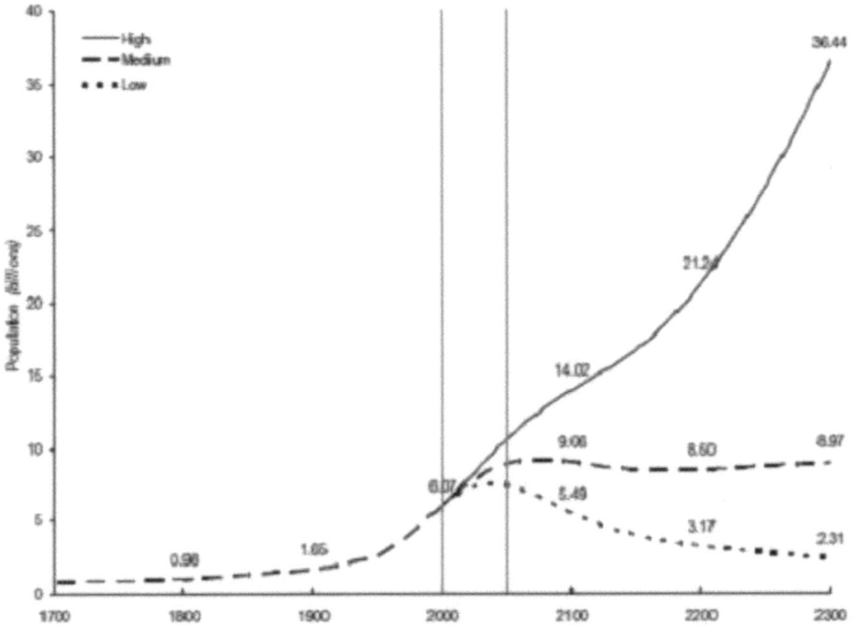

Figure 1. World population estimation and Prediction 1700th – 2300th, in reference [1].

With factories more closely connected to markets and the larger population of potential consumers, many more people could afford consumer goods. This stimulated sales, providing more jobs, increased production, and lower prices. With business booming, companies developed new products, triggering a virtual explosion of new technological advances, inventions, and consumer products in the latter 1800's. All these advances led to a higher standard of living, which further increased the consumer market, starting the process all over again.

The first step most countries took to industrialize was to build railroads to link coal to iron deposits and factories to markets. Once a transportation system was in place, factory building and production could proceed. By 1900. railroads had virtually revolutionized overland transportation and travel, pulling whole continents tightly together (both economically and politically), helping create a higher standard of living, the modern consumer society, and a proliferation of new technologies.

From the start, industrialization meant the transformation of countries' populations from being predominantly rural to being predominantly urban. By 1850. Britain had become the first nation in history to have a larger urban than rural population, and London had become the largest town in the world.

These early industrial cities created problems in three areas: living conditions, working conditions, and the social structure. First of all, cities built so rapidly were also built shoddily. Tenement houses were crammed together along narrow streets, poorly built, and incredibly crowded.

But in second half of 19th century, the standard of living of the common people improved, they had money to buy goods. Sales and profits led to more production and jobs for more people, who also now had money to spend. This further improved the standard of living, leading to more sales, production, jobs, and so on, all of which generated the incentive to create new products to sell this growing consumer market. It was the age of progress.

Steam powered ships reduced travel time at sea much as the steam locomotives did on land since ships were no longer dependent on tail winds for smooth sailing.

By 1900, the automobile, powered by the internal combustion engine, was ushering in an age of fast personal travel that took individuals wherever and whenever they wanted independently of train schedules. In 1903. the internal combustion engine also allowed human beings to achieve their dream of powered flight. The sky was now the limit, and even that would not hold up, as the latter twentieth century would see flights to the moon and beyond.

Fuelling these new developments were new sources of energy. Petroleum powered the automobile, while natural gas was used extensively for lighting street lamps. Possibly most important of all was electricity, which could be transmitted over long distances and whose voltage could be adapted for use by small household appliances. Among these was Thomas Edison's light bulb, providing homes with cheaper, brighter, and more constant light than the candle ever could provide.

The 19th century was the age of electricity. For the development of electric vehicles is important 1800. when for the first time Allessandro Volta (Italian) produces an electrical power from a battery made of silver and zinc plates. After many other more or less successful attempts with relatively weak rotating and reciprocating apparatus the Moritz Jacobi created the first real usable rotating electric motor in May 1834 that actually developed a remarkable mechanical output power. His motor set a world record which was improved only four years later. On 13 September 1838 Jacobi demonstrates on the river Neva an 8 m long electrically driven paddle wheel boat, in [3]. The zinc batteries of 320 pairs of plates weight 200 kg and are placed along the two side walls of the vessel. The motor has an output power of 1/5 to 1/4 hp (300 W). The boat travels with 2,5 km/h over a 7,5 km long route, and can carry a dozen passengers. He drives his boat for days on the Neva. A contemporary newspaper reports states the zinc consumption after two to three months operating time was 24 pounds.

In 1887 Nikola Tesla (Serbian, naturalized US-American) files the first patents for a two-phase AC system with four electric power lines, which consists of a generator, a transmission system and a multi-phase motor. Presently he invention the three-phase electric power system which is the basis for modern electrical power transmission and advanced electric motors. The inventor for the three-phase power system was Nikola Tesla, see reference [4]. But, the highly successful three-phase cage induction motor was built first by Michael Dolivo-Dobrowolsky in 1889.

2. Beginning of the EV development

The first attempt of electric propulsion was made on railways in the first half of the 19th century. It was not about cars, but as a locomotive fed by batteries, it is reasonable that this is considered a forerunner of the current prototype electric vehicles.

Robert Davidson (Scottish) also developed electric motors since 1837th in [5]. He made several drives for a lathe and model vehicles. In 1839. Davidson manages the construction of the first electrically powered vehicle. In September 1842. he makes trial runs with a 5-ton, 4,8 m long locomotive on the railway line from Edinburgh to Glasgow. Its electromotor makes about 1 hp (0,74 kW) and reaches a speed of 4 mph (6,4 km/h); a vehicle could carry almost no payload. Therefore, the use of the vehicles was very limited. Gaston Plante found a suitable battery pack in 1860th year, enabling the commercialization of electric vehicles.

At the world exhibition in Berlin 1879th years, Siemens has demonstrated the first practical electric vehicle applicable for, for example, a small electric battery tractor on rails, which was able to pull three small carriages full of people. Motor has had almost all the characteristics of today's motors for electric traction.

Already in 1881st year after on the streets of Paris was driven tricycle powered from lead-acid batteries. A year later, a horse power-drawn tram with electric propulsion was rebuilt, so that up to 50 passengers could be driving these carriages without horses. Several years later, Thomas Edison had constructed a little better first electric vehicle with nickel-alkaline batteries that are powered electric vehicle with nominal power of 3,5 kW. Immediately afterwards, the electric bus was built as well.

In England J.K.Starley constructed in 1888th the small electric vehicle [6]. Several years later, on 1893. Bersey constructed a postal vehicle and a passenger vehicle with four seats using a battery brand Elwell - Parker.

Since then, efforts are continuing, especially in America. According to some sources, the first electric vehicle in the United States was constructed by Fred M.Kimball 1888th, from Boston. In commercial use, the vehicle began to produce the first company Electric Carriage and Wagon Co. of Philadelphia, which has produced a vehicle 1894th, and the 1897th New York City has delivered a number of electric taxis. Another company, Pope Manufacturing Co. from Hartford, began producing electric vehicles 1897th years and has evolved considerably. Company produced 2.000 taxis as well as buses and electric trucks. However, they did not have great commercial success.

The first small batch production of EV had began in 1892. in Chicago. These vehicles had been very cumbersome but even so had a very good pass by customers also. They had carriages of look like (Figure 2), with large wheels, no roof, with eaves that protected passengers from rain and sun. They were used for trips, in order to perform some business, and even as a taxi to transport more passengers. Passenger's EV had the engine up to several kilowatts, which were allowed at the maximum speed of about 20 km / h, and cross a distance over a hundred kilometers on a single charge of batteries. Series DC electric motors were used, usually. Batteries have a high capacity, as far as 400 Ah, and voltages up to 100 V. Proportion of battery weight, compared to a fully loaded vehicle with passengers, was over half, which allowed so many autonomous movement radius.

Figure 2. First EV,s were possible to cross up to 100 km, moving with speed below 20 km/h.

The first production of small batch EV had began in 1892. in Chicago. These vehicles had been very cumbersome but even so had a very good pass by customers also. They had look like of carriages (figure 2), with large wheels, no roof, with eaves that protected passengers from rain and sun. They were used for trips, in order to perform some business, and even as a taxi to transport more passengers. Passenger's EV had the engine up to several kilowatts, which were allowed at the maximum speed of about 20 km/h, and cross a distance over a hundred kilometers on a single charge of batteries. Series DC electric motors were used, usually. Batteries have a high capacity, as far as 400 Ah, and voltages up to 100 V. Proportion of battery weight, compared to a fully loaded vehicle with passengers, was over half, which allowed so many autonomous movement radius.

In Europe, the first real electric vehicle was constructed by the French and Jeantaud Raffard in 1893rd. Electric motor power was 2,2 to 2,9 kW (3-4 hp), a battery capacity of 200 Ah was placed behind and had a weight of 420 kg.

In 1894, five electric vehicles participated in the first automobile race held from Paris to Rouen, a distance of 126 km. One steam vehicle won, from manufacturers De Dion.

The first race of motor vehicles was won by electric. Five vehicles with internal combustion engine and two cars with electro propulsion were racing on the road, which consisted of five sections, each one mile long (1.609 m). The winners in all five sections were electric with an average speed of 43 km/h.

Bright moment for electric vehicles in Europe was the 1899th, when on the May 1, an electric vehicle in the form of torpedo, called James Contente or "dissatisfied" reference [7], reached a speed of 100 km/h. Electric vehicle weight about 1.800 kg and was constructed by Belgian Camille Jenatzy, in [8].

The next world record speed was achieved a few years later with the vehicle which had a gasoline engine and electric vehicles were never more able to develop greater speed than vehicles with internal combustion engine.

Figure 3. Electric vehicles named Jamais Contente, which in 1899. reached previously unimaginable speed of over 100 km/h.

Waldemar Junger in 1899. first patented alkaline battery in the world. In the summer 1900th he demonstrated its capacity before the wondering audience of professionals. One battery is kept at the Waverly American Run car with which the inventor was able to drive around Stockholm in an electric vehicle for about 12 hours and with whom he went 92,3 miles (148,5 km) before the battery was discharged.

Given the fact that at the end of the 19th and early 20th century EV were moving at low speeds when the power required for handling the air resistance is negligible, the power obtained from batteries was mainly used for handling the rolling resistance, which is generally small. On the other hand, less power drain causes battery operation with a higher efficiency level so a large quantity of batteries loaded allowed a relatively large radius of movement.

http://dx.doi.org/10.5772/51771

3. The development of EV in 20th century

The twentieth century has been a century of change. It has been a century of unprecedented world population growth, unprecedented world economic development and unprecedented change in the earth's physical environment.

From 1900 to 2000, world population grew from 1,6 billion to 6,1 billion persons, about 85 per cent of the growth having taken place in Asia, Africa and Latin America.

In 1900, about 86 per cent of the world populations were rural dwellers and just 14 per cent were city dwellers, but by 2000, the share of the world population living in rural areas had declined to 53 per cent, while the number of urban-dwellers had risen to 47 per cent, in [9]. By 2030, over three fifths of the world will be living in cities. Virtually all the population growth expected during 2000-2030 will be concentrated in the urban areas of the world.

The enormous expansion in the global production of goods and services driven by technological, social and economic change has allowed the world to sustain much larger total and urban populations, and vastly higher standards of living, than ever before. For example, from 1900 to 2000, world real GDP increased 20 to 40 times, while world population increased close to 4 times and the urban population increased 13 times.

The first motor show held in New York 1901st was shown 23 and 58 steam electric and petrol cars were presented together. At the beginning of this century were used three types of motor vehicles with internal combustion engines that used: gasoline, steam or electricity. Statistics show that in 1900. from 8.000 cars driven on the roads in America, 38 % were powered by electricity. Almost equally, the third of the total number of vehicles, at the time was powered to electric power, steam vehicles and vehicles with internal combustion engines.

The car with the internal combustion engine has received increasing popularity due to its ease of charging, mobility, speed and autonomy, although the electric vehicle was still kept. Electric vehicles are especially favored by women, whom thought of the car with petrol as dirty and difficult to drive, and in the same time those looked like the features for which they were more preferred by men, driven by passion for the sport.

Defect of those electric vehicles then has been relatively short range between charges. In the late 19th century, the specific energy in the battery pack was about 10 Wh/kg. Already in the early 20 century, this value improved to the level of 18 Wh/kg, which would amount to only a decade later to 25 Wh/kg. In addition, the charging stations were not sufficiently widespread, although the situation began to improve in the early 20th century. However, sources of oil found in that period caused the low price of gasoline and the advancement of technology in the production of internal combustion engines has created the conditions for rapid progress on these cars. Therefore, the development of electric vehicles remained on the sidelines.

Studebaker developed in 1905 five models of electric traction, using the same chassis. From 1900. to 1915. year a hundred manufacturers of electric vehicles appeared. In 1904. about a third of U.S. vehicles were produced with electro propulsion. In 1912. about 10.000 electric vehicles were produced, of which about 6.000 as passenger's vehicles and 4.000 for the transportation of goods. The total traffic had approximately 20.000 vehicles to transport people and about 10.000 for freight transport. 1913th twenty companies manufacturing electric vehicles produced about 6.000 electric cars and trucks.

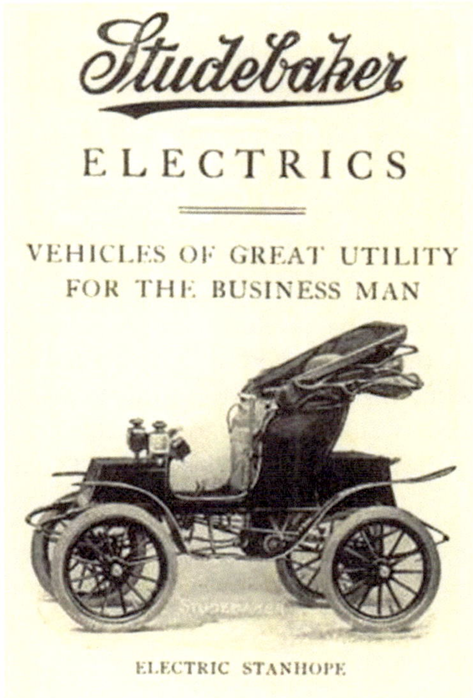

Figure 4. The external appearance of the first EV in early 20th century.

In the rally ride in the long run, from Beijing to Paris 1907th, gasoline cars definitely won over steam and electricity vehicles.Wide publicity made Dey and Harry Staymez invention in 1915. Their electric car instead of the differential had motor that was designed in the way that the rotor and stator, each connected to one half of axle, were able to turn in relation to one another. Thus, power shared between the two axles was able to turn at different speeds when cornering. Upon driving on downhill, the electric motor was turning into a dynamo serving as brakes and converting mechanical energy into electrical energy.

One passenger electric vehicle in 1917. crossed the distance from Atlantic City to New York (200 km) at an average speed of 33 km per hour.

In the twenties of this century in Germany, France and Italy, electric vehicles were designed mainly for special purposes, where it did not require more speed and autonomy. Stigler from Milan, a company specialized in electric products, constructed in 1922 more then one car with electric drive power of 4.5 kW (6 KS) and battery capacity of 250 Ah, which could speed up to 25 km/h to cross 100 km without recharging.

Before and after World War II many electric vehicles were on the streets of America, Western Europe and South Africa. The Last Car Show in America where a new type of electric vehicle was shown was in 1923. year.

1930 was the year when the appearance of the Fords model T, for some time, marked the dissolution of the companies that produced electric vehicles.

Soon after t interest in electric vehicles was lost, even in Europe and the success of the vehicles with internal combustion engine was triumphant. The performance of electric vehicles compared to internal combustion vehicles was fairly weak. The problem of batteries that were heavy and inefficient remained unresolved. Performance of the car made for special purposes, with a short radius of movement, could not be accepted for cars that could compete with gasoline powered ones.

World War II re-emphasized in the foreground electrical traction. For convenience in the normal production some vehicles were transferred to vehicles with electro propulsion. In Italy, you could have seen the car Fiat 500 (old Topolino), accumulator battery-powered weighing over 400 kg, as well as the bigger vehicles s with batteries stored in the engine and trunk space. During this period was specially designed and manufactured in a number electric Peugeot VLV. These vehicles have an advantage over the vehicles with internal combustion engines due to significantly lower maintenance costs and longer service life, making them seem more economical for exploitation.

After World War II, electric traction has remained largely reserved for special transportation and the smaller vehicles that are commonly used in the city.

3.1. Early development of drive systems

In the first EV were mostly used serious DC motors with a simple speed control solutions. In these electric motors are the excitation coil and the inductive coil connected to the serious so that the current that passes through the inductors passes through the excitation coil. This means that is in the great parts of range machine, until it comes into part of the saturation, flux is proportional to the loaded current. Only at higher loads and currents when the magnetic material enters the saturation, there is no proportionality between the magnetic flux and current, because the increase in current does not produce increase in flux.

For the operation of the serious DC motors are characteristic the great changes in flux with the load. Electric motor speed is changed in wide limits as a function of load change.

At idle load current are small and the excitation flux, so there is a risk of engine ran. Therefore, the engine should never be put into operation, under full power, without at least 20 – 30 % rated load.

At idle, load current is small as the excitation flux, so there is a risk of electromotor over speed. Therefore, the engine should never be put into operation under full voltage without at least 20 – 30 % rated load.

Speed regulation of DC electromotor can be making by changing the supply voltage or by load changing. Because DC electromotor has feature, that torque increase with the increasing of load and rotation speed falling, these electromotor are sometimes called traction. DC electromotor can be very hardly move in the regenerative mode and only if we make a re-connection of the winding.

- Start-up with additional resistance

Additional resistance is connected into the serious with a driving motor and thus lowers the voltage at the ends of the motor and reduces the starting current. With more resistance, which allows successively excluding, it is possible step-shaped voltage and speed regulator. This is a wasteful method with a low degree of usefulness.

- Commissioning and speed control via contactors (controller)

Relatively inexpensive and efficient method, but not enough good for regulation of electric vehicle speed. The necessary condition is that the voltages of all electric sources have to be equal, so appropriate involvement of the switches can get the basic voltages on the electric motor.

Additional regulation of the speed of rotation of electric motors can be done by additional rheostat for the step by step decreasing of flux, by which the speed increases and the torque decreases. Former methods of starting the electric motor and speed control of the vehicle were less quality but good enough to move the EV with relatively low speeds. In addition, there were certain losses in the resistor for speed control of electric motors and did not provide recuperative braking.

3.2. The first oil crisis

Since 1869, US crude oil prices adjusted for inflation averaged 23,67 $ per barrel (1 barel = 159 l) in 2010 dollars compared to 24,58 $ for world oil prices. Fifty percent of the time prices U.S. and world prices were below the median oil price of 24,58 $ per barrel.

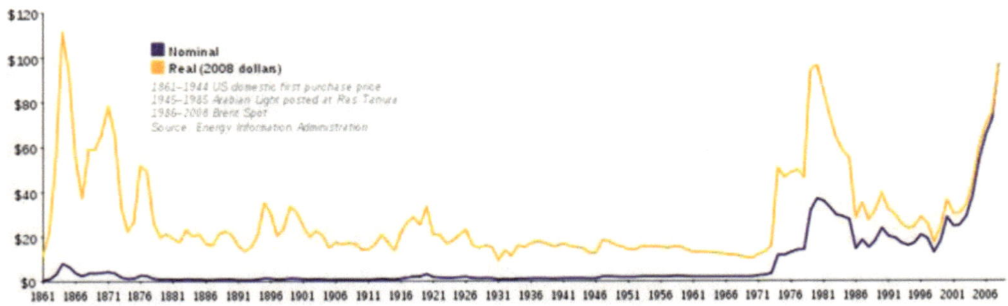

Figure 5. Long-term oil prices, 1861-2008 (orange line adjusted for inflation, blue not adjusted). Due to exchange rate fluctuations, the orange line represents the price experience of U.S. consumers only, in [10].

If long-term history is a guide, those in the upstream segment of the crude oil industry should structure their business to be able to operate with a profit, below 24,58 $ per barrel half of the time. The very long-term data and the post World War II data suggest a "normal" price far below the current price.

From 1948 through the end of the 1960s, crude oil prices ranged between 2,50 $ and 3,00 $. The price oil rose from 2,50 $ in 1948 to about 3,00 $ in 1957. When viewed in 2010 dollars, a different story emerges with crude oil prices fluctuating between 17 $ and 19 $ during most of the period. The apparent 20 % price increase in nominal prices just kept up with inflation.

From 1958 to 1970, prices were stable near 3,00 $ per barrel, but in real terms the price of crude oil declined from 19 $ to 14 $ per barrel. Not only was price of crude lower when adjusted for inflation, but in 1971 and 1972 the international producer suffered the additional effect of a weaker US dollar.

OPEC was established in 1960 with five founding members: Iran, Iraq, Kuwait, Saudi Arabia and Venezuela. Two of the representatives at the initial meetings previously studied the Texas Railroad Commission's method of controlling price through limitations on production. By the end of 1971, six other nations had joined the group: Qatar, Indonesia, Libya, United Arab Emirates, Algeria and Nigeria. From the foundation of the Organization of Petroleum Exporting Countries through 1972, member countries experienced steady decline in the purchasing power of a barrel of oil.

Throughout the post war period exporting countries found increased demand for their crude oil but a 30 % decline in the purchasing power of a barrel of oil. In March 1971, the balance of power shifted. That month the Texas Railroad Commission set proration at 100 percent for the first time. This meant that Texas producers were no longer limited in the volume of oil that they could produce from their wells. More important, it meant that the power to control crude oil prices shifted from the United States (Texas, Oklahoma and Louisiana) to OPEC. By 1971, there was no spare production capacity in the U.S. and therefore no tool to put an upper limit on prices.

A little more than two years later, OPEC through the unintended consequence of war obtained a glimpse of its power to influence prices. It took over a decade from its formation for OPEC to realize the extent of its ability to influence the world market.

In 1972, the price of crude oil was below 3,50 $ per barrel. The Yom Kippur War started with an attack on Israel by Syria and Egypt on October 5, 1973. The United States and many countries in the western world showed support for Israel. In reaction to the support of Israel, several Arab exporting nations joined by Iran imposed an embargo on the countries supporting Israel. While these nations curtailed production by five million barrels per day, other countries were able to increase production by a million barrels. The net loss of four million barrels per day extended through March of 1974. It represented 7 percent of the free world production. By the end of 1974, the nominal price of oil had quadrupled to more than 12,00 $.

Any doubt that the ability to influence and in some cases control crude oil prices had passed from the United States to OPEC was removed as a consequence of the Oil Embargo. The extreme sensitivity of prices to supply shortages, became all too apparent when prices increased 400 percent in six short months.

From 1974 to 1978, the world crude oil price was relatively flat ranging from 12,52 $ per barrel to 14,57 $ per barrel. When adjusted for inflation world oil prices were in a period of moderate decline. During that period OPEC capacity and production was relatively flat near 30 million barrels per day. In contrast, non-OPEC production increased from 25 million barrels per day to 31 million barrels per day.

In 1979 and 1980, events in Iran and Iraq led to another round of crude oil price increases. The Iranian revolution resulted in the loss of 2,0-2,5 million barrels per day of oil production between November 1978 and June 1979. At one point production almost halted.

The Iranian revolution was the proximate cause of the highest price in post-WWII history. However, revolution's impact on prices would have been limited and of relatively short duration had it not been for subsequent events. In fact, shortly after the revolution, Iranian production was up to four million barrels per day.

In September 1980, Iran already weakened by the revolution was invaded by Iraq. By November, the combined production of both countries was only a million barrels per day. It was down 6,5 million barrels per day from a year before. As a consequence, worldwide crude oil production was 10 percent lower than in 1979.

The loss of production from the combined effects of the Iranian revolution and the Iraq-Iran War caused crude oil prices to more than double. The nominal price went from 14 $ in 1978 to 35 $ per barrel in 1981.

3.3. Renaissance of EV

In the seventies began the renaissance of EV. Fixed price of oil, which is less and less available, and the problems associated with its production and transport, leads to renewed interest in electric vehicles. At that time, it seemed that the coal and oil reserves would exhaust quickly, predicted at the beginning of the third millennium, so the world began to think about the "energy conservation". In addition, ongoing technical advances made with high quality and effective solutions of speed regulator for electric motor, lighter batteries and lighter materials for the body.

After 1970. environmental problems and oil crises increased the actuality of electric vehicles. Especially in the United States the interest of the citizens awoke who have acquired a habit to use widely electric vehicles for golf courses, for airports, for parks and fairs. According to some sources, one third of vehicles intended for driving on gravel roads were with electric traction. So there was a need to develop a new industry.

1974 Sebring - Vanguard began producing electric vehicles on the lane. City Car with two-seat, weighs 670 kg, and an electric voltage 48 V, 2,5 kW power only, achieved a maximum speed of 45 km/h. With an improved variant of this operation the maximum speed of 60 km/h was accomplished. The vehicle exceeded up to 75 kW with a single charge of batteries and the cost was about 3.000 US$. Only between the 1974th and 1976. about 2.000 of these vehicles was produced 1974. Copper Development Association Inc. made a prototype electric passenger vehicle. Although it used lead-acid batteries, it could develop a top speed of 55 mph (90 km/h), and could go over 100 mph (161 km/h with one battery charge at a speed of 40 mph (65 km/h).

Among the achievements of the General Motors company at the time was the GM 512 vehicle designed for drive in urban areas that are closed for classic cars. These are two types of small passenger vehicles with a carriage-body constructed partly of glass resin, but one is with pure electro propulsion and the other is a hybrid. Basic data on pure electric version are: weight 560 kg, the engine of 6 kW, a maximum speed of 70 km/h. With a 150 kg lead acid batteries could be run without charge from 50 to 70 km. It was supplied even with an air conditioning.

The largest exhibition of electric vehicles ever made till then, EV Expo 78, in [11], was held in Philadelphia. Expo displayed more than 60 electric vehicles with prices from 4.000 $ to as much as 120.000 $.

The first electric vehicle, General Motors, a prototype car with four seats cost 6.000 $. It was planned as a second family vehicle.

Secondly there is an electric vehicle Electric Runabouth Copper, who is a manufacturer of Copper Development Association Inc. said that it can be produced for 5.000 $. The vehicle mass of 950 kg, with four seats, made of fiberglass, had a top speed around 110 km/h could not move without charge to 130 km before its battery runs out of battery. It has a 10 kW electric motor that could, in one-hour mode, it delivers up to 15 kW and ups eliminates up to 22 %. Weight of batteries was about 380 kg.

Figure 6. A typical city car (City Car) with two seats, weighs only 670 kg had a top speed of 28 mph (45 km/h) and radius of movement up to 65 km.

Most EV were relatively modestly equipped, but the Electric Car Corporation of Michigan, he believed the first luxury electric vehicle called the Silver Volt. The prototype of this five-seat EV has achieved a top speed of movement 110 km/h had a radius of 160 km between charges the battery. Silver Volt owned air conditioning and was sold for about 15.000 $.

Some companies also produce and display luxury EV priced up to 120.000 $ the most expensive ever built passenger car of this type.

Figure 7. Copper City Electric Car Runabouth power 15 kW made on the basis of cooperation for the use components of Renault R5.

The majority of EV is driven by a conventional lead-acid batteries that are found even 1868th years and are still the mainstay of the vehicles. But the lead-acid batteries have also already been the primary limiting factor for the development of EV. Pointed out that at least 40 million vehicles in the U.S., a total of 110 million, can be electrically driven second family vehicle as meeting the ecology and urban and suburban driving conditions. However, lead batteries and still remain a limiting factor in EV that time.

Laden vehicle **total weight 1.134kg**		**Empty vehicle** **Curb weight 934 kg**
46,3 %	Body 542 kg	56,1 %
27,3 %	Batteries 310 kg	33,2 %
8,8 %	Electric propulsion 100 kg	10,7 %
13,2 %	2 passengers 150 kg	
4,4 %	Payload 50 kg	

Table 1. Percentage distribution of the reconstructed mass of the vehicle YUGO-E when it is empty and loaded.

From this period, the EV was largely rebuilt vehicles from the existing series production vehicle with the drive IC. And with a maximum weight of lead acid batteries, the performances of these cars were quite limited. As an example, the percentage distribution of the reconstructed mass of the vehicle can serve example of the reconstructed vehicle YUGO-E when it is empty and loaded, reference [12, 13].

1. Body YUGO - E	Type of vehicle passanger
dimensions 3,49*1,542*1,392 m	Empty vehicle weight 934 kg
Useful load 2 persons + 50 kg	drive front-wheel
Brakes disk, front and back	Control over the rack
2. Direct current electric motor	Power 6,3 kW
Voltage 72 V	Rated current 113 A
Number of revolution 2.800 min^{-1}	Weight 38 kg
3. battery	Type traction
Total voltage 72 V	Capacity (20h) 143 Ah
Pieces 6	Total Weight 294 kg
4. Voltage regulator	Type transistor chopper
Current limit 180 A	Voltage drop at current of 100 A 0,7 V
Undervoltage disconnection 48 V	Weight 4 kg
5. Battery charger	Battery charger characteristic IUUo
Voltage 72 V	Current 18 A
Power 1.800 W	Weight 38 kg
6. DC / DC converter	Type with galvanic isolation
Output voltage 13,5 V	Maximum output currant 22,2 A
power 300 W	Weight 2 kg

Table 2. Technical data of electric drive Yugo-E, in [13].

3.4. Impact of the development of power electronics on the development of EV

The invention of the transistor in 1948 revolutionized the electronics industry. Semiconductor devices were first used in low power level applications for communications, information processing, and computers. In 1958, General Electric developed the first Tyristor, which was at that time called SCR, in [14]. Since around 1975, more turn-off power semiconductor elements were developed and implemented during the next 20 years, which have vastly improved modern electronics. Included here are improved bipolar transistors (with fine structure, also with shorter switching times), Field Effects Transistors (MOSFETs), Gate Turnoff Thyristors (GTOs) and Insulated Gate Bipolar Transistors (IGBTs).

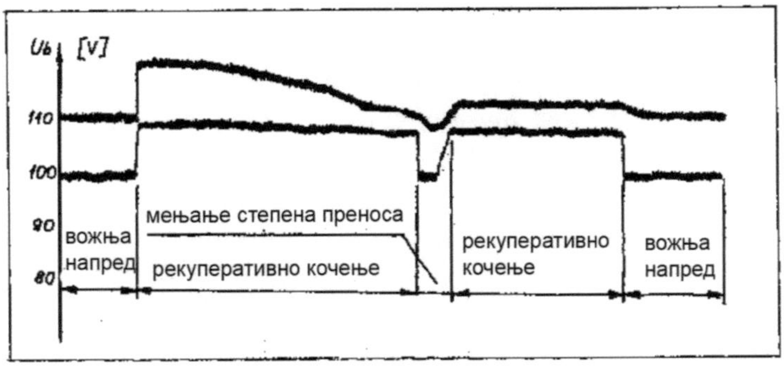

Figure 8. Change the battery voltage during DC recuperative braking in [15, 16].

Although they initially made Chopper with thyristors, later almost exclusively were made with transistors. The main difference is that Chopper with thyristors operates up to several hundred Hz, and the power transistors and up to several tens of kHz. For use the EV used Chopper with mutual influence (for lowering and raising the voltage), because this type of chopper allows propulsion and recuperative or regenerative braking drive motors. In this way it is possible to drive DC generator machine brake or braking to convert mechanical energy into electrical energy in [17].

It is well known, there are two modes of operation of electric vehicles. In the electric motor drive mode, in the operation is step down chopper and the average voltage on the electric motor is less then battery voltage. In the electric braking mode, in the operation is step up chopper, so the less voltage of the electric motor supply battery on higher level voltage and on that way there is recuperative braking.

3.5. End of the 20th century

Late 20th century contributed to an even greater exacerbation of conditions around the EV application. Scientists have become aware that environmental pollution is becoming larger, the emission of exhaust gases and particles affect climate change and that non-renewable energy sources under the influence of high demand and exploitation are becoming more expensive and slowly deplete.

Technology is certainly a double edged sword that has also created new problems such as pollution, overpopulation, the greenhouse effect, depletion of the ozone layer, and the threat of extinction from nuclear war. It has also been used to give us prosperity our ancestors could never have dreamed about. Whether it is ultimately used for our benefit or destruction is up to us and remains in the balance

In 2010, the world's population reached 6,9 billion persons in [18]. It is expected to attain 9,3 billion in 2050 and 10,1 billion by the end of the century. The proportion of the population living in urban areas grew from 29 per cent in 1950 to 50 per cent in 2010. By 2050, 69 per cent of the global population, or 6,3 billion people, are expected to live in urban areas. The atmospheric concentration of carbon dioxide (CO2), the main gas linked to global warming,

http://dx.doi.org/10.5772/51771

has increased substantially in the course of economic and industrial development. CO2 emissions are largely determined by a country's energy use and production systems, its transportation system, its agricultural and forestry sectors and the consumption patterns of the population. In addition to the impact of CO2 and other greenhouse gases on the global climate, the use of carbon-based energy also affects human health through local air pollution. Currently, CO2 emissions per person are markedly higher in the more developed regions (12 metric tons per capita) than in the less developed regions (3,4 metric tons per capita) and are lowest in the least developed countries (0,3 metric tons per capita). Industrial and household activities as well as unpaved roads produce fine liquid or solid particles such as dust, smoke, mist, fumes, or smog, found in air or emissions. Protracted exposure to Particulates is detrimental to health and sudden rises of concentration may immediately result in fatalities. Concentration of particulate matter in the air of medium and large cities is inversely correlated with the level of development.

Ownership of passenger cars has increased considerably worldwide and the transportation of goods and services by road has intensified. Rising demand for roads and vehicles is associated with economic growth but also contributes to urban congestion, air and noise pollution, increasing health hazards, traffic accidents and injuries. Motor vehicle use also places pressure on the environment, since transportation now accounts for about a quarter of the world's energy use and half of the global oil consumption, and is a major contributor to greenhouse gas emissions. In the more developed regions there are more than 500 motor vehicles per 1000 population. In the less developed regions this ratio is only 70 vehicles per 1000 population, but it is increasing more rapidly than in the more developed regions.

Energy generated by the combustion of fossil fuels and biomass often results in air pollution, affecting the health of ecosystems and people. This type of combustion is also the main source of greenhouse gases and rising atmospheric temperatures.

However, in the late 20th century has made improvements in electric drives. Quality inverters are designed with the ability to control the voltage and frequency, enabling the use of induction motors to drive the EV in [19]. Asynchronous (induction) motor is simpler, lighter, more efficient and robust than DC motors. Despite all that, its price is considerably lower than the DC motor. Maximum speed is increased by 50% to 150% of maximum speed DC motor which is limited because of problems with commutation. The efficiency of induction motors is from 95% to 97%, and is higher than that of DC motor from 85% to 89% for DC motors. Inverters are power converters that convert the DC voltage alternating current, the required frequency and amplitude [20, 21].

4. Start of the 21st century

The unprecedented decrease in mortality that began to accelerate in the more developed parts of the world in the nineteenth century and expanded to all the world in the twentieth century is one of the major achievements of humanity. By one estimate, life expectancy at birth increased from 30 to 67 years between 1800 and 2005, leading to a rapid growth of the population: from 1 billion in 1810 to nearly 7 billion in 2010, in [22].

With the growth of population in the world there is a need to increase transportation of people, goods and raw materials as a prerequisite for the growth of production and consumption and the standard of living. This constant growth is natural and expected process of development of civilization and one of the most important indicators of development of society and humanity so that today a life without road traffics considered unthinkable.

Big boost for electric vehicle development was given by the developed countries where air pollution is receiving alarming values.

In cities with large population, and where there is a big environmental pollution, the city authorities have taken some steps to the special places provided for movement and recreation citizens to reduce air pollution. In places where there are a large number of urban populations, city governments often support the eco-drive vehicles.

First of all vehicles are required city services that are moving in the streets intended for pedestrians, such as travel or vehicle inspection. In addition, various types of tourist vehicles moving at pedestrian areas or in city parks. Then, various kinds of utility and delivery vehicles that work in limited areas such as rail bus stations or airports.

In order to significantly reduce oil consumption and pollution in the world that creates traffic especially in big cities it is necessary to make the transition from today's cars with internal combustion engines to electric drives. Given the poor performance of EV on the market there are fewer of these vehicles, although almost all major manufacturers of passenger vehicles operate on the development of these vehicles.

Although scientist Nikola Tesla wrote and discussed the use of EV with the alternate (induction) engine until 1904. in [23], when the EV is already contained in the traffic in the United States a decade ago founded the company bearing his name, Tesla Motors, which is producing very interesting and modern sports EV.

EV "Tesla Roadster" is a sport, the first serial built car that used lithium-ion battery in [24], and the first one which had a radius greater than 320 km on a single charge.

The vehicle has a length of 3.946mm, 1.851mm width and a curb weight is 1.234 kg. Useful load is for 2 persons, and the weight of batteries is 450 kg. The AC drive motor has a power 185 kW and a maximum speed of rotation 14.000 min^{-1}. Voltage Li-ion battery is a 375 V and capacity 145 Ah. Charger of the rechargeable battery is inductive (contactless). The vehicle can travel up to 231mile (372 km) in city driving with standard EPA testing procedure. Speed of 60 mph (97 km/h) can be achieved only by 3,9 s, top speed is electronically limited to 125 mph (201 km/h). This vehicle has made the largest radius of movement on single charge EV batteries 311 miles (501 km). Electricity consumption is only 145 Wh per kilometer of road travelled.

Mass production of this vehicle was started in early 2008. year. Despite the crisis that is evident and the prices of over 100,000 USD in the beginning of sales, has so far sold more than 1000. pieces of this vehicle in [25].

http://dx.doi.org/10.5772/51771

Figure 9. Tesla Roadster electric car of the firm Tesla Motors.

On The development of modern EV worked both large and small manufacturers of motor vehicles. EV still has significant problems arising from low-volume production so that these vehicles are still expensive and thus less attractive. In the first place it is air-conditioning for passengers and a relatively small possibility of storing electricity in batteries. The necessity of development of plant components specially developed for series production will be affected by the low price of these components. Great stimulus to the occurrence of EV on the World Fair is given by Far eastern markets provide producers in [26, 27], which also made a series of large vehicles substantially at lower prices and affordable to most buyers in developed countries.

4.1. Hybrid Vehicle (HV)

Oil prices value on world markets in spring 2008. exceeded 100 $/barrel, with previous analyzes have designated this value as the marginal cost of EV use. Oil prices reached a value of 147 $/barrel in early July 2008., and shortly thereafter dropped to a value of only 40 $/barrel, it is nowday stabilized at value around 100 $/barrel.

One of the objectives of the new plan, which President Obama has described as "historic", is to replace the existing complex system of federal and state laws and regulations on exhaust emissions and fuel economy. Announcing the plan in [28], President Obama said that "the status quo is no longer acceptable," as it creates dependency on foreign oil and contributes to climate change. Effects of new measures will be as if from the roads in America 177 million vehicles have been removed and that the state saves as much oil as in 2008. was imported from Saudi Arabia, Venezuela, Libya and Nigeria.

Since then it speeds up the development and improvement of a mostly EV batteries or "power tank" which the vehicle carries. Parallely is working on improving the use of EV which now can be used for some applications, as well as the use of HV.

Not finding the opportunity to meet the existing types of EV driving habits with conventional drive vehicles, and vehicles with conventional drive to meet certain environmental requirements, motor vehicle manufacturers have come to the medium solution, so called. hybrid drive. If the hybrid has a higher capacity battery that can be recharged via connection to an external source and distribution network, then it is a "plug in" hybrid vehicle (PHV)

HV makes real breakthrough in terms of reducing consumption of fossil fuels, as well as in terms of environmental benefits, and improving air quality in cities, which is encouraged by governments in some western countries. Using PHV reduces smog emissions established in the cities, in [29].

Although PHV will never become a "zero-emission vehicles" (ZEV) due to their internal combustion engine, the first PHV which appeared on the market reduce emissions by one third to half in [30], and is expected from more modern models to reduce emissions even more.

There are several types of applications in hybrid drive vehicles. Common to all is that a shorter time in the city center, vehicle can move with the electric drive as an environmentally clean and then to aggregate that includes the IC engine that runs at the optimal point of operation. In this way the HV has minimal emissions and minimal consumption of petroleum products.

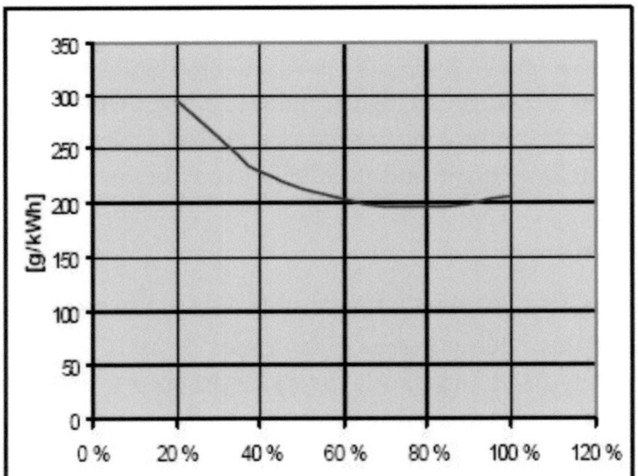

Figure 10. Diagram of the specific consumption of diesel engine as a function of maximum continuous power, [31, 32].

HV has two drives, and practically unlimited radius of movement. In the regime of pure electric drive with modest performance with maximum speed of 80 km/h small autonomous movement of about 80 km radius, but because of that the hybrid drive doubles the speed and radius becomes practically unlimited. Because the two types of power, HV is about 35 % more expensive than the equivalent of cars with internal combustion engine, but to create habits of drivers, some states stimulated by reducing taxes for these vehicles.

The general conclusion is that is a positive step towards the introduction of environmental drive vehicles. However, since no definitive solution is found, experiments with pure electric and hybrid solutions carried out, as well as various types of technical drive solutions. Despite the turbulent development of EV and HV, some experts believe that vehicles with ICE will dominate for more 15 years, but even after that will not disappear in [33].

The main reason for the production and purchase of hybrid vehicles down to fuel economy in city driving, but are often cited and highlight information on saving energy and reducing pollution in [34, 35]. Best-selling HV Prius in [36], has a fuel-efficiency of 51 mpg (21,7 km/l) in the city and 48 mpg (20,4 km/l) on the open road. Typically, in our present data on consumption per 100km distance traveled, so that consumption in the city is 4,6 l/100 km and on the open road is about 4,9 l/100 km.

4.2. Plug in EV (PEV)

EV with batteries still have a small market share in the sale and use of cars, but different types of EV, especially the Army, that made significant progress. This was especially favored new legislation announced by the U.S. administration.

It is known that the EV motor vehicle was powered by an electric motor fed from an electrochemical power sources. Often, an electric vehicle (EV) is called the zero vehicle emissions (ZEV), because it emits no harmful particles into the atmosphere. In the older literature, for EV use the terms electric vehicle (EM) or autonomous electric vehicle (AEV) [37].

The basic components of the EV are battery pack as a "reservoir of power" and drive electric motor with speed regulator.

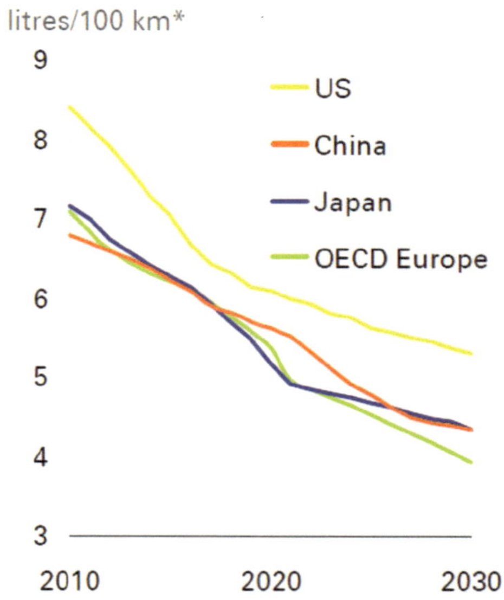

Figure 11. Experts' forecasts of consumption of hybrid vehicles by 2030. in [38].

If someone install aggregate in the EV that has a combustion engine and generator, we get a hybrid variant of EV and then it is always possible when driving or when necessary to recharge the battery. With this solution the drive gets slightly higher consumption of oil products in long-distance driving and slightly lower performance with the drive in vehicles with internal combustion engine. But, in the city center, when the internal combustion engine is not in operation, the car behaves ecologically and uses less oil derivatives per kilometer of road vehicles then vehicle with internal combustion engine.

Hybrid vehicles are vehicles in which exists a combination of internal combustion engines (gasoline or diesel) and electric drive, but have limited features of the electric drive mode and can be supplemented from the power grid.

"Plug in" HV are vehicles that can move a distance of 15 to 60 km with a charged battery pack and then the batteries need to be supplemented from the power grid or by combustion engines. Often embedded computer determines the optimal conditions to charge.

The main differences between HV and "Plug in" HV Prius becomes obvious if one looks at the range or increase the radius of the vehicle in electric mode, approximately 2 km (Prius) to 23,4 km (PHV), in [39].

	Prius PHV	**Prius HV**
Dimensions (length/width/height)	4.460/1.745/1.490 mm	←
Curb weight	1.490 kg	1.350 kg
Seats	5 persons	←
Maximum engine power	60 kW (82 KS)	←
Maximum power of the entire system	100 kW (136 KS)	←
Storage energy	Li-ion battery (5,2 kWh)	NiMH batterya (1,3kWh)
Engine Displacement / maximum power	1.797cc / 73kW (99hp)	←
Fuel consumption PHV	57,0 km/l	–
Fuel consumption HV	30,6 km/l	32,6 km/l
EV range	23,4 km	around 2 km
EV top speed	100 km/h	55 km/h
Electrical energy efficiency	6,57 km/kWh	–
Battery recharge time	About 100 min. (200V) about 180 min. (100V)	–

Table 3. Technical characteristics comparison of the hybrid Prius"Plug in" hybrid vehicles and Prius hybrid vehicles) in [39].

In addition, it is improved specific fuel consumption in the hybrid mode. Studies have shown that in Japan, 90 % of drivers exceed the average daily distance below 50 km and 60 km and 75 in the EU and the U.S. respectively. In this case, the expected cost of vehicles greatly influences the price of electricity which during the day in Japan is about 20 cents/kWh and late at night around 8 cents/kWh. It should be noted that the average price of electricity in Serbia amounts to only 5 EU cents/kWh.

The best-selling hybrid car in the U.S. "Toyota Prius", has the highest demand when fuel prices rise. The state encourages the producer price of 6.400$, in [40], so that the standard model sells for just 21.610 US$). The fuel economy of this vehicle is 48 mpg (4,9 l/100 km) in city driving and 45 mpg (5,2 l/100 km) on the open road. Translated into fuel consumption per 100 km is 5,2 l/100 km in city driving and 4,9 l/100 km on the open road.

Large oil producers, such as BP33, consider that in future, up to 2030. PHV will be dominant, primarily due to a reduction in fuel consumption per kilometer of the road, figure 11.

5. Factors that influence the further development of the EV

Transport in cities today is based on other petroleum derivatives. With today's technical solutions existing EV's does not have enough energy so that it can achieve a radius of movement and performance competitive with internal combustion powered vehicles. On the other hand, the absence of exhaust emissions and low noise make the EV attractive for some specific purposes such as short trips with frequent stops in which vehicles with internal combustion engines would have inefficient work.

In addition to high economic dependence on oil and oil products, is a common problem and protecting the environment, reducing emissions and greenhouse gases. It is anticipated that, due to technology development, energy consumption in production systems, despite the larger volume of production in the coming years largely be stagnant.

There are several factors that influence the development of EV:

- Growth in world population and transportation needs
- Energy demand in the world
- Crude oil as an energy source
- Pollution and global warming
- World production and consumption
- Efficiency of electric drives

5.1. The growth in world population and transportation needs

As the main means of mass transportation, cars with internal combustion engines marked the twentieth century. However, the consequences of this form of mass transportation are a large amount of harmful exhaust substances that pollute the environment. Finding alternative energy sources that would move the vehicle could solve this problem. One possible solution is EV.

	Country	Number of vehicles
01	China	13.897.083
02	Japan	8.307.382
03	Germany	5.552.409
04	South Korea	3.866.206
05	Brazil	2.828.273
06	India	2.814.584
07	US	2.731.105
08	France	1.922.339
09	Spain	1.913.513
10	Mexico	1.390.163

Table 4. Production of passenger cars in the world's 2010th in [41].

The world in 2010. year, according to OICA in [41], produced 58,305,112 passenger vehicles used to transport passengers. China topped the list with almost 24% of produced cars followed by Japan, Germany and South Korea. Despite the large car manufacturers for which she is known in the world, the U.S. ranks only seventh in the world,

5.2. Energy demand in the world

Population growth in the world and general technical advances cause a growing need for all types of energy. Percentage of growth energy use needs in the world is greater than the percentage of population growth. Today, more than half, or 56 % of the world's energy consumed in the U.S., Japan and the European Union. As these countries are relatively poor in energy resources, they represent the largest energy importers.

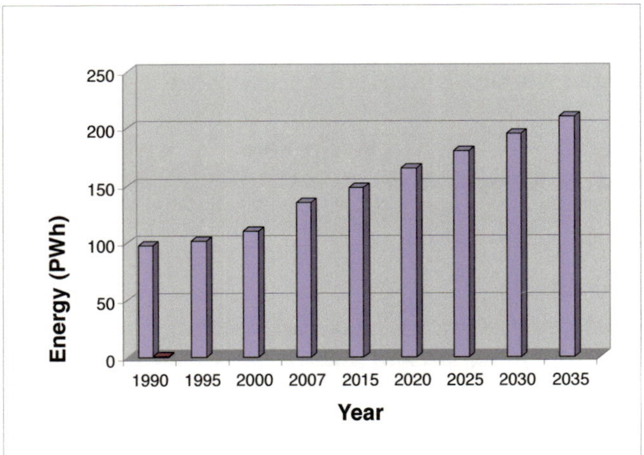

Figure 12. Consumption or total primary energy in the world since 1990. to the date and forecast till 2035.

The statistical overview of the total consumption of primary energy in the world since 1.990. to date, as well as forecast till 2.035. years is shown in figure 12 and expressed in PWh. in [42].

Estimates are that due to increasing consumer demands, and especially because of increasing demands for the transportation of goods and people, energy demand increased by about 1.5 to 2 % per annum. It is believed that in the period from 2000. to 2050. The demand for energy will be more than doubled.

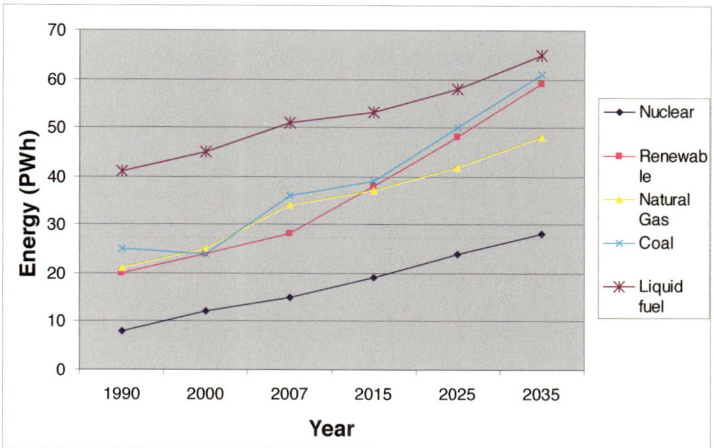

Figure 13. Types of suitable monitoring of energy in the world in the period since 1990. year to date and forecast by 2035.

The different energy sources in total or primary energy consumption in the world in the same period and forecast up until 2035. is presented in figure 13. This balance includes oil, natural gas, solid fuels, nuclear energy and renewable energy sources with heat recovery lost during combustion of other fuel types. Weaker energy sources, such as wood, biomass and other sources in these considerations are not taken into account.

It may be noted that the share of nuclear 'energy significantly increases and the prediction indicate that, despite all the concern and dissatisfaction of the "green" this type of energy will be exploited more and more. There are expectations that all types of renewable energy products and exploit all the more. Although these sources are currently produced per unit of energy even more expensive than others, it is believed that in the future primarily due to new technologies and mass production price significantly reduced.

Coal remains the main source of energy. Consumption and production of natural gas is increasing. Production of hydropower is poor because the share of water flows in the production of electricity is utilized enough.

5.3. Oil as an energy source

Although the share of oil in total primary energy percentage decreases, production, consumption of oil is generally increasing. There are opposing tendencies: on the one hand, increased daily transport of people and goods, while the second reduction of imported energy, environment and the negative economic balance. Over 97 % of fuel consumed in the transport sector, U.S. in [43], is based on oil, and this represents about two-thirds of the total national oil consumption. Although the specific consumption of liquid fuels in vehicles since 1970. The steadily declining, population growth and the length of distance traveled per capita is increasing and contributing to the total consumption of liquid fuels for transport.

And if efforts are made to find new sources and new facts indicate that this type of energy is slowly decreasing and scientists expect that for some time all sources of energy will dry up.

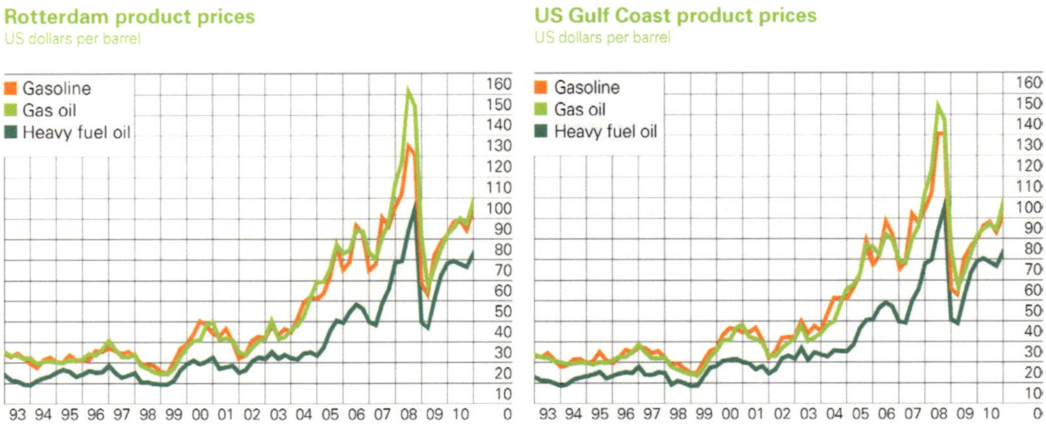

Figure 14. Prices of petroleum products on the market in Rotterdam in [44]. since 1993. expressed in U.S. $ per barrel.

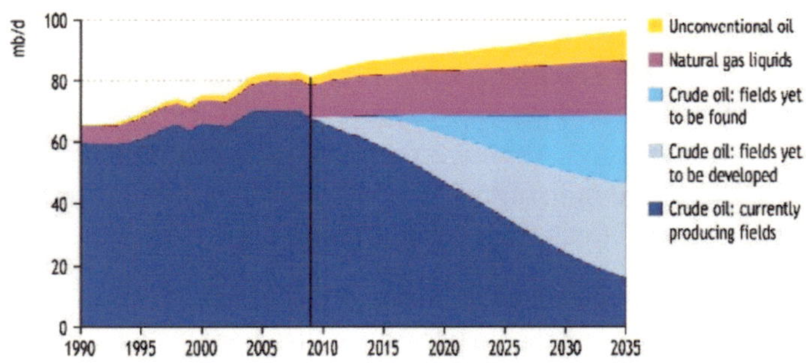

Figure 15. Forecast of global production of liquid fuels by 2035.

According to a statistical review of BP (British Petroleum) [44], in 2.011., figure 14 shows the increase in prices of petroleum products in Rotterdam since 1993. expressed in U.S. dollars per barrel.

Forecast of production of petroleum products in the world by 2.035. year, according to the Energy Information Administration (EIA) in [42], is shown in figure 15. We hope to discover new oil fields, and activate the existing drain current, so that the next 25 years, production of crude oil will mainly keep the existing values. Expected to increase consumption of natural gas and non-conventional liquid fuels. At the same time certain redistribution of the consumption of liquid fuels will be made. Expected increase in consumption of liquid fuels for transport and to a lesser extent for other consumers.

Taking into account today and proven preset fossil fuel reserves can be estimated that up to half of the century the transport sector and transport of energy resources was largely satisfied, but certainly not after the 2050th year, if only with today's fuel reserves appeared a new energy crisis, in [45].

5.4. Environmental pollution and global warming

Modern transport has contributed to overall economic progress but also caused problems and environmental pollution, traffic congestion and problems of energy supply - particularly in times of energy crisis.

Air pollution by burning fuel in motor vehicles becomes the most important global issue, especially in urban areas worldwide. Emission of pollutants originating from motor vehicles caused by the level of traffic, possibility of roads and weather conditions. Pollutants from the exhaust system of motor vehicles reach the atmosphere and are dependent composition, and fuel volatility.

In terms of impact on global atmospheric pollution and problems associated with it, the most important effect is the increase in global mean temperature. From the standpoint of global warming the greatest danger represents carbon dioxide, an unavoidable component of the combustion products of petroleum products, in [46].

Human activities in the past two centuries have been based on the large use of hydrocarbons to obtain the necessary energy. Therefore, the amount of "greenhouse gases" in the atmosphere has increased and is expected to lead to increase in average global temperature.

In addition to air pollution in violation of the environment and space as a significant natural resource waste oils are participating, as well as uncontrolled release of oil, in [47]. to contaminate surface and groundwater.

In contrast to the natural greenhouse effect, an additional effect caused by human activity contributes to global warming and may have serious consequences for humanity. Earth's average surface temperature has increased by about 0,6 °C in [48], only during the twentieth century.

In addition, if we can not take any steps toward limiting emissions of greenhouse gases in the atmosphere, concentrations of carbon dioxide by 2100. can be expected to reach values between 540 and 970 million particles of the volume. This concentration of carbon dioxide is leading to global temperature increase between 1,4 and 5,8 °C by the end of this century.

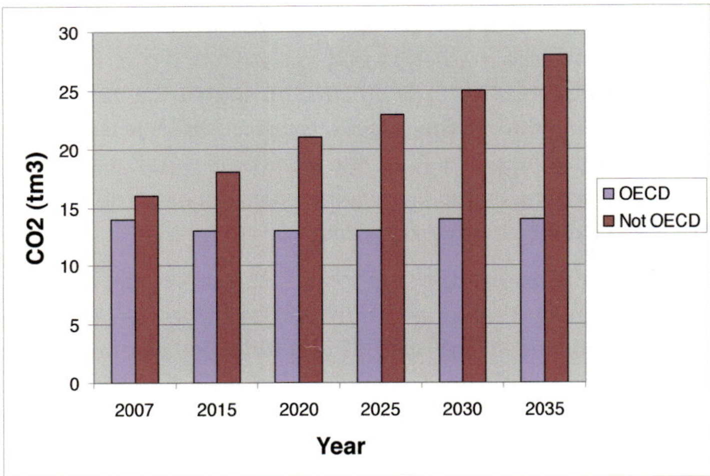

Figure 16. Forecast comparison of carbon emissions in the period since 2007. until 2035. The OECD countries and other countries.

The temperature rise of this magnitude would also have impacted on the entire Earth's climate, and would be manifested trough the frequent rainfall, more tropical cyclones and natural disasters every year in certain regions, or on the other hand, in other regions such as long periods of drought, which would overall have a very bad effect on agriculture. Entire ecosystems could be severely threatened extinction of species that could not be fast enough to adapt to climate change.

In order to reduce air pollution from vehicles and to make more economical cars in the fight against global warming and reducing dependence on oil in the U.S. are preparing new standards for reducing automobile emissions and reduce consumption of fossil fuels. The intention of the U.S. administration is that these measures by 2016. reduce he emissions from vehicles by 30 %. Under the new standards for passenger vehicles, fuel consumption must be reduced to a level of 35,5 miles/gallon (6,62 l/100 km) in [49]. It is expected that new proposals for new vehicles in the average rise in price by about 1,300 $ in 2016. year. It should be noted that the U.S. is the largest automobile market in the world with about 250 million registered vehicles

5.5. World production and consumption of electric energy in the World

A necessary precondition for economic development and growth of each country and the region is safe and reliable electricity supply. Electricity consumption per capita is highest in the Nordic countries (to a maximum of 24,677 kWh, Iceland) and in North America. Almost half of EU countries have nuclear power plants so that in France and Lithuania almost 75 % of electricity is obtained from nuclear power plants in [50].

The growth and forecast growth of electricity production in the world and the total energy consumption in the period 1990 - 2035, according to the Energy Information Administration (EIA) is shown in Figure 17.

Base for observation of this comparison was taken 1990. year. It may be noted that the real growth of electricity consumption in the period since 1990. to 2006. is 59 % and overall energy consumption 36 %. Forecasted growth in electricity consumption by 2025. amounts to 181 % and overall energy consumption 95 %.

Production and consumption of electricity for years has a steady growth of around 3.3 % per year. Normal for middle-income countries has a slightly higher growth. Electricity production is obtained mostly by burning solid fuel 40 % and natural gases about 20 %. About 16 % of electricity obtained from hydropower and only slightly less, 15 % from nuclear power plants. Less than 10 % is obtained from petroleum.

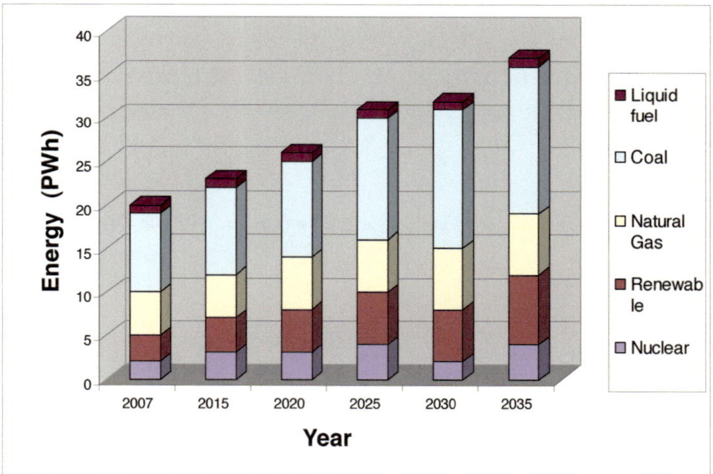

Figure 17. The share of energy in electricity generation in the world since 1971. to 2001, [50] Last few decades, the share of electricity derived from nuclear power plants have increased considerably and from hydro has declined, although the total growth in electricity production obtained from hydropower continued. It is believed that the near future will experience significant increase in production of electricity from nuclear power plants, to a lesser extent from natural gas, and later also from renewable sources.

5.6. Efficiency of electric drives

Efficiency of electric vehicles was marked several times when lead-acid batteries were used. It can be divided into two parts: the degree of usefulness in the charging and discharging the batteries.

Batteries with a charger efficiency of 85 % conditioned that 15 % of the total power dissipated in heat, all for process for charging batteries or refill the tank "of electricity." Charging process is followed by the inevitable losses, so that for certain conditions and the charge current was 82 %. This creates a loss of primary energy by 15,3 %. This implies that already in the charging of batteries about 30 % of the total electrical energy is converted into losses.

The process of discharging the battery is quite complex. How discharge current overcome five-hour discharge current and they belong to one-hour mode current to or even lower, there is a significant drop in efficiency. For example. one-hour discharge mode, discharge current is about 3,7 times higher than the five-hour, and a level of efficiency is 0,65. In discharge mode for 0,5 h, discharge current is about 5,5 times higher and the efficiency is only 0,45. In the tested vehicle we had a 45-minute discharge mode in which the utilization rate of 0,56, so that the primary energy from the power grid consumes an additional 30.7%. Practically, this much power is necessary to drive electric cars and overcoming all resistance to traction.

Assembly drive motor and voltage regulator exceeds the value of the degree of utilization of 94 % with the direction of growth, regardless of whether the DC or AC powered. For these components not more than 7 % is lost of electricity drawn from the power grid. Transmission along with the transmission gear has high efficiency of about 96 %, so that the components of the electric drive consumes only 1,5 % of primary energy.

Taking into account all the losses in transport of the electricity from the power grid to power the drive wheels of the vehicle may be test requirements for electric vehicles Yugo – E, in [51] obtain overall efficiency:

$$\eta = \eta_{pa} \bullet \eta_{a1} \bullet \eta_{a2} \bullet \eta_r \bullet \eta_{em} \bullet \eta_t = 0,85 \bullet 0,82 \bullet 0,65 \bullet 0,94 \bullet 0,96 = 0,41 \quad (1)$$

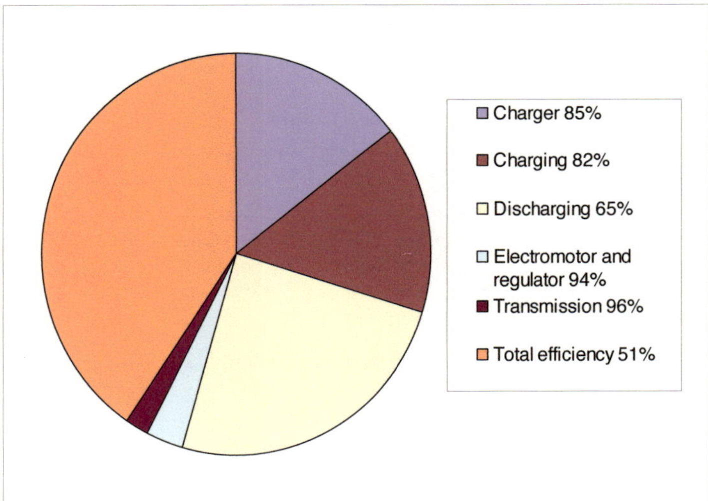

Figure 18. Diagram of losses and efficiency of electric vehicles.

The efficiency of primary energy is much better than machines with conventional drive. Useful power is consumed in four parts and to overcoming of resistance: frictional, wind (aerodynamic), climb and acceleration. Computer data indicates that at a constant speed on flat road of 60 km/h, about 60 % of output used for overcoming the friction force, and about 40 % to overcoming aerodynamic drag.

In order to analyze the total energy efficiency level of the energy source to the wheels of the vehicle, it is necessary to bear in mind the following:

- The efficiency of exploitation from the mine of natural fuels (fossil fuel or nuclear energy),
- Electricity production and
- The network transport.

Efficiency of electricity production can vary widely. According to European measurements, ranges from 39 % for plants with coal production to 44 % for power plants with natural gas, or the average value of 42 %. Combined cycle power plant with natural gas can reach the level of efficiency over 58 %. If we multiplied the average value of 42 % by the transfer efficiency of 92 %, the sources of efficiency of the reservoir of 38 % is obtained. Battery charger recharges the battery, and transmission losses in the electric motor give the utility of the reservoir of energy to the wheels of 65-80 %. Thus the total utility from the source to the wheels is from 25 to 30 %.

Exploitation of natural fuel and transport network are dependent of the type of energy but have an average efficiency of about 92 %. Together with the losses in transport and processing of getting the total level of efficiency from source to reservoir of about 83 %. But the internal combustion engine is only 15-20 % of energy into useful work. Thus the total utility of the source to the wheels is 12 to 17 %.

Energy efficiency is extremely important information on the consumption of electricity from power grid to travel kilometer of the road. It is obtained as the ratio of distance traveled per unit of electricity consumed. Measurements have been made in Serbia, in [51, 53]. driving a constant speed along a straight road in the hilly city driving. The results showed that the energy efficiency of a flat open road is about 5,1 km/kWh, while in the hilly city driving about 4,5 miles/kWh. The specific energy consumed, defined as the ratio of electrical energy from the power grid per unit distance traveled, or as the reciprocal of the energy economy, is on a flat open road below 0,2 km/kWh in the hilly city driving around 0,22 km/kWh.

	ICE	EV
From source to reservoir	83 %	38 %
From the reservoir of energy to the wheels	15–20 %	65-80 %
Total: From the source to the wheels	12–17 %	25-30 %

Table 5. The current level of utility vehicles with ICE and the EV, in [52].

6. Problems and Prospects "energy reservoir"

Development and implementation of future EV largely depend on the technical characteristics of the components of the drive. It is difficult to change established habits of drivers in the world, with the expectation from a motor vehicle to transport them quickly from one location to another. The main disadvantage of EV is in the battery pack and that they still can not accumulate more than 200 Wh/kg energy. If compared to liquid fuels about 12.000 Wh/kg, this very fact means that the tank cars with conventional internal combustion engine, which weighs about 40 kg can store approximately 480 kWh of energy in modern Li ion battery heavy around 300 kg only about 60 kWh electricity.

Promising system Li-air batteries with 1.700 Wh/kg will be able to fully provide the comparative characteristics of the EV and to thereby make the transition to a completely pure EV.

It is interesting to note that the investigation of an aluminum-air battery has started several decades ago because of the high energy potential, because of the opportunities for quick replacement of worn out mechanical anode and the economy, in [54]. It was worked on the development of aluminum-air battery with the anode of aluminum which is alloyed with small amounts of alloying components and a neutral aqueous solution of sodium chloride NaCl as the electrolyte in [55]. The prototype battery achieved 34/39 W/kg specific power and specific energy of 170-190 Wh/kg, the optimal current density between 50 and 100 mA/cm^2, which at the present level of development of chemical power sources is a battery of exceptional quality. The lack of battery life is relatively high cost of components which are used for alloying aluminum anode.

The energy density of gasoline is 13.000 Wh/kg, which is shown as "a theoretical energy density" (Figure 19). The average utilization rate of passenger cars with IC engine, from the fuel tank to the wheels, is about 13 % in US, so that "useful energy density" of gasoline for vehicles use is around 1.700 Wh/kg. It is shown as "practical" energy density of gasoline. The efficiencie of autonomous electric propulsion system (battery-wheels) is about 85 %. Significantly improvement of current Li-ion energy density of batteries is about 10 times, which today is between 100 and 200 Wh/kg (at the cellular level), could make that electric propulsion system be equated with a gasoline powered, at least, to specific useful energy. However, there is no expectation that the existing batteries, as Li-ion, have ever come close to the target of 1,700 Wh/kg.

Oxidation of 1 kg of lithium metal, releases about 11.680 Wh/kg, which is slightly lower than gasoline. This is shown as a theoretical energy density of lithium-air batteries. However, it is expected that the real energy density of Li-ion batteries will be much smaller.

The existing metal-air batteries, such as Zn-air, usually have a practical energy density of about 40-50 % of its theoretical energy density. However, it is safe to assume, that even fully developed Li-air cells will not achieve such a great relationship, because lithium is very lightweight, and therefore, the mass of the battery casing and electrolytes will have a much bigger impact.

http://dx.doi.org/10.5772/51771

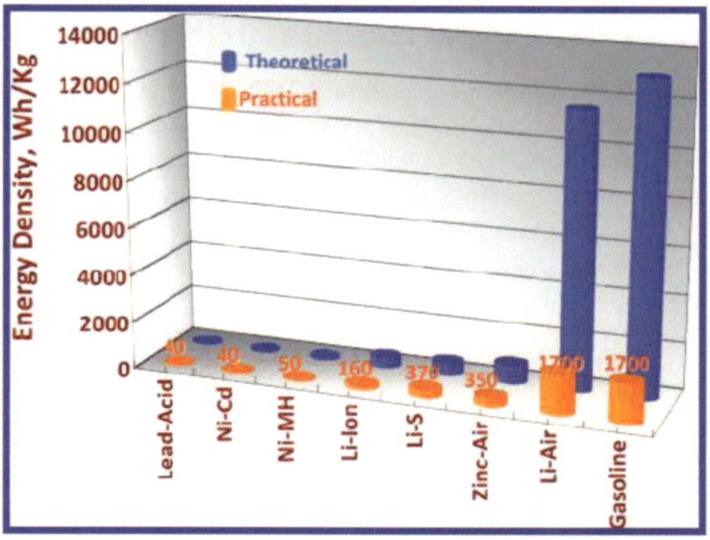

Figure 19. Energy density of different types of batteries and gasoline in [56].

Fortunately, the energy density of 1700 Wh/kg for a fully charged battery pack fits only 14.5 % of the theoretical energy content of lithium metal. It is realistic to expect, achieve mint of such energy density, at the cellular level, considering the intense and long team's development in [54]. Energy density of complete batteries is only a half of density, realizedat the cellular level.

It is interesting to mention, that the significant results in development this type of battery are achieved in the laboratories of the Institute of Electrochemistry ICTM and the Institute of Technical Sciences SASA, where they were working on development of aluminum-air battery with the aluminium anode alloyed with small amounts of alloying components and the neutral aqueous solution NaCl, as the electrolyte in [57]. The prototype of such batteries, had achieved a power density of 34/39 W/kg, and energy density of 170-190 Wh/kg, by optimal current density between 50 and 100 mA/cm^2.

Volumetric energy (in Wh/l) in the storage batteries is an important feature of the design considerations also. This requirement is the best expressed by condition that there is a maximum capacity of 300 dm^3 (family car) for battery pack and auxiliary systems. A driving range of 500 miles (800 km) requires that the reservoir of energy, store energy of 125 kWh (with power consumption of 250 Wh/km), so that the volume of 300 dm^3 is limiting specific gravity of the battery pack, including space for air circulation, must not be less than 0,5 kg/dm^3.

Power density: While Li-air systems imply an extremely high energy density, their power density (measured in W/kg of batteries weight) is relatively low. The prototype of Li-air cells achieves current density, in average 1mA/cm^2, which is insufficient and is expecting significantly increase of the current density for at least 10 times. One way to achieve the required power density is the creation of a hybrid electric drive system, where a small, high power battery, for example, on the basis of Li-ion technology, would provide the power in short periods of high demand, such as it is acceleration. Supercapacitors could be used instead of these batteries.

Duration: The current Li-air cells show a possibility of full charge cycles, only about 50, with less capacity loss. Future research efforts must be directed towards improving the accumulated capacity in multiple discharges. In addition, the total number of charge cycles and discharge do not mean to be very large, due to the high energy capacity of Li-ion cells. For example, a battery, designed for duration of 250.000 km, and projected to cross the EV radius of movement of 800 km, should be charged only 300 times (Full cycle equivalent) in [58]. It is necessary to keep in mind that a lot of air will go through the battery during operation, and even a short-term accumulation of moisture, can be harmful to duration.

Safety: EV batteries will be, especially in the beginning of the application, complying with extremely high safety standards, even more strictly than at gasoline car.

Price: Design requirements of high-capacity battery for the drive EV are quite strict, but they are quite well defined. They will serve as guidelines for the scientific research, conducted on the Li-air battery system. Batteries for EV power have been just carrying out the transition from nickel metal hydride to Li-ion batteries, after years of researching and developing. Transition to the Li-ion batteries should be viewed in terms of a similar development cycle. It is known that, the price of each product, decreases with increasing mass production. It is expecting that the EV prices will decline, because of falling down prices of Li-air batteries, including the price of EV. However, support to introduction of new vehicles in traffic would be systematically addressed.

Battery types	Energy density Wh/kg/ Wh/litar	Specific power W/kg	Number of rechar. cycles	Energy efficiency	Self disch. for 24 hours	Duration years	Price US$/kWh
PbO	40/60-75	180	500	82 %	1 %	2,5-4	100-150
NiCd	50/50-150	150	1.350	72,5 %	5 %		
NiMH	70/140-300	250-1000	1.350	70,0 %	2 %	5-7	300-500
Li-ion	125/270	1800	1.000	90,0 %	1 %	5-10	"/"/1000
Li-ion polymer	200/300	"/3000	-	-	-		
NaNiCl (Zebra)	125/300	-	1.000	92,5 %	0 %		

Table 6. Characteristics of different types of batteries.

Accommodation of batteries as a power source, for vehicles with electric drive, is a big problem also depending on technological solution of batteries. As it can be seen, in table 6 in [59], lead-acid batteries have a low energy, per unit mass and volume and a relatively small number of charge cycles. In contrast, modern Li-ion batteries and NaNiCl, have significant energy capacity, with a larger number of charges and are of a stable voltage. However, the latter ones are sensitive to warming and may have an energy loss up to 7,2 %.

Battery duration should be, always, taken into account, when their price is consideration. The duration depends on several factors, such as how often the vehicle is in use and how many times the batteries have been filled up. In table 6, there are data on duration expectancy of certain batteries types and price per unit of energy.

7. Conclusion

It can be concluded that the future and the past belong to the EV. Nevertheless, new sources of liquid fuels are still to be found, their exploitation is more expensive and there is less of it in the world. In addition, it is necessary to preserve oil as a resource to other industry where you can not find an alternative. On the other hand, electricity is usually sufficient. If in the meantime renewable energy booms, the possibility of its cheap production will open. This means that, in addition to the environmental, economic and conditions for wider use of electric vehicles will gain.

Almost all the problems related to the production of EV technology are sufficiently well resolved, with high efficiency. The biggest problem is the electrical energy storage. Fuel cells, electrochemical sources, supercapacitors, or new sources that could be made sufficiently compact and inexpensive, would allow in the near future, the transition from vehicles that use liquid fuels to electric vehicles.

It is likely that the transition from internal combustion vehicles to EV won't be quick. Still these ones are inferior and can not meet potential customers in all circumstances. Battery development has made great progress but still not enough. In addition, if the battery problem will be solved, there are still many problems that need to be better addressed. Some of these problems will resolve themselves, as prices fall with the increased production, but others, supporting the introduction of new vehicle traffic will be much harder to resolve spontaneously.

So far EV's are more expensive than existing and have certain restrictions of applications you still can not replace the existing vehicles of most vehicle owners in the world. In order to create habits of the driver for the purchase and use of EV, economically strong countries are introducing incentive funds for the EV and HV, which gives definite results. First, there are certain financial incentives for the purchase of the vehicle. In addition, the purchase of EV are not paying taxes, in the cities parking is free for them, vehicles do not pay a toll and in the cities they can move in traffic bands reserved for public transport vehicles. The most important thing is to develop a refilling station for batteries which often offer free recharge EV.EV should not be that expensive investment, especially in large-scale production. So far, the most expensive and also less than perfect for use in EV its battery. Therefore, the most intensive scientific research carried out exactly in this area.

In a situation of permanent oil price increases and increased air pollution, especially in cities, two solutions to the problem occurred.

In accordance with the statements of U.S. President, U.S. moved in the direction of energy efficiency and savings in transportation of petroleum products. This means that it is headed in the direction of HV use with the aim to reduce consumption of the average U.S. vehicle to 6,62 l/100km. Although the U.S. made the extremely popular EV Tesla Roadstar, more U.S. government supports all major car manufacturers to start producing HV.

At the same time as the major importers of oil turned to the study and making Plug in EV or pure EV. First who did it is Germany ahead of the EU, but also China and other countries.

8. Nomenclature

AC-Alternating current

BP-British Petroleum

DC-Direct current

EIA-Energy Information Administration (EIA)

EU-European Union

EV-Electric Vehicle

HV-Hybrid vehicle

Li-air-Lithium- air

Li-ion-Lithium- ion

OIC-AInternational Organization of Motor Vehicle

OPEC-Organization of Petroleum Exporting States

PHV-Plug-in Hybrid

IC-Internal combustion engine

UN-United Nations

ZEV-Zero Emissions Vehicle

Acknowledgements

This work was financially supported by the Ministry of Education and Science Republic of Serbia through projects TR 35041, TR 35042 and TR 35036

Author details

Zoran Nikolić[1*] and Zlatomir Živanović[2]

*Address all correspondence to: zoran.nikolic@itn.sanu.ac.rs

1 Institute of Technical Sciences of the SASA, Belgrade, Serbia

2 University of Belgrade, Institute of Nuclear Sciences VINCA, Belgrade, Serbia

References

[1] UN Department of Economic and Social Affairs. (2004). World Population to 2300. New York, 240, *ST/ESA/SER.A/236/*, http://www.un.org/esa/population/publications/, accessed 1 July.

[2] The Flow of History. (2012). *Railroads and Their Impact (c.1825-1900)*, http://www.flowofhistory.com/, accessed 1 July.

[3] Fredzon, I. R. (1958). Sudovbie elektromehanizmbi. *Gosudarsvenoe soioznoe izdetelbstvosudostroitelbnoi promibišlennosti*, Leningrad, 5.

[4] Tesla, N. (1888). Electro magnetic motor. *U.S. Patent 381968*, May 1.

[5] Kordesch, K. (1978). The electric automobile. *Union Carbide Corporation Battery ProductDivision*, Ohio.

[6] Loeb, D. (1987). *EV Focus*, 1(13).

[7] Larminie, L., & John, Lowry. (2003). *Electric Vehicle Technology Explained*, John Wiley & Sons Ltd., UK, 2.

[8] Latham, H, & Nicholas, M. Choppers set for take-off. *Electric & Hybrid Vehicle Technology*.

[9] United Nations. (2001). Department of Economic and Social Affairs ST/ESA/SER.A/202, *Population Division, Population, Environment and Development, The Concise Report*, New York, 44-45, http://www.un.org/spanish/esa/population/, accessed 1 July 2012.

[10] James, L. Williams. (1996). *Oil Price History and Analysis*, RG Economics, Copyright ©, 2012, London.

[11] Pothier, D. (1978). Electric cars: The future is now. *The Philadelphia Inquirer, Tuesday*, 3, 4.

[12] Nikolić, Z., Dakić, P., Marjanović, S., & Pavlović, S. (1997). *Some features of the reconstructed electric vehicles Yugo-E for Electric distribution company Elektrodistribucija*, 25, Beograd, 3, 266-280.

[13] Nikolić, Z., Marjanović, S., & Dakić, P. (1997). Some results of electric vehicle YUGO-E testing. Beograd. *Proceedings of the conference NMV 97*, 53-56.

[14] Gutzwiller, F. W., et al. (1960). *GE Silicon Controlled Rectifier Manual*, Liverpool, New York, General Electric Co.

[15] Nikolić, Z. (1981). Some experiences with electromobile. Kragujevac. *Proceedings of the conference "Science and motor vehicles 81"*, A01-1 up to A01-14.

[16] Nikolić, Z. (2010). *Electric vehicle in world and in our country*, Institute Goša, Belgrade, 312.

[17] Grafham, D. R., & Hey, J. C. (1972). *SCR Manual* (Fifth Edition), General Electric Company, USA.

[18] Urban Population. (2011). *Development and the Environment*, United Nations, Bernan Assoc, 978-9-21151-488-9, http://www.unpopulation.org.

[19] Murphy, J.M.D., & Turnbull, F.G. *DC-AC Invertor Circuits, Power Electronic Conrol of AC motors*, Pergamon Press, 101-184.

[20] Rashid, M. H. (1988). *Power Electronics Circuits, Devices and Applications*, Prentice-Hall international Inc.

[21] Guzinski, J., & Abu-Rub, H. (2010). Sensorless induction motor drive for electric vehicle application. *International Journal of Engineering, Science and Technology*, 2(10), 20-34, http://www.ijest-ng.com, accessed 1 July 2012.

[22] UN Department of Economic and Social Affairs,. (2012). *Population Facts*, http://www.un.org/en/events/populationday/, accessed 1 July.

[23] Tesla, N. (1904). Electric Autos, Special Correspondence. accessed 1 July 12. *Manufacturers' Record*, December 29, http://www.tfcbooks.com/tesla/.

[24] Van Amburg, B. (1998). Transforming transportation. *Electric & Hybrid Vehicle Technology*, 98, 196-199.

[25] Magda, M. (2007). Tesla roadster. *Electric & Hybrid Vehicle Technology International*, annual (04-08).

[26] Motavalli, J. (2009). An Electric Car With Chinese Roots. *The New York Time*, June 9.

[27] Coda performance. (2012). http://www.codaautomotive.com/, accessed 1 July.

[28] Garden, R. (2009). *Remarks by the President of national fuel efficiency standards*, http://www.whitehouse.gov/the_press_office/Remarks-by-the-President-on-national-fuel-efficiency-standards/, May 19, accessed 1 July 2012.

[29] Kampman, B., Leguijt, C., Bennink, D., Wielders, L., Rijkee, X., De Buck, A., & Braat, W. (, January). Green Power for Electric Cars. *Development of policy recommendations to harvest the potential of electric vehicles*, Delft, CE Delft, 20.

[30] Bagot, N. (1997). It,s good, but is it enough? *Electric & Hybrid Vehicle Technology*, 97, 22-23.

[31] Marintek Neste generasjon innenriksferjer- optimalt fremdriftssystem. (2001). *Marintek,report MT23 A01-008*.

[32] Caterpillar. (1997). Engine performance. *2526B Marine propulsion*.

[33] Akihira, W. (1997). The Automobile of Tomorrow: Toyota,s Approach. *A Toyota quaterly rewiew* [100].

[34] Zivanovic, Z, Diligenski, Dj, & Sakota, Z. (2009). The Application of Hybrid Drive Technology in City Buses. Belgrade. *XXII JUMV International Automotive Conference Science & Motor Vehicles*, Proceedings on CD, 1-15.

[35] Nikolic, Z., Filipovic, Z., & Janjusevic, Lj. (2011). State of development of the electric and hybrid vehicles, energetic and ecological aspect of applications. *Industry*, 34(4), 267-292.

[36] United States Council For Automotive Research Llc. (2012). www.uscar.org/freedom-car, accessed 1 July.

[37] Annual Energy Outlook 2011 with projections to 2035. (2011). *EIA, U.S. Energy Information Administration, DOE/EIA-0383*, accessed 1 July 2012, http://www.eia.gov/forecasts/aeo/.

[38] Despić, , et al. (1973). *Prospects and problems of development of electric vehicle and traffic*, Institute of the technical sciences of the SASA, Belgrade.

[39] Abe, S. (, December). Development of Toyota Plug-in Hybrid Vehicle. *Journal of Asian Electric Vehicles*, 8(2), 1399-1404.

[40] Anderson, J. (2011). Kiplinger. *Best Used Cars*, May 13, http://autos.yahoo.com/news/best-used-cars-2011.htmlaccessed, 1 July 2012.

[41] International Organization of Motor Vehicle Manufacturers. (2012). *via NationMaster*, http://www.nationmaster.com/graph/ind_car_pro-industry-car-production, accessed 1 July.

[42] International Energy Outlook. (2010). *EIA, U.S. Energy Information Administration, DOE/EIA-0484*, www.eia.gov/oiaf/ieo/, accessed 1 July 2012.

[43] World energy Outlook. (2010). International Energy Agency. *Key Graphs, OECD/IEA*, http://www.iea.org/statist/index.htm, accessed 1 July 2012.

[44] BP Energy Outlook 2030. (, January). London, accessed 1 July 2012.

[45] Medium-term markets. (2011). *Overview*, International Energy Agency, OECD/IEA, http://www.iea.org, accessed 1 July2012.

[46] Pucar, M., & Josimović, B. (2002). The influence of transport on energy and environmental crisis and some possible solutions to these problems. Žabljak. *Proceedings of the meeting: Road and Environment*, 27-34.

[47] Talberth, J., & Posner, S. (2010). *Ecosystem Services and the Gulf Disaster*, World Resources Institute, July 7, http://www.wri.org/stories/2010/07/ecosystem-services-and-gulf-disaster, accessed 1 July 2012.

[48] Metz, B., Davidson, O., Bosch, P., Dave, R., & Meyer, L. (2007). *Climate Change Mitigation*, Cambridge University Press, New York, USA.

[49] Garden, R. (2009). *Remarks by the President of national fuel efficiency standards*, http://www.whitehouse.gov/the_press_office/Remarks-by-the-President-on-national-fuel-efficiency-standards/, May 19, accessed 1 July 2012.

[50] United nations Economic commission for Europe. (2012). http://www.unece.org, accessed 1 July.

[51] Kampman, B., Leguijt, C., Bennink, D., Wielders, L., Rijkee, X., De Buck, A., & Braat, W. (, January). Green Power for Electric Cars. *Development of policy recommendations to harvest the potential of electric vehicles*, Delft, CE Delft, 20.

[52] Nikolić, Z. (1996). *Report on the electric vehicle*, Institute of the technical sciences of the SASA, Belgrade, 87-90.

[53] Nikolić, Z., & Bilen, B. (1997). Electric vehicles in traffic conditions. Beograd. *Proceedings of the meeting: "Rational energy consumption in a wide"*, 617-624.

[54] Despić, A. R., Dražić, D. M., Zečević, S. K., & Grozdić, T. D. (1976). Problems in the use of high-energy-density aluminium-air batteries for traction. *Power sources*, 6, 361-368.

[55] Despić, A. R., & Milanović, P. D. (1979). *Aluminium-air battery for electric vehicles, Recueil des travaux de L*, Institut des Sciences techniques de L,Academie Serbe des Sciences et arts, 12(1), 1-18.

[56] Latham, H., & Nicholas, M. (1999). Choppers set for take-off. *Electric & Hybrid Vehicle Technology In4ernational*, 99(110), 112.

[57] Advanced hybrid architecture. (2009). *Electric & Hybrid Vehicle Technology International*, 99-100.

[58] Metz, B., Davidson, O., Bosch, P., Dave, R., & Meyer, L. (2007). *Climate Change Mitigation*, Cambridge University Press, New York, USA.

[59] *Next generation lithium ion batteries for electrical vehicles Edited by Chong Rae Park*, Published by In-Teh, Vukovar, Croatia.

Chapter 3

Electric Vehicles – Consumers and Suppliers of the Electric Utility Systems

Cristina Camus and Tiago Farias

Additional information is available at the end of the chapter

http://dx.doi.org/10.5772/51911

1. Introduction

Electric vehicles (EVs) have been gaining attention in the last few years due to growing public concerns about urban air pollution and other environmental and resource problems. The technological evolution of the EVs of different types: Hybrid electric vehicles (HEV), battery electric vehicles (BEV) and plug-in hybrid electric vehicles (PHEV), will probably lead to a progressive penetration of EV´s in the transportation sector taking the place of vehicles with internal combustion engines (ICEV). The interesting feature of EVs (only available for BEVs and PHEVs) is the possibility of plugging into a standard electric power outlet so that they can charge batteries with electric energy from the grid.

While a large penetration of plug-in EVs is expected to increase electricity sales, extra generation capacity is not needed if the EVs are recharged at times of low demand, such as overnight hours. EVs, as a local zero emissions' vehicle, could only provide a good opportunity to reduce CO_2 emissions from transport activities if the emissions that might be saved from reducing the consumption of oil wouldn´t be off-set by the additional CO_2 generated by the power sector in providing for the load the EVs represent. Therefore, EVs can only become a viable effective carbon mitigating option if the electricity they use to charge their batteries is generated through low carbon technologies.

In a scenario where a commitment was made to reduce emissions from power generation, the build-up of large amounts of renewable power capacity raises important issues related to the power system operation (Skea, J, et al., 2008), (Halamay et al., 2011), as a result, power system operators need to take measures to balance an increasingly volatile power generation with the demand, and to keep the system reliability. To perform these actions, the SO (system operator) needs to access active and reactive power reserves which are either contractually established with the power generators or traded in the ancillary system market

(Estanqueiro, A. et al., 2010). These requirements represent an extra cost for the system which might adequately quantify the negative effect of the variability and uncertainty of each renewable generation technology.

Practically speaking, there are additional external costs of integrating renewable inflexible generation in the power systems, namely in terms of backup capacity, needed to balance power generation and demand when the renewable generation is lower than forecasted, and some kind of storage or demand shift, needed to integrate excesses of renewable generation, especially likely to occur in the off-peak periods.

In this context, electric vehicles can bring techno-economical advantages for the electric power system because of their great load flexibility and increase the system storage capacity. In fact, EVs are parked 93% of their lifetime, making it easy for them to charge either at home, at work, or at parking facilities, hence implying that the time of day in which they charge, can easily vary and, furthermore, for future energy systems, with a high electrification of transportation, Vehicle to Grid (V2G), where the EV works also as an energy supplier, can offer a potential storage capacity and use stored energy in batteries to support the grid in periods of shortage (Kempton and Tomic, 2005). Although each vehicle is small in its impact on the power system, a large number of vehicles can be significant either as an additional charge or a source of distributed generating capacity.

While the aggregate demand for electricity is increasing, decentralized power generation is gaining significance in liberalized electricity markets, and small size electricity consumers are becoming also potential producers. Prosumer is a portmanteau derived by combining the word producer, or provider, with the word consumer. It refers to the evolution of the small size passive consumer towards a more active role in electricity generation and the provision of grid services.

This chapter is concerned with studying how the electric vehicle can work as a "prosumer" (producer and consumer) of electricity. The benefits to the electric utilities and the costs of services provided by EVs in each type of power market will be addressed. The potential impacts of the EVs on the electricity systems, with a great amount of renewable sources in the generation mix will be studied with a focus on the additional power demand and power supply an EV can represent, the role of a new agent on the power market – The EV aggregator – and the economic impacts of EVs on electric utilities.

The analysis of the impact on the electric utilities of large-scale adoption of plug-in electric vehicles as prosumers will be illustrated with a real case study.

Many studies regarding battery electric vehicles and Plug in hybrids have been, and continue to be performed in different countries. In the US, for instance, the capacity of the electric power infrastructure in different regions was studied for the supply of the additional load due to PHEV penetration (Kintner-Meyer et al., 2007) and the economic assessment of the impacts of PHEV adoption on vehicles owners and on electric utilities (Scott et al., 2007). Other studies (Hadley, 2006) considered the scenario of one million PHEVs added to a US sub-region and analyzed the potential changes in demand, impacts on generation adequacy, transmission and distribution and later the same analysis was extended to 13 US regions with the inclusion of GHG estimation for each of the seven scenarios performed for each region (Hadley, 2008). The ability to schedule both charging and very limited discharging of PHEVs could significantly increase power system utilization. The evaluation of the effects of optimal PHEV charging, under the assumption that utilities will indirectly or directly control when charging takes place, providing consumers with the absolute lowest cost of driving energy by using low-cost off-peak electricity, was also studied (Denholm and Short, 2006). This study was based on existing electricity demand and driving patterns, six geographic regions in the United States were evaluated and found that when PHEVs derive 40% of their miles from electricity, no new electric generation capacity was required under optimal dispatch rules for a 50% PHEV penetration. A similar study was made also by NREL (National Renewable Energy Laboratory) but here the analysis focused only one specific region and four scenarios for charging were evaluated in terms of grid impact and also in terms of GHG emissions (Parks et al., 2007). The results showed that off-peak charging would be more efficient in terms of grid stress and energy costs and a significant reduction on CO_2 emissions was expected thought an increase in SO_2 emissions was also expected due to the off peak charging being composed of a large amount of coal generation. Studies made for Portugal (Camus et al., 2011) of the impacts in load profiles, spot electricity prices and emissions of a mass penetration of EV showed that reductions in primary energy consumption, fossil fuels use and CO_2 emissions of up to 3%, 14% and 10% could be achieved by year 2020 in a 2 million EVs' scenario, energy prices could range 0.9€ to 3.2€ per 100 km according to the time of charging (peak and off-peak) and the electricity production mix. A recent report (Grunig M. et al., 2011) that analyzed the EV market for the next years concluded that, the market penetration of EVs will remain fairly low compared to conventional vehicles. The estimation based on several government announcements, industry capacities and proliferation projects sees more than five million new Electric Vehicles on the road globally until 2015 (excluding two- and three-wheelers), the majority of these in the European Union. The main markets for Electric Vehicle are in order of importance the EU, the US and Asia (China and Japan). Some further target markets like Israel and the Indian subcontinent are also expected to evolve. In the long term, the share of EVs will most likely increase as additional countries adopt technologies and initiate projects.

The first description of the key concepts of V2G appeared in 1996, in an article (Kempton and Letendre, 1996) written by researchers at the University of Delaware. In this report the approach was to describe the advantages of peak power to be supplied by EDVs connected to the grid. Further work from the same researchers was continued (Kempton and Letendre, 2002) and the possible power services provided for the grid by vehicles were increased by the analysis of spinning reserve and regulation. The formulation of the business models for V2G and the advantages for a grid that supports a lot of intermittent renewable were described specially for the case of wind power shortage (Kempton and Tomic, 2005a; Kempton and Tomic, 2005b). The use of a fleet for providing regulation down and up was studied and how the V2G power could provide a significant revenue stream that would improve the economics of grid-connected electric-drive vehicles and further encourage their adoption were evaluated (Tomic and Kempton, 2007). The potential impact of renewable generation on the ancillary service market, with a focus on the ability of EVs to provide such services via demand response (DR) and V2G were analyzed. The document also presents a revenue model that incorporates potential scenarios regarding EV adoption, electricity prices, and driver behavior. The output of the model determines the overall revenue opportunity for aggregators who plan to provide DR-EV (Leo M. et al., 2011), although, there is a significantly large market for these services, the limited revenue opportunity for aggregators on a per car basis is unlikely to be compelling enough to justify a business model. According to a recent report from Pike Research (Gibson B., Gartner J., 2011), EVs compete with traditional generation sources as well as emerging technologies, such as stationary battery storage, for revenue from ancillary services such as frequency regulation and demand response.

2. Electricity generation

Electricity generation faces nowadays a greater number of challenges related to reliability, sustainability and security of supply. The use of renewable resources in power generation has been adopted in most OECD countries as an answer to the climate change problems originated by the burning of fossil fuels in the traditional thermal plants to supply the ongoing increase in electricity demand.

In this section, a description of the electric power systems demand is done emphasizing its evolution along a day and seasonal profile, the different technologies available for power generation are also presented, their main features and when and how each of them produces and the emissions associated with electricity production from thermal units are also addressed in this section. A description of the renewable sources, identifying the factors that influence the value of each renewable technology for the power system is done. These factors include the variability, uncertainty, complementarities with other sources and with the demand and implications for reserve requirements. The impacts of EVs recharge in the typical load profiles will be assessed and also the effects of EVs working as electricity suppliers.

2.1. Electricity demand and supply

Electric power systems are designed to respond to instantaneous consumer demand. One of the main features of power consumption is the difference in demand along the day hours, the week days and seasons. This evolution along the day, with a valley during the night that represents about 60% of the peak consumption, has great financial consequences with the need of having several power plants that are useless and an underutilized network during the night.

To supply this load, there are a different set of technologies, from renewable sources (hydro, solar, wind, biomass and waves) to conventional thermal units (natural gas, coal, fuel oil and nuclear).

These different technologies, with different load factors (ratio of average load to capacity), supply the system in different periods and power levels. There are mainly two types of power plants in the electric system: base load or peak power plants. Base load plants are used to meet some of a given region's continuous energy demand, and produce energy at a constant rate, usually at a low cost relative to other production facilities available to the system. Peak power plants are used few hours a year only to fulfill the peaks at higher unit energy prices.

The intermittent renewable sources like the hydro run-of-river and wind are not included in this definition as they are not controllable, but have to be included in the power supply with the highest priority according to the energy and climate policies established (EC, 2009) so they can be considered as base load power plants.

Sometimes the renewable production has an average production profile that works in opposition with demand. Fig. 1 shows as an example, the average production profile of wind power in Portugal verified in year 2010.

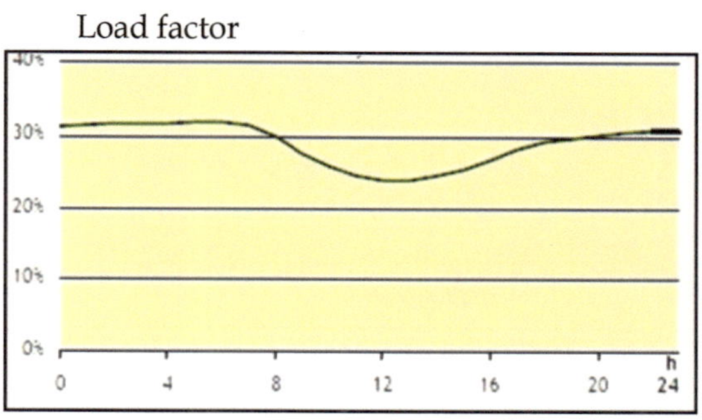

Figure 1. Average wind power profile for year 2010 in Portugal (REN, 2011)

In fact, it has been observed along the years that the wind power production has in average this same profile, with more power production during the night hours.

This situation is even sharper in summer months. In Fig. 2 are the average power produced by wind, solar and small hydro in July 2011 in Portugal.

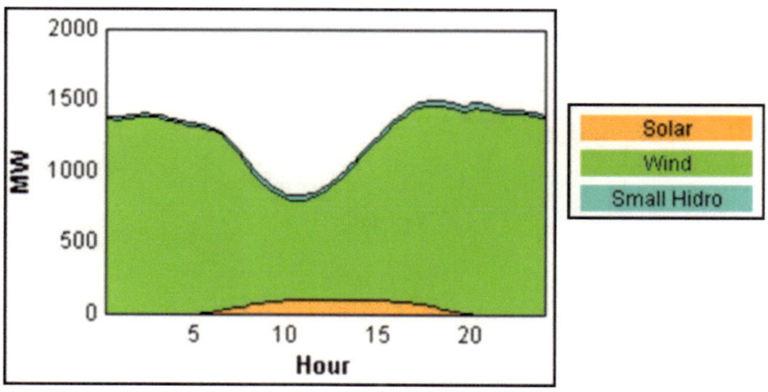

Figure 2. Average power production profile of the renewable sources in July 2011 in Portugal

In summer months, the renewable production is lower when the demand is higher.

For this same case study, Fig. 3 shows the July average power profile with the production technologies. The lowest renewable production level coincides with the peak consumption.

This situation gives the opportunity for electric vehicles contribution for levelling the power consumption diagram and allowing the penetration of more renewable production, by increasing the load during the night hours and supplying the system at the peak hours (Fig. 4).

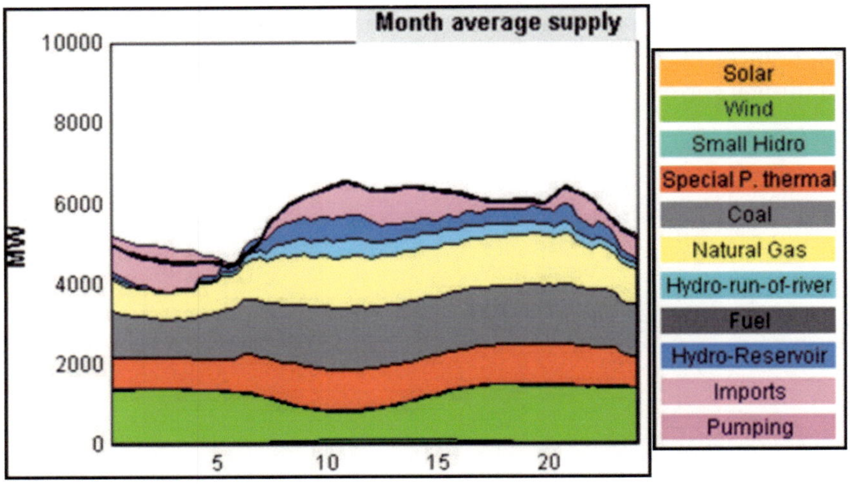

Figure 3. Average load profile with production technologies in July 2011 for Portugal.

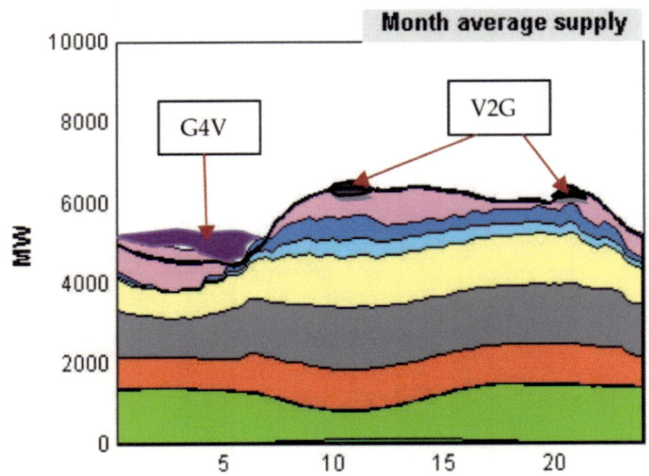

Figure 4. Example of the effect EVs can produce in the electricity demand profile as consumers and suppliers of electricity through G4V (Grid for Vehicle) and V2G respectively.

2.2. The main technologies for electricity generation and the merit order

As described in the previous sub-section, there are many technologies available for electricity production.

The aim of a power plant in a power system is to supply the load in an economical, reliable and environmentally acceptable way. Different power plants can fulfill these requirements in different ways. Different power plants have different characteristics concerning how they can be controlled in the power system. When operating a power system, the total amount of electricity that is provided has to correspond, at each instant, to a varying load from the electricity consumers. To achieve this in a cost-effective way, the power plants are usually scheduled according to marginal operation costs, also known as merit order. Units with low marginal operation costs will operate almost all the time (base load demand), and the power plants with higher marginal operation costs will be scheduled for additional operation during times with higher demand. Wind power plants as well as other variable sources, such as solar and tidal sources, have very low operating costs. They are usually assumed to be zero therefore these power plants are at the top of the merit order. That means that their power is used whenever it is available.

In parallel with marginal operation costs of the power plants are the environmental costs, nowadays assessed by the GHG emissions, they represent. In Table 1, are the average emission rates considered for the typical thermal power plants to compute the GHG emissions from power generation. Those average values can increase if the power plants are subjected to many start-up cycles.

Technology	**Emission rate (kg/MWh)**		
	CO_2	**NO_x**	**SO_2**
Coal	900	2.8	6.3
Fuel	830	3.9	4.5
Nat gas (Comb. Cycle)	360	0.13	0
Cogeneration (N.Gas)	600	0.5	

Table 1. Emission rates considered for the thermal power plants for GHG emission computation (EDP, 2008).

Summarizing, we can dispose of flexible plants, where the power output can be adjusted (within limits), and inflexible plants, where power output cannot be adjusted for technical or commercial reasons. Examples of flexible and inflexible power plants are in Table 2.

As mentioned, the output of the inflexible power plants is treated as given when optimizing the operation of the system.

Not all the flexible power plants can be used the same way to adjust to power demand. The hydro plants with reservoir are the more flexible. Thermal units must be "warmed up" before they can be brought on-line, warming up a unit costs money and start-up cost depends on time unit has been off. There is the need to "balance" start-up costs and running costs. For example a Diesel generator has a low start-up cost but a high running cost, while a Coal plant has a high start-up cost and a low running cost.

Flexible Plants	Inflexible Plants
Coal-fired	Nuclear
Oil-fired	Run-of-the-river hydro
Open cycle gas turbines	Renewable sources (wind, solar,...)
Combined cycle gas turbines	Combined heat and power (CHP, cogeneration)
Hydro plants with storage	

Table 2. Available power plants for electricity generation.

2.3. The renewable sources

The percentage of renewable production depends on the location (the endogenous resources available) and the energy policy of the local economy.

Many sources of renewable energy, including solar, wind, and ocean wave, offer significant advantages such as no fuel costs and no emissions from generation. However, in most cases these renewable power sources are variable and non-dispatchable. The utility grid is already able to accommodate the variability of the load and some additional variability introduced by sources such as wind. However, at high penetration levels, the variability of renewable power sources can severely impact the utility reserve requirements.

For instance, at low penetration levels, the variable output of wind power plants is easily absorbed within the variability of the load. However, as the penetration level increases, the added variability of the wind resource can cause greater ramp-rates, greater inter-hour variability, and greater scheduling error. This ultimately increases the amount of generation the system operators must hold in reserve (i.e., the reserve requirement) to accommodate the unplanned excursions in wind generation.

2.3.1. Wind, solar, and wave generation characteristics

Wind power is now a very mature and established renewable resource throughout the world. However, other renewable power sources such as solar (PV or concentrating/thermal) and ocean wave energy also have significant potential. Each of these renewable power sources can be described by three major characteristics.

1st –Variable. The output power of a large-scale wind, solar, or wave power plant varies over time. The vast majority of the time, the variability from one minute to the next is very small, and even the hourly variation is usually small. However, on occasion the output of a large plant, as high as several hundred MW, may go from full output to low production or vice versa over several hours (Fig. 5);

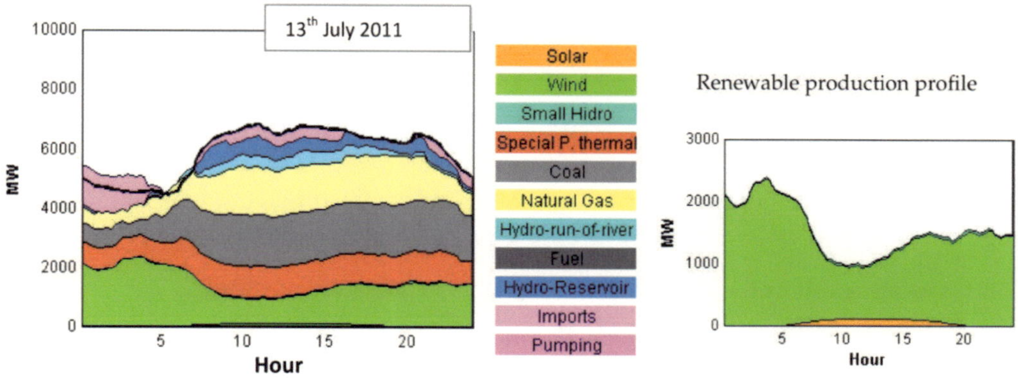

Figure 5. Example of the consumption and production on the 13th July 2011 in Portugal where, in less than 5 hours, a loss of more than 1000MW in renewable production occurred.

2nd –Non-dispatchable. As implemented now, the system operator has very limited control of the output of large scale renewable generation. In general, the operator must deal with whatever the renewable generation outputs are in much the same manner as dealing with the load. Therefore it is common in the analysis of the impact of renewable power generation to subtract its contribution from the load: renewable power generation appears as a negative load;

3rd – Energy source. Due to the non-dispatchable nature of wind, solar, or wave, they generally have a relatively low capacity credit. That is, they do not make a significant contribution to the power requirements of the grid for planning purposes. However, each Joule of energy converted by a renewable source is one Joule saved for "traditional" generation, such as coal. Therefore, renewable energy sources can make a significant impact on the energy requirements of the grid.

2.3.2. Adequacy of renewable production with power demand

The variable, non-dispatchable nature of wind, wave, and solar has a significant impact on the utility reserve requirements. Analyzing the effect of these renewable energy sources on the reserve requirements provides a meaningful and concrete method of characterizing the variability of a given renewable energy source, including its short and long-term correlation with the load.

In order to balance generation with load on a minute-by-minute, hourly, or daily basis, the variability of both the generation and the load must be examined.

With renewable resources like wind, solar, and ocean wave, forecasting of the available generation can present a particular challenge, which, while having a large impact on the hourly or daily reserve requirements, often has less of an impact on the intra-hour requirements. Given the focus on reserve requirements, it readily becomes apparent that a clear understanding of the different types/timescales of reserves is necessary.

Three different timescales are currently used to calculate reserve requirements.

The first, regulation, is defined as the difference between the minute-to-minute power generation/load and the 10-minute average power generation/load. This timescale accounts for small changes in power demand or supply that can be readily met through Automatic Generation Control (AGC) via spinning reserves.

The second timescale of interest, following, is defined as the difference between the 10-minute average power generation/load and the hourly average power generation/load.

This timescale accounts for larger changes in the power demand or supply.

The final timescale, imbalance, is defined as the difference between the hourly average power generation/load and the forecasted generation/load for that hour. The imbalance component of the reserve requirements is directly impacted by the accuracy and frequency of the forecasted generation/load. With the large increase in wind power generation, the imbalance component of the reserve requirement is forecasted to grow rapidly.

In order to calculate imbalance reserve requirements, the scheduled or forecasted power must be determined for both the renewable resource and the load.

2.3.3. Energy storage needs

Reliability is an important feature of power systems. A reliable power system implies that there is always enough generating capacity to satisfy the power demand. In reality this aim can only be achieved to a certain security level. As the installation of power plants is a long process, future power portfolios and their ability to cover the demand must be assessed in advance. The contribution of wind power to the availability of generating capacity becomes important with increasing wind penetration. The capacity value of wind power is therefore identified for future, potentially large wind power penetration levels.

Capacity value designates the contribution of a power plant to the generation adequacy of the power system. It gives the amount of additional load that can be served in the system at the same reliability level due to the addition of the unit. It is a long established value for conventional power plants. Over recent years similar values have been calculated for wind power. A higher correlation between wind and load will lead to higher capacity values. In the case of low correlation between wind and load, there will be need of more storage capacity to respond to renewable and load in-balances.

The additional requirements and costs of balancing the system on the operational time scale (from several minutes to several hours) are primarily due to the fluctuations in power output generated from wind. A part of the fluctuations is predictable for 2 h to 40 h ahead. The variable production pattern of wind power changes the scheduling of the other production plants and the use of the transmission capacity between regions.

This will cause losses or benefits to the system as a result of the incorporation of wind power. Part of the fluctuation, however, is not predicted or is wrongly predicted. This corresponds to the amount that reserves have to take care of.

The economic, social and political costs of failing to provide adequate capacity to meet demand are so high that utilities have traditionally been reluctant to rely on intermittent resources for capacity. Dimensioning the system for system adequacy usually involves estimations of the LOLP (loss of load probability) index. The risk at system level is the probability (LOLP) times the consequences of the event. For an electricity system, the consequences of a blackout are large, thus the risk is considered substantial even if the probability of the incident is small.

The loss of load expectation (LOLE) is a measure of system adequacy and nominates the expectation of a loss of load event. The required reliability of the system is usually in the order of one larger blackout in 10–50 years.

Since no generating plant is completely reliable, there is always a finite risk of not having enough capacity available. Variable sources may be available at the critical moment when demand is high and many other units fail. Fuel source diversity can also reduce risk.

3. Electricity market

In terms of the economic model, the electricity industry has evolved from a vertically integrated state-owned monopoly company (not subjected to the normal rules of competition) to a liberalized market where generators and consumers have the opportunity to freely negotiate the purchase and sale of electricity. In this section the typical electricity markets are described and the more adequate markets for EVs are addressed.

3.1. Electricity market structure

Electric power systems include power plants, consumers of electric energy and transmission and distribution networks connecting the production and consumption sites. This interconnected system experiences a continuous change in demand and the challenge is to maintain at all times a balance between production and consumption of electric energy. In addition, faults and disturbances should be cleared with the minimum effect possible on the delivery of electric energy.

Power systems comprise a wide variety of generating plant types, which have different capital and operating costs. When operating a power system, the total amount of electricity that is provided has to correspond, at each instant, to a varying load from the electricity consumers. To achieve this in a cost-effective way, the power plants are usually scheduled according to marginal operation costs, also known as merit order. Units with low marginal operation costs will operate almost all the time (base load demand), and the power plants with higher marginal operation costs will be scheduled for additional operation during times with higher demand. Wind power plants as well as other variable sources, such as solar and tidal sources, have very low operating costs. They are usually assumed to be zero therefore these power plants are at the top of the merit order. That means that their power is used whenever it is available.

The electricity markets operate in a similar way, at least in theory. The price the producers bid to the market is slightly higher than their marginal cost, because it is cost-effective for the producers to operate as long as they get a price higher than their marginal costs. Once the market is cleared, the power plants that operate at the lowest bids come first.

If the electricity system fails the consequences are far-reaching and costly. Therefore, power system reliability has to be kept at a very high level. Security of supply has to be maintained both short-term and long-term. This means maintaining both flexibility and reserves that are necessary to keep the system operating under a range of conditions, also in peak load situations. These conditions include power plant outages as well as predictable or uncertain variations in demand and in primary generation resources, including intermittent renewable sources.

3.2. Base load power

Base-load power is the "bulk" power generation that is running most of the time. Base-load power is typically sold via long term contracts for steady production at a relatively low price and can better be provided by large power plants because they last longer and cost less per kWh.

3.3. Peak power

Peak power is used during times of predictable highest demand. Peak power is typically generated by power plants that can be switched on for shorter periods, such as gas turbines and hydro plants with reservoir. Since peak power is typically needed only a few hundred hours per year, it is economically sensible to draw on generators that are low in capital cost, even if each kWh generated is more expensive.

3.4. Spinning reserve

Spinning reserves are supplied by generators set-up and ready to respond quickly in case of failures (whether equipment failure or failure of a power supplier to meet contract requirements). They would typically be called, say, 20 times per year; a typical duration is 10 min but must be able to last up to 1 h (spinning reserves are the fastest-response and highest-value component of the more general electric market for "operating reserves"). Operation reserves include several types of reserves in place to respond to short-term unscheduled demand fluctuations, or generator/other system failure. Operating reserve represents generators that can be started or ramped up quickly. There are several categories of operating reserves, often referred to as ancillary services.

Quick-start capacity includes combustion turbines and hydroelectricity, while spinning capacity represents other partly loaded fossil and/or hydroelectric plants. The introduction of wind power into a grid can increase these operation-reserve requirements, due to the variability in wind generation.

3.5. Balancing

Balancing or regulation is used to keep the frequency and voltage steady, they are called for only one up to a few minutes at a time, but might be called 400 times per day; Spinning re-

http://dx.doi.org/10.5772/51911

serves and balancing are paid in part for just being available, a capacity payment per hour available; Base-load and peak are paid only per kWh generated.

The variability in wind generation precludes wind from contributing fully to the reserve margins required by utilities to ensure continuous system reliability.

Planning reserves ensure adequate capacity during all hours of the year. Typical systems require a "peak reserve margin" of 10%-18%. This means a utility must have in place 10%-18% more capacity than their projected peak power demand for the year. This ensures reliability against generator or transmission failure, underestimates of peak demand, or extreme weather events.

Due to the resource variability of wind generation, only a small fraction of a wind farm's nameplate capacity is usually counted toward the planning reserve margin requirement. In fact, as wind penetrates further into an electric grid, this "capacity credit" for wind generally declines, especially if the wind farms are developed near each other, i.e. if their output is well correlated.

3.6. The aggregator

To access to the electricity market means, among other aspects, to have access to the so called "market prices". Under this concept, an EV does not have individually the capacity to access to the electricity market, as each quantity of energy produced is insignificant when compared with the regular power players'.

There arises then a new element for the interconnection between the micro-generation and the electricity market, that it can be called by "commercial agent" or "aggregator". The commercial agent or aggregator adds a set of small power producers so that they can became, in a certain way, a fair concurrent in the market by the fact of dealing with a substantial quantity of energy. Under the point of view of the aggregator, there is also the possibility of dealing either with energy generation and/or energy consumption to maximize the economic value of the EV to the consumer and at the same time revenue to the aggregator, it is almost certain that the charging and discharging vehicle will be done in order to allow the vehicle to be charged with the lowest-cost electricity, and also allows the vehicle to provide high-value ancillary services. EVs could be connected to the power system through the aggregator that sells the aggregated demand of many individual vehicles to a utility, regional system operator, or a regional wholesale electricity market. The idea is that EVs respond intelligently to real-time price signals or some other price schedule to buy or sell electricity at the appropriate time so that the vehicles would be effectively "dispatched" to provide the most economical charging and discharging.

4. Electric vehicle as a consumer and supplier of electricity

Given the nature and physical characteristics of EVs, their integration into the grid is performed at the distribution voltage level. Such an interconnection allows each EV to be plugged into the grid to get the energy to charge up the battery. The EVs, when aggregated in sizeable numbers, constitute a new load that the electricity system must supply. However, an EV can be much more than just a simple load given that bi-directional power transfers are possible once the interconnection is implemented. Indeed, the integration allows the deployment of EVs as a generation resource as well as a storage device for certain periods of time when such deployment aids the system operator to maintain reliable operations in a more economic manner. We refer to the aggregated EVs as a generation/storage device in this case. The entire concept of using the EVs as a distributed resource – load and generation/storage device–by their integration into the grid is known as the vehicle-to-grid (V2G). Under this concept, the EVs become active players in grid operations and play an important role in improving the reliability, economics and environmental attributes of system operations. Such benefits include the provision of capacity and energy-based ancillary services, the reduction of the need for peakers and load levelization.

4.1. Electric vehicle modeling

Electric vehicles constitute a variety of vehicle types with different battery capacities, vehicle ranges, and vehicle drive trains. Such differences are important to the electric industry because of their influence on daily vehicle electricity consumption. The common characteristic of EVs and PHEVs is that they require a battery, which is the source of all or part of the energy required for propulsion. For EVs, the original energy consumption unit in kWh and the energy consumption per unit distance in kWh/km is generally used to evaluate the vehicle energy consumption. The battery energy capacity is usually measured in kWh and the driving range per battery charge can be easily calculated.

A typical electric vehicle (EV) traction battery system consists of a chain of batteries connected in a series, forming a battery pack with nominal voltages ranging from 72 to 324 V and capable of discharge/charge rates of several hundred amperes.

As vehicles, EVs are not always stationary and, therefore, may be dispersed over a region at any point in time. In a moving state, EVs may be used for commuting purposes or, possibly for longer trips – if the battery capacity is large or if the EV is a PHEV.

For the EVs used for commuting, we can view, therefore, that the vehicles are idle an average of 22 h a day. We note that as the commuting distance is smaller than the potential range of the EVs, not all the energy in the batteries is consumed by the commute. We may see each EV as a potential source of both energy and available capacity that can be harnessed by the grid in addition to supplying the load of the EV to charge up the battery.

In addition to the storage capacity, there are some other aspects of interest in characterizing the batteries. A critically important one is the state of charge (s.o.c.) of the batteries. It is defined as the ratio of the energy stored in a battery to the capacity of the battery. It varies from 0 when the battery is fully discharged to 1– often expressed in percentages as a variation from 0% to 100% – when the battery is fully charged and provides a measure of how much energy is stored in the battery. The s.o.c. typically decreases when energy is withdrawn from the battery and increases when energy is absorbed by the battery. Thus, for a day during which the EV owner goes to work in the morning, parks the EV, goes back home in the late afternoon and then plugs the EV for charging during the night, the s.o.c. will evolve along a pattern illustrated in Fig. 6.

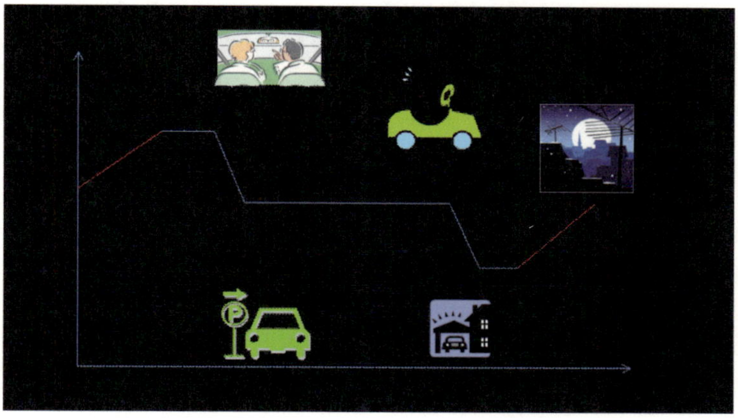

Figure 6. S.o.c. evolution for an EV along a typical working day with only home charging

Batteries release energy more easily when their s.o.c. is high or more exactly above a tolerance level. We stipulate 60% to be the tolerance level in the examples of this work. When the s.o.c. is lower than 60%, a more appropriate utilization of this battery is for energy absorption. If the battery releases energy, then the EV acts as a supply-side resource. If it absorbs energy, the EV acts as a demand-side resource. We can view the battery store present supply-and demand-side resources as a function of the s.o.c. The diagram in Fig. 7 summarizes this information.

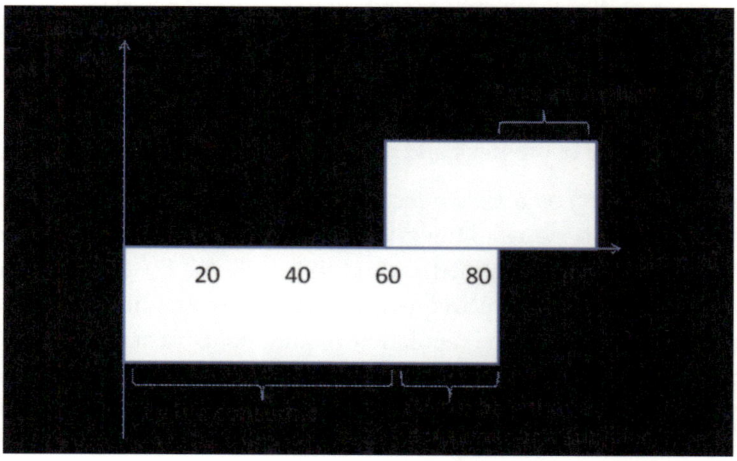

Figure 7. Relation between the s.o.c. and the function of the EV

The frequent switching of the s.o.c. may cause a decrease in battery storage capability which is defined as the battery degradation.

4.2. EVs aggregation

The battery storage of an individual EV is too small to impact the grid in any meaningful manner. An effective approach to deal with the negligibly small impact of a single EV is to group together a large number of EVs – from thousands to hundreds of thousands. The aggregation, then, can impact the grid both as a load and a generation/storage device.

The basic idea behind such aggregation is the consolidation of the EVs, so that together they represent a load or a resource of a size appropriate to exploit economic efficiencies in electricity markets. The Aggregator is a new player whose role is to collect the EVs by attracting and retaining them so as to result in a MW capacity that can impact beneficially the grid. The size of the aggregation is indeed the key to ensuring its effective role. In terms of load, an aggregation of EVs represents the total capacity of the batteries, an amount in MWs that constitutes a significant size and allows each EV to benefit from the buying power of a large industrial/commercial customer. There are additional economic benefits that accrue as a result of the economies of scale. The aggregated collection behaves as a single decision maker that can undertake transactions with considerably lower transaction costs than would be incurred by the individual EV owners. So, the aggregated entity can make purchases – be it electricity, batteries or other services – more economically than the individual EV owners can and can pass on the savings to each EV owner. As a resource, the aggregated EVs constitute a significant capacity that may beneficially impact the operations of a system operator. The SO deals directly with the Aggregator, who sells the aggregated capacity and energy services that the collection of EVs can provide. The Aggregator's role is to effectively collect the distributed resources into a single entity that can act either as a generation/storage device capable of supplying capacity and energy services needed by the grid or as a controllable load to be connected to the ESP to be charged in a way so as to be the most beneficial to the grid. It is the role of the Aggregator to determine which EVs to select to join the aggregation and to determine the optimal deployment of the aggregation. A single aggregation may function either as a controllable load or as a resource, as depicted in Fig. 8.

The charging of the EVs introduces a new load into the system. For every SO, the load has a typical daily shape formed of on- peak and off-peak periods as described in section 2.

The EV aggregation can act as a very effective resource by helping the operator to supply both capacity and energy services to the grid. To allow the operator to ensure that the supply– demand equilibrium is maintained around the clock, the EV aggregation may be used for frequency regulation to control frequency fluctuations that are caused by supply–demand imbalances. The shape of the regulation requirements varies markedly from the on-peak to the off-peak periods. We define regulation down as the absorption of power and regulation up as the provision of power. A battery may provide regulation up or regulation down service as a function of its s.o.c. Depending on its value for each EV in the aggregation, the collection maybe deployed for either regulation up or regulation down at a point in time. Resources that provide regulation services are paid for the capacity they offer.

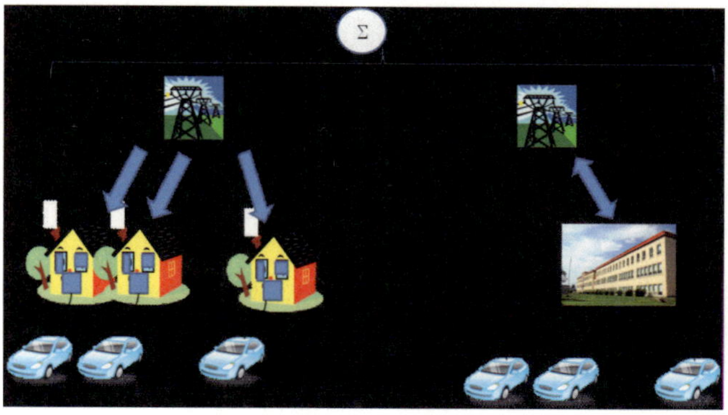

Figure 8. EVs working as load and as supplier of electricity

4.3. The adequate electricity markets for EVs

EVs, with their fast response and low capital costs, appear to be a better match for the quick-response, short-duration, electric services, such as spinning reserves and balancing. The equivalent of those markets in the Portuguese Electric sector, are secondary and tertiary regulation (REN, 2012).

Spinning reserves are paid for by the amount of time they are available and ready even though no energy was actually produced. If the spinning reserve is called, the generator is paid an additional amount for the energy that is actually delivered (e.g., based on the market-clearing price of electricity at that time). The capacity of power available for 1 h has the unit MW-h (meaning 1MW of capacity is available for 1 h) and should not be confused with MWh, an energy unit that means 1MW is flowing for 1 h. These contract arrangements are favorable for EVs, since they are paid as "spinning" for many hours, just for being plugged in, while they incur relatively short periods of generating power.

Regulation or balancing, also referred to as automatic generation control (AGC) or frequency control, is used to fine-tune the frequency and voltage of the grid by matching generation to load demand. Some markets split regulation into two elements: one for the ability to increase power generation from a baseline level, and the other to decrease from a baseline. These are commonly referred to as "regulation up" and "regulation down", respectively. Compared to spinning reserves, it is called far more often, requires faster response, and is required to continue running for shorter durations.

4.4. Estimation of costs and revenues for vehicles owners

4.4.1. Estimation of revenues for vehicles owners

Calculating revenue for vehicle owners depend on the market that V2G power is sold into. Equation 1 can be used for markets that pay for available capacity and for energy (Kempton and Tomic, 2005a).

$$r = p_{cap} P t_{plug} + p_{el} R_{d-c} P t_{plug} \tag{1}$$

Where r is the total revenue [€], p_{cap} is the market price for capacity [€/kW-h], P is the contracted capacity available less or equal to P_{V2G} [kW], t_{plug} is the time the EV is plugged in and available [h], p_{el} is the price of electricity for the plugged in hours [cents/kWh], $R_{d\text{-}c}$ is the dispatch to contract ratio given by $E_{disp}/(P.t_{plug})$.

Capacity payments are an important part of revenue and compensation for energy delivered generally nets out taking into account the energy that must be purchased to charge the vehicle and the cost of batteries depreciation. Furthermore, to compute energy payments, a profile of grid services provided by the vehicle must be defined.

In Portugal, the average capacity prices for regulation between 2007 and 2011 and for the first months of 2012 were shown in Table 3.

Year	Capacity Price	Power range	Regulation [€/MW]	
	[€/MW-h]	[MW]	up	down
2007	18.5	188	45.7	32.7
2008	21.4	158	63.5	42.2
2009	28.9	197	48.5	21.5
2010	27.2	290	53.8	13.5
2011	27.8	286	73.7	12.6
2012	36.3	291	64,9	24,4

Table 3. Average prices and capacity for regulation services in Portugal (REN, 2012)

In Fig. 9 is depicted the annual average regulation band evolution and the weighed unit capacity price. The average power range for regulation has increased in the last 5 years, representing from 2.8% of average power in 2007 till 5% in 2011.

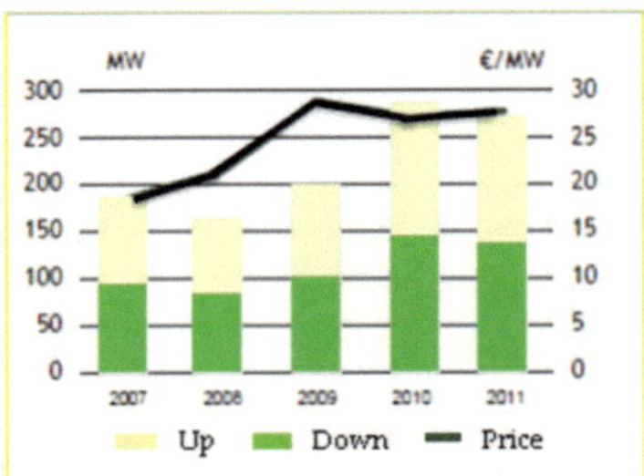

Figure 9. Evolution of regulation band and unit capacity price (REN, 2012)

http://dx.doi.org/10.5772/51911

Looking at the percentage of wind power production in the same 5 years, it increased from 9.3% in 2007 to 18% in 2011. It can be assumed that the increase of intermittent power sources like wind, in the electricity generation mix, leads to an increase of need of power band reserves to assure the same level of system reliability.

In Fig. 10 it is depicted the evolution of capacity installed and energy production in Portugal among the different technologies.

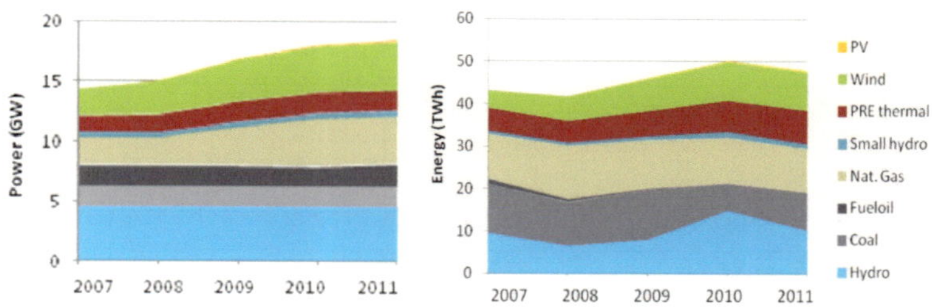

Figure 10. Evolution of capacity installed in the different technologies (left-side) and of annual production from the different technologies (right-side) (REN, 2012)

The increase needs for ancillary services (spinning reserves and regulation) had been fulfilled by the dispatchable technologies in the proportion described in Table 4.

Years	2007	2008	2009	2010	2011
Hydro	67%	39%	18%	28%	27%
Coal	10%	3%	16%	12%	14%
Nat.gas	23%	58%	66%	60%	59%

Table 4. Evolution of the contracted power band among the dispatchable available technologies in Portugal (REN, 2012)

From 2007 till 2011 the power band has increased in 100 MW. To fulfil this 100MW needs, about 30000 EVs at a 3.5 kW each should be plugged. If only 20% of total EVs were available to supply this service, 140000 EVs should be necessary (3% of the total actual light duty fleet). For instance considering the average prices occurred in the first 2012 months and depicted in Table 5.

Capacity	Power	Regulation [€/MW]			
Price	range	Valley		off-valley	
[€/MW-h]	[MW]	up	down	up	down
36,3	291	64,9	24,4	61,3	35

Table 5. Average prices and capacity for regulation services in Portugal in 2012 (REN, 2012)

An EV can expect to achieve a daily revenue of 2.3 € for providing ancillary services to the power grid (Table 6).

	Capacity	Energy	
		valley	Peak
p_{cap} [cents/kW-h]	3.6		
P [kW]	3.5		
t_{plug} [h]	16	9	7
E_{disp} [kWh/day]		3.0	3.0
P_{elup} [cents/kWh]		6.5	6.1
P_{eldown}[cents/kWh]		2.4	3.5
R_{c-d}		0.10	0.12
Revenue [€/day]	2.02	0.13	0.14

Table 6. Expected daily revenues for an EV that provides ancillary services in Portugal

4.4.2. Estimation of costs for vehicles owners

The cost of V2G is calculated from purchased energy, wear and capital cost. The energy and wear for V2G are those incurred above energy and wear for the primary function of the vehicle, transportation. Similarly, the capital cost is that of additional equipment needed for V2G, but not for driving. The general formula for annual cost is (equation 2):

$$c = c_{en} E_{need} + c_{ac} \tag{2}$$

c is the total cost per year [€], c_{en} the cost per energy unit produced for V2G [€/kWh], E_{need} is the electric energy needed to be dispatched in the year [kWh] considering the conversion's efficiencies (equation 3).

$$E_{need} = E_{disp} / \eta_{conv} \tag{3}$$

c_{ac} is the annualized capital cost for additional equipment needed for V2G including also the cost of equipment degradation (wear) due to extra use for V2G (equation 4).

$$c_{ca} = c_d + c_c \frac{d}{1-(1+d)^{-n}} \tag{4}$$

c_d represents the annual costs of battery degradation, c_c, the capital cost of extra equipment, d the discount rate and n the investment's life time.

The costs for battery degradation depend on the cycling regimes. As V2G extra cycling would increase battery replacement and additional cost for that should be taken into account. For example considering that a lithium-ion battery could have a 3000 cycle life time (Tomic and Kempton, 2007) at a 100% of discharge and could last almost 10 years with less than a daily charge, an extra shallow, 4% cycling for regulation services occurring in average 10 times in a day it would shorten batteries life in 40% so that after 6 years they should have to be replaced. To compare investments with different life times we use the annuity method (equation 5).

$$c_d = c_{bat}\left(\frac{d}{1-(1+d)^{-nl}} - \frac{d}{1-(1+d)^{-n2}}\right) \tag{5}$$

c_{bat} is the cost of battery and n_1 and n_2 are the expected life times without and with V2G.

The estimated costs for EVs' owners for providing ancillary services are depicted in Table 7.

	Costs
Bat Cap [kWh]	16
E_{disp} [kWh]	900
η_{conv}	0,8
E_{need} [kWh]	1125
c_{en} [cents/kWh]	7
c_{bat} [€/kWh]	700
c_d [€/yr]	754
c_c [€]	500
d [%]	8%
n_1 [yr]	10
n_2 [yr]	6
c_{ca} [€/yr]	828
c[€/yr]	907

Table 7. Expected annual costs for an EV that provides ancillary services in Portugal

4.4.3. Estimation of financial results for vehicles owners

In this way, estimates for annual profits for EVs' owners, as a result of capacity payments providing regulation capacity could be computed considering the values in table 8.

	Revenue	Costs
Capacity [€/yr]	605	
Reg. Up [€/yr]	43	
Reg. down [€/yr]	40	
sav. in recharge [€/yr]	63	
c [€/yr]		907

Table 8. Estimation of costs and revenues for V2G providing regulation services in Portugal

As V2G is connected at low voltage this regulation service should be purchased by a distribution company that could act as an aggregator to provide enough regulation power to sell in the power markets subjected to the prices shown in Table 5.

We consider that the EVs provide regulation during valley and off-valley hours. During valley hours they are mainly used for charging for further use (for driving and for grid support) but also could provide regulation up and down during this time. For providing regulation up and down we considered the vehicles are plugged-in daily during at least 7 off-valley hours and 9 valley hours. If the vehicles offer this service for 300 days per year a total of 605€ could be earned only for providing. If an average energy of 3.0 kWh is supplied daily to the grid, an annual revenue of 43.2€ could be expected for regulation up and a total revenue of 40€ for regulation down plus 63€ in savings for recharging (due to energy input). Unfortunately, under the described assumptions, total annual costs exceed total revenues in 156€. This loss is very sensitive to battery degradation, if we consider n_2=7 years instead of n_2=6 years, total annual costs decrease to 635€ and a result of 116€ could be obtained (318€ with n_2=8, table 9).

	n_2=6	n_2=7	n_2=8
Total revenue [€/yr]	751	751	751
Total costs [€/yr]	907	635	433
Result [€/yr]	-156	116	318

Table 9. Estimation of costs and revenues for V2G for different assumptions of battery life

It should not be forgotten that, it is the aggregator that trades directly to the grid for offering regulation services with V2G and works with the market prices showed in table 5 and so a percentage of the earnings should go to the aggregator. Considering a 4% revenue for the aggregator services, the EV's owner profits could range from -156€ to 305€.

5. Conclusion

Electric vehicles (EVs) and plug-in hybrid electric vehicles (PHEVs), which obtain their fuel from the grid by charging a battery, are set to be introduced into the mass market and expected to contribute to oil consumption reduction. PHEVs and EVs can also provide a good opportunity to reduce CO_2 emissions from transport activities if the electricity they use to charge their batteries is generated through low carbon technologies. In addition to the environmental issue, EVs bring techno-economical challenges for utilities as well, because EVs will have great load flexibility as they are parked 93% of their lifetime, making it easy for them to charge either at home, at work, or at parking facilities, hence implying that the time of day in which they charge, can easily vary. EV aggregations can act as controllable loads that contribute to level the off-peak load at night or as generation/storage devices that can provide up and down regulation service when the vehicles are parked.

This chapter described how the electric vehicle can work as a "prosumer" of electricity. The benefits to the electric utilities and the costs of services provided by EVs in each type of power market were addressed, the role of a new agent on the power market – The EV aggregator – and the economic advantages for EVs owners considered the Portuguese energy market as a case study.

There are still many doubts about the life time of EV batteries and battery degradation when proving V2G. Global costs are very sensitive to battery costs and degradation assumptions so that profits can range from -155€/yr to 305€/yr considering respectively 40% to 20% in batteries life range reduction due to V2G supply.

The pressure to generate electricity from endogenous low carbon resources in the majority of the countries makes naturally transport electrification a solution to lower emissions and fossil fuels use from the transportation sector. On the other hand, the increasing of intermittent renewable sources in the power systems, forces the increase of the regulation power band in order to assure the same level of reliability to the power system which would increase the power installed and fixed costs to the power system.

EVs can be a benefit to the environment by reducing emissions and noise in the cities while, at the same time, by providing ancillary services to the power grid, reduce the investments and operation costs in thermal generation and allows the integration of more renewable production. To provide a 100 MW of band power a total of 30000 EVs at a 3.5 kW each should be plugged-in. If only 20% of total EVs are available to supply this service, 140000 EVs should be necessary which corresponds of 3% of the total actual light duty fleet in the Portuguese case study.

Acknowledgment

The authors would like to acknowledge FCT- Fundação para a Ciência e Tecnologia through the national project Power demand estimation and power system impacts resulting of fleet penetration of electric/plug-in vehicles (MIT-Pt/SES-GI/0008/2008).

The authors also thank the REN – Portuguese Energy Networks for supplying up-to-date and valuable data concerning electric power consumption and generation.

Author details

Cristina Camus and Tiago Farias

Polytechnical Institute of Lisbon - Instituto Superior de Engenharia de Lisboa, Technical University of Lisbon - Instituto Superior Técnico, Portugal

References

[1] Camus C., Farias T., Esteves J., Potential impacts assessment of plug-in electric vehicles on the portuguese energy market, Elsevier Energy Policy 39, Sep. (2011). 5883-5897.

[2] Denholm, P., & Short, W. (2006). An Evaluation of Utility System Impacts and Benefits of Optimally Dispatched Plug-In Hybrid Electric Vehicles. Technical Report NREL/TP- October 2006.10.2172/888683 , 620-40293.

[3] Duvall, Dr., & Mark, S. (2006). Plug-in hybrid electric Vehicles Technology Challenges", EPRI Sep 2006

[4] EC (2009). Directive 2009/28/EC of the European Parliament and of the Council of 23 April 2009 on the promotion of the use of energy from renewable sources

[5] EDP(2008), Sustainability Report available at http://www.edp.pt/pt/aedp/unidadesdenegocio/producaodeelectricidade/Pages/CompromissoAmbiental.aspx

[6] EPRI (2006). Comparing the Benefits and Impacts of Hybrid Electric Vehicle Options for compact sedan and sport utility vehicles, EPRI,1006892. July 2002

[7] ERSE (2011). http://www.erse.pt/pt/electricidade/tarifaseprecos/Paginas/default.aspx.last accessed 11/01/15)

[8] Estanqueiro, A., Mateus, C., Pestana, , & , R. Operational " Experience of Extreme Wind Penetrations" in. T. Ackerman, 9th International Workshop on Large-Scale Integration of Wind Power into Power Systems and Transmission Networks for Offshore Wind Power Plants. Québec, CA, (2010). Available at: http://repositorio.lneg.pt/bitstream/10400.9/1172/1/Paper_w159_Estanqueiro.pdf. , 371-383.

[9] Gibson, B., & Gartner, J. (2011). Vehicle to Grid Technologies Applications for Demand Response, Vehicle to Building, Frequency Regulation, and Other Ancillary Services: Market Analysis and Forecasts, Pike Research, 2011, 2G

[10] Grünig, M., Witte, M., Marcellino, D., Selig, J., & van Essen, H. Impact of Electric-Deliverable Vehicles. An overview of Electric Vehicles on the market and in development Delft, CE Delft, April (2011).

[11] Hadley, Stanton. W. (2006). Impact of Plug-in Hybrid Vehicles on the Electric Grid. October 200610.2172/974613

[12] Hadley, Stanton. W. (2008). Potential Impacts of Plug-in Hybrid Vehicles on Regional Power Generation, ORNL/TM. January 2008.10.2172/932632

[13] Kempton W., Letendre S., Electric Vehicles as a new power source for Electric Utilities, Elsevier Science , Vol. Dec (1996). (3), 157-175.

[14] Kempton W., Letendre S., The V2G Concept: A New For Model Power? Connecting utility infrastructure and automobiles, Public utilities fortnightly 140(4): 16-26, Feb 2002

[15] Kempton, W., & Tomic, J. (2005a). Vehicle-to-grid power fundamentals: calculating capacity and net revenue. J. Power Sources , 144-2005.

[16] Kempton, W., & Tomic, J. (2005b). Vehicle-to-grid power implementations: from stabilizing the grid to supporting large-scale renewable energy, J. Power Sources , 144-2005.

[17] Kintner-Meyer Michael, Kevin Schneider, Robert Pratt, (2007). Impacts assessment of plug-in hybrid vehicles on electric utilities and regional u.s. power grids", Part 1: technical analysis, Pacific Northwest National Laboratory (a), November, 2007.

[18] Leo M., Kavi K., Anders H., Moss B, Ancillary Service Revenue Opportunities from Electric Vehicles via Demand Response, School of Natural Resources and Environment University of Michigan, April 2011.

[19] Parks, K., Denholm, P., & Markel, T. (2007). Costs and Emissions Associated with Plug-in Hybrid Electric Vehicle Charging in the Xcel Energy Colorado Service Territory. Technical Report NREL/TP- May 200710.2172/903293 , 640-41410.

[20] REN, (2012). REN- Energy Networks of Portugal- Technical reports, available on line at http://www.centrodeinformacao.ren.pt/PT/Paginas/CIHomePage.aspx,last time accessed at 6/03/2012).

[21] Scott, Michael. J., Kintner-Meyer, Michael., Elliott, Douglas. B., & Warwick, William. M. ((2007).), Impacts assessment of plug-in hybrid vehicles on electric utilities and regional u.s. power grids: part 2: economic assessment. Pacific Northwest National Laboratory (a), November, 2007.

[22] Skea, J., Anderson, D., Green, T., Gross, R., Heptonstall, P., & Leach, M. (2008). Intermittent renewable generation and maintaining power system reliability", Generation, Transmission & Distribution, IET January 2008 Page(s):82- 89, 2(1)

[23] Tomic, J., & Kempton, W. Using fleets of electric-drive vehicles for grid support. J. Power Sources ((2007). doi:10.1016/j.jpowsour.2007.03.010

Chapter 4

Energy Efficiency of Electric Vehicles

Zoran Stevic and Ilija Radovanovic

Additional information is available at the end of the chapter

http://dx.doi.org/10.5772/55237

1. Introduction

In this chapter, the most important possibilities for increasing energy efficiency of electric vehicles would be considered, regarding energy savings accumulated in the vehicle itself and increasing the range of performances of the cars with given initial resources. Some of the possibilities that should provide such a progress nowadays are:

- Using energy under braking
- Using waste heat energy
- Additional supply by solar cells
- Improved mechanical energy transmission system
- Improved cars shell design
- Increasing of efficiency of power convertors
- Special design of electric engines
- Using supercapacitors, fuel cells and new generation batteries
- Route selection on the criterion of minimum consumption in real time
- Parameter monitoring inside and outside of the vehicle and computerized system control with optimization of energy consumption

Today, the problem of energy becomes so important that an entire industry is turning towards clean, renewable energy (solar energy, wind energy, etc.). Prototypes of hybrid vehicles with the announcement of mass production scheduled for the near future have become everyday occurrence. In addition, many cars are designed to use only electricity as motive power, which reduces emissions to zero.

Photo cells in a glass roof generate electricity, even at lower intensity of solar radiation; this current operates using a fan in a vehicle. In this way the vehicle interior has a constant supply of fresh air and pleasant temperatures (up to 50% lower), although the motor vehicle is turned off so that fuel economy is evident. The solar roof is only the beginning, while the development of city cars is going toward solar vehicles prototype.

A solar vehicle is an electric vehicle powered completely or significantly by direct solar energy. Usually, photovoltaic (PV) cells contained in solar panels convert the sun's energy directly into electric energy. The term "solar vehicle" usually implies that solar energy is used to power all or part of a vehicle's propilsion. Solar power may be also used to provide power for communications or controls or other auxiliary functions.

Another concept that has been developing over the years is a kinetic energy recovery system, often known simply as KERS. KERS is an automotive system for recovering a moving vehicle's kinetic energy under braking. The recovered energy is stored in a reservoir (for example a flyeheel or a batterry or supercapacitor) for later use under acceleration. Electrical systems use a motor-generator incorporated in the car's transmission which converts mechanical energy into electrical energy and vice versa. Once the energy has been harnessed, it is stored in a battery and released when required. The mechanical KERS system utilizes flywheel technology to recover and store a moving vehicle's kinetic energy which is otherwise wasted when the vehicle is decelerated. Compared to the alternative of electrical-battery systems, the mechanical KERS system provides a significantly more compact, efficient, lighter and environmentally-friendly solution. There is one other option available - hydraulic KERS, where braking energy is used to accumulate hydraulic pressure which is then sent to the wheels when required.

Development of new components, improved connections and electric engine control algorithms allow increase of efficiency of power convertors, therefore electric engine itself, to the maximum theoretical limits. New generation improvements of electric engine system has an impact on price, however investment quickly pays off during operating.

Major efforts are invested in the development of high energy density batteries with minimum ESR. Also, current research show that fuel cells have reached needed performances for commercial use in electric vehicles. Supercapacitors that provide high power density increase the acceleration of vehicle as well as collecting all the energy from instant braking, therefore improvements of the characteristics of power supply are made.

Modern electric vehicles have full information system that has constant modifications and does monitoring of inside and outside parameters in order to achieve maximum energy savings. Except for smart sensors, it is highly important to process GPS signals and route selection on the criterion of minimum energy consumption.

By combining these technologies, concepts and their improvements, we are slowly going towards energy-efficient vehicles which will greatly simplify our lives in the future.

2. Electrical losses reduction in EV

2.1. Energy efficiency of the converters

Increasing of the energy efficiency of the convertors can be achieved by optimizing their configuration and control, as well as choosing the adequate component. Converter configuration depends on the type of the electric motor (DC or AC), possible recovery energy braking, drive dynamics etc.

For DC motor supply there are mostly used chopper voltage reducers, so they will be considered here. Figure 7 shows simplified presentation of the chopper supply of a DC motor. Chopper is shown as ideal breaker controlled by voltage (U_{up}), so it can control switching on (T_{ON}) and switch-off (T_{off}) exiting voltage (U_{do}). For all four quadrant operation transistor bridge as shown in fig. 1 can be used [1].

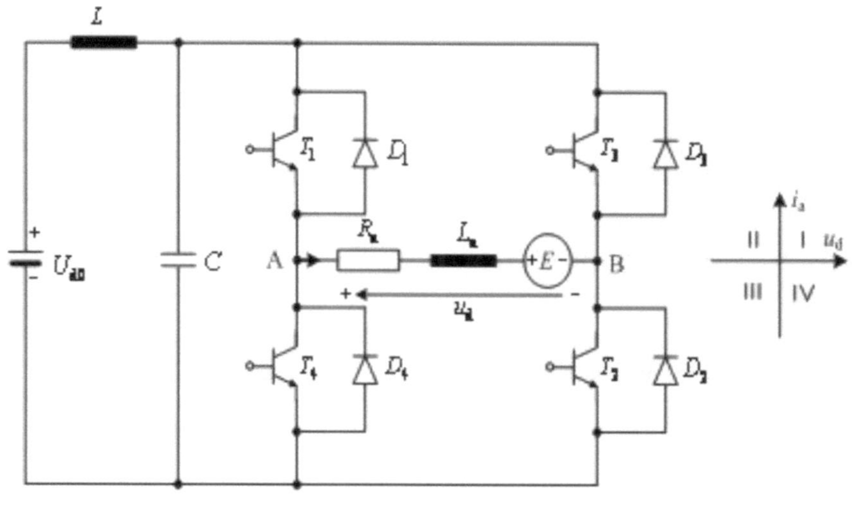

Figure 1. Transistor Bridge

By switching on transistor pairs T_1-T_2 or T_3-T_4 positive or negative polarity of motor voltage u_d is provided. To close motor current at null or reverse polarization, diodes D_1 to D_4 are provided. Converter part of the AC drive of the vehicle consists of the inverters, regulators and control set. The inverter is part of the drive inverter that inverts DC voltage to AC voltage necessary waveform to ensure the required control electric motors. Three-phase inverter consists of three inverter bridges with two switching elements in each bridge, therefore, a total of six switches. By controlling the moments of switching of the particular switches, and by controlling the length of their involvement, the appropriate waveforms at the output of inverter are achieved.

General modern circuit for speed regulation of DC motor is shown in figure 2. Reference rotary speed W_{ref} is set and also maximum armature current I_{amax} and their actual values are monitored and also brought into regulator which outputs present command values for excitation actuators and inductor [1-2].

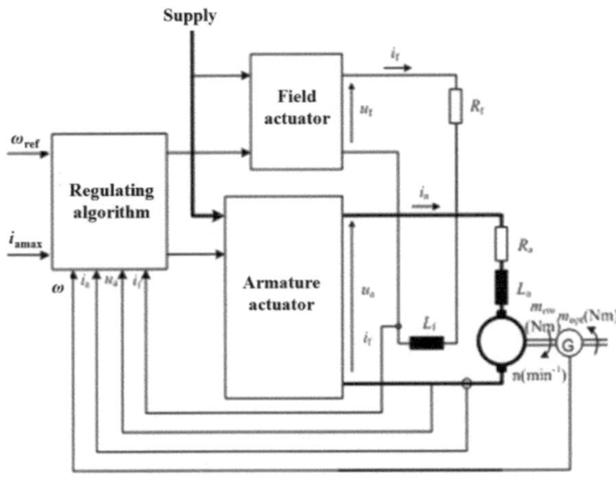

Figure 2. Circuit for speed regulation of DC motor with independent field

Out of base range (for speeds above nominal) method of reduced field is used so among basic values excitation current, i_f, is monitored. Apart from classic PID action, regulating algorithm comprises other tasks (actuator command input adaptation, change of regulating method in accordance with the given speed, alarms etc.). Standard way of regulating DC drives, cascade regulation, consists of two feedbacks: internal – current and external – speed.

Asynchronous motor at constant frequency and amplitude of supply voltage rotor speed depends of load torque, which requires complicated governing algorithms in case when precise speed control and/or position. This phenomenon is a consequence of principle of asynchronous motor, and it is electromagnetic induction, which requires difference in between rotor speed and rotary magnetic field generated by stator to create electromagnetic torque. Electronics that creates algorithms mentioned was expensive earlier and such a use of asynchronous motors was difficult, but today with cheaper electronics components and use of microprocessors for regulating algorithms they are more often used.

Figure 3 represents block-diagram of regulated drive for AC motor. Depending on use and requirements, some of feedbacks and regulators can be left out. Power block (converter + motor) has two input and five output values. Input (command) parameters are effective polyphase supply voltage U_d and frequency Ws. Output (regulated) values are motor current Is, flux w, position O, rotary frequency w and torque me. Each of those has proper regulator in negative feedback, in order as shown in figure 10.

http://dx.doi.org/10.5772/55237

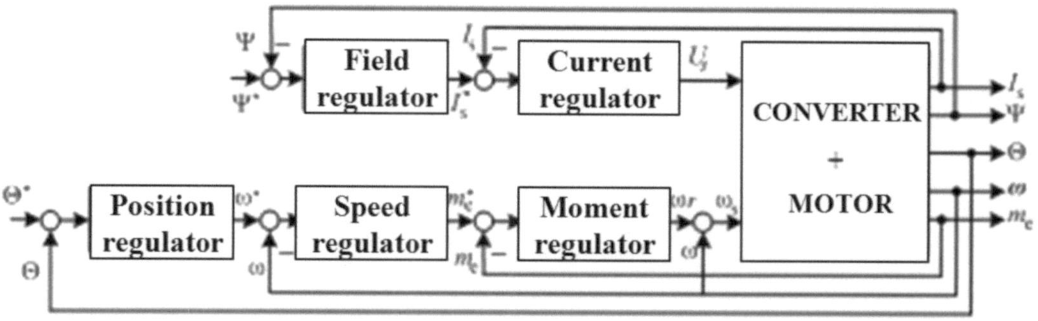

Figure 3. Block diagram of AC motor regulator

Regulation (close-loop control) comprises control with negative feedback, or feedbacks, by means of which, by means of measuring regulated parameters and comparing with required (reference) parameters those values, is acted upon command parameters, so it is automatically achieved ahead defined values of controlled values [1-2].

There may be a large energy saving by selecting the suitable power switching elements, which development is in high prosperity. As switch elements in the inverters and choppers high-power bipolar transistors, MOS (Metal Oxide Semiconductor) transistors or IGBTs (Insulated Gate Bipolar Transistor) are used. High-power bipolar transistors have very low collector-emitter resistance in the conducting state, while their control must provide sufficient supply base, it is required a relatively high power for control. On the other hand, MOS transistors have very high input resistance, and to control them it is just enough to provide the appropriate value of the voltage between the gate and source. Therefore the MOS transistor control current is almost zero and there is no power dissipation in the control circuit. Lack of MOS transistors is relatively high resistance in ON state. IGBT belongs to the family BiMOS transistors and combines these fine qualities of high-power bipolar and MOS transistors [2].

Development of multi-axis distributed control systems where sensors, actuators and controller are distributed across networks. System features system synchronized control and high speed serial communications using fiber optic channels for noise immunity. In addition, communication protocols have been developed that monitor data integrity and can sustain operation in the event of a temporary loss of communication channel. Engineers can design a system to meet exact customer requirements (fig. 4) [2].

In this way, the optimization of the drive by the criteria of the dynamics and energy efficiency, while following the user's request. For supply of certain components, particularly in hybrid vehicles, high power supplies of constant current or current impulses are needed. Precise management and optimization of such sources today is exclusively microprocessor controlled [3].

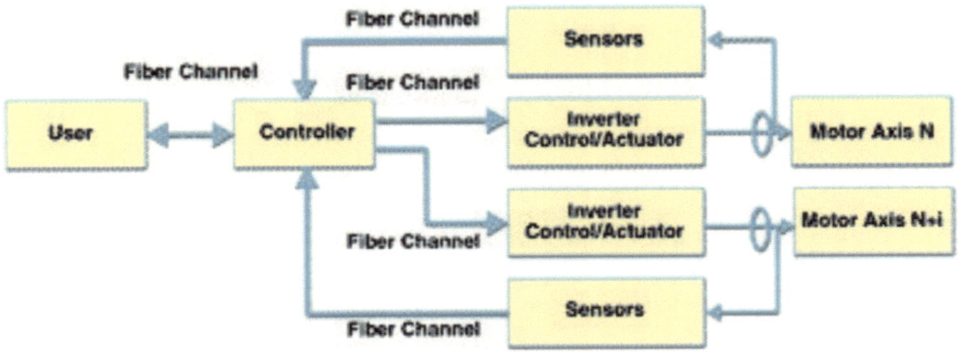

Figure 4. Distributed control

2.2. Energy efficiency of the electric motor

The electric motor is the most important part of the electrical drive and the last link in the chain of energy conversion. DC motors because of their good qualities, control of the rotation speed and control of the torque, for a long time have been irreplaceable part of the controlled electric motor drives. In recent years, thanks to the advanced control techniques, asynchronous motors take place of the DC motors in regulated drives because of its good properties (robustness, lower maintenance requirements and their appliances in explosive environments, which are especially important in the case of hybrid vehicles).

Electric motor drive is designed and optimized starting from the known parameters of the engine. The latest methods for minimizing the power losses in real-time by reducing the level of flux does not require knowing of all engine parameters, and can be applied to asynchronous motor drives with scalar and vector control. Optimization of efficiency of asynchronous motors is based on adaptive adjustment of flux levels in order to determine the optimum operating point by minimizing losses [2].

Losses due to higher harmonics have to be taken into account when determining the degree of efficiency of the entire drive. The voltage at the output of the inverter is considered ideal sinusoidal in the case of control structure developing and the produced effects of higher harmonics are subsequently taken into account.

Any well-designed controller for optimization should meet the following requirements [2]:

- to determine the optimal operating point for each speed and each load torque of the defined areas of work;
- the duration of the optimization process is as short as possible;
- to have a minimum number of sensors required;
- to be easy to use;
- that it can be applied to any standard electric motor drive;

- that it can be applied to any type of engine if the only known data are on motor nameplate;
- to demonstrates a high degree of robustness in the case of disruption load torque;
- Demonstrates a high degree of robustness in the case of motor parameter variations.

Beside the standard electric motors, solutions specially made for EV are developing. Therefore, switched reluctance motor (SRM) is gaining much interest as a candidate for electric vehicle (EV) and hybrid electric vehicle (HEV) electric propulsion for its simple and rugged construction, ability of extremely high-speed operation, and insensitivity to high temperatures. However, because SRM construction with doubly salient poles and its non-linear magnetic characteristics, the problems of acoustic noise and torque ripple are more severe than these of other traditional motors. Power electronic technology has made the SRM an attractive choice for many applications. The SRM is a doubly salient, singly excited synchronous motor. The rotor and stator are comprised of stacked iron laminations with copper windings on the stator, as shown in Fig. 5 [4]. The motor is excited with a power electronic inverter that energizes appropriate phases based on shaft position. The excitation of a phase creates a magnetic field that attracts the nearest rotor pole to the excited stator pole in an attempt to minimize the reluctance path through the rotor. The excitation is performed in a sequence that steps the rotor around.

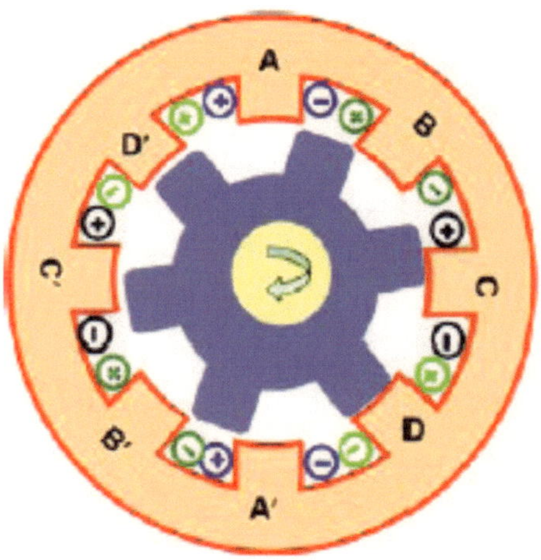

Figure 5. A switched-reluctance motor with 8 stator poles and 6 rotor poles

The SRM is similar in structure to the stepping motor, but it is operated in a manner that allows for smooth rotation. Because there are no permanent magnets or windings on the rotor, all of the torque developed in the SRM is reluctance torque. While the SRM is simple in principle, it is rather difficult to design and develop performance predictions. This is due to the nonlinear magnetic characteristics of the motor under normally saturated operation.

The special design of electric motors used in direct-drive vehicles where the engines are installed in each wheel. This will be discussed more in the mechanical transmission part.

2.3. Supercapacitors vs. accumulator batteries and fuel cells

Supercapacitors are relatively new type of capacitors distinguished by phenomenon of electrochemical double-layer, diffusion and large effective area, which leads to extremely large capacitance per unit of geometrical area (in order of multiple times compared to conventional capacitors). They are taking place in the area in-between lead batteries and conventional capacitors. In terms of specific energy (accumulated energy per mass unity or volume) and in terms of specific power (power per mass unity or volume) they take place in the area that covers the order of several magnitudes. Supercapacitors fulfill a very wide area between accumulator batteries and conventional capacitors taking into account specific energy and specific power [1]. Batteries and fuel cells are typical devices of small specific power, while conventional capacitors can have specific power higher than $1MW/dm^3$, but at a very low specific energy. Electrochemical capacitors improve batteries characteristics considering specific power or improve capacitors characteristics considering specific energy in combination with them. In relation to other capacitor types, supercapacitors offer much higher capacitance and specific energies [5-6].

Accumulator batteries and low temperature fuel cells are typical devices with low specific power, where conventional capacitors may have specific power over $1MW/dm^3$, but at very low specific energy. Electrochemical capacitor can improve characteristics of batteries in terms of specific power and improve properties of capacitors in terms of specific energy when they are combined with them [7].

The principal supercapacitor characteristic that makes it suitable for using in energy storage systems (ESS), is the possibility of fast charge and discharge without lost of efficiency, for thousands of cycles. This is because they store electrical energy directly. Supercapacitors can recharge in a very short time having a great facility to supply high and frequent power demand peaks [8].

2.3.1. Supercapacitor caracterization

Electrochemical investigation methods are widely used for characterization of different kinds of materials, as well as of the processes in systems where the electrochemical reactions take part. There is a series of well known methods, but some new methods from electrotechnical area have been introduced [9-10]. So, first of all it was given an overview of the standard electrochemical methods and parameters, beginning with potential measurement and simple methods such as chronopotentiometry and chronoamperometry, till electrochemical impedance spectroscopy [10]. The last named method is adapted for systems containing large capacitances that became actually with appearance of electrochemical supercapacitors. New methods are Dirac voltage excitation and Dirac current excitation. Measurement system described here allows application of electrochemical methods, as follows: measuring open circuit potential, chronopotentiometry, chronoamperometry, galvanostatic method, potentiostatic method, Dirac voltage excitation, galvanodynamic method, cyclic voltammetry and electrochemical impedance spectroscopy [10-11].

2.3.2. Supercapacitors as a function of increasing energy efficiency of EV

Most strict requirements are related to supercapacitors applying in electric haulage, i.e. for vehicles of the future. Nowadays, batteries of several hundred farad capacitance are with working voltage of several hundred volts have been produced. Beside great capacitance and relatively high working voltage, these capacitors must have great specific energy and power (because of limited space in vehicle). Considering their specific power, they have great advantage in relation to accumulator batteries, but, on the other side, they are incomparably weaker considering specific energy. Hence, ideal combination is parallel connection of accumulator and condenser batteries. In an established regime (normal drawing) vehicle engine is supplied from accu-battery, and in the case of rapidly speeding, from supercapacitor. Very important is the fact that in the case of abrupt breaking, complete mechanical energy could be taken back to system by converting into electrical energy only in presence of supercapacitor with great specific power [1].

In Figure 6 the scheme of an electrical drive vehicle in which supercapacitor is used for energy storage and so-called regenerative breaking is presented.

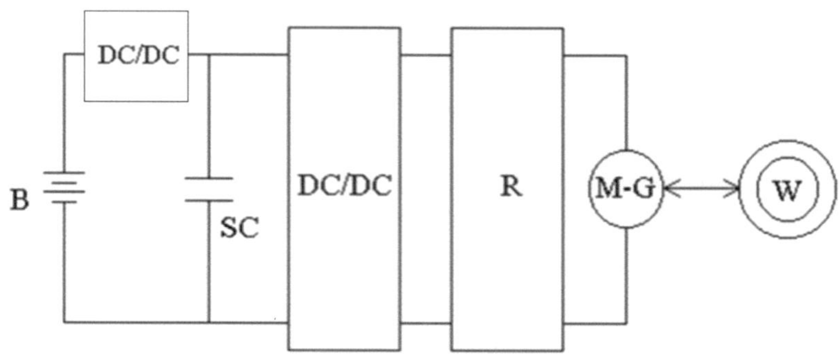

Figure 6. Scheme of electrical drive vehicle with supercapacitor with possibitlity for using breaking energy; B – one-way voltage source, SC – supercapacitor; DC/DC – direct voltage converter; R – regulator; M-G – engine – generator (depending on working regime; W – drive wheels

2.3.3. Supercapacitors in regulated electrical drives

Regulated electrical drives are more than 30% of all electric drives. They are developing quickly and present to constructors stricter and stricter speed regulation (and position) and torque. From energy point of view it is desirable their more participation, since optimal speed setting or required can lead to reduction of used energy [1].

DC source voltage is performed by means of DC-DC converter (chopper). Figure 7 shows principal scheme of such a system.

To provide breaking, or to dissipate braking energy that cannot be returned to the network through diode rectifier, it is required to have braking device with transistor T and resistor R.

Input voltage U_{do} is filtered by simple LC filter and brought to the chopper input that regulates mean value of output voltage U_d.

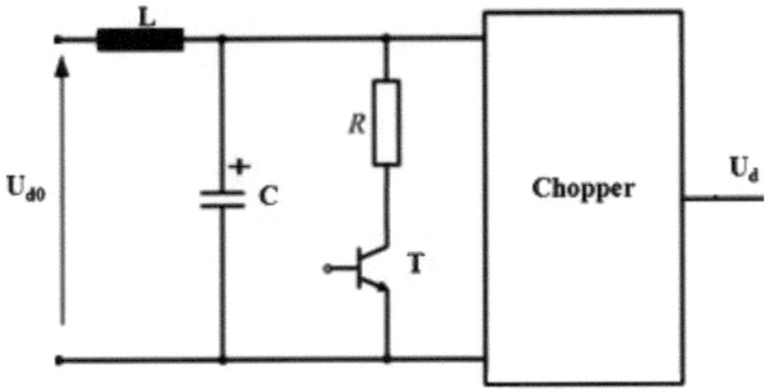

Figure 7. Principal scheme of chopper supply with supercapacitor

2.3.4. Accu batteries

Nowadays, there are a great amount of standard batteries that can be used for EV, however every single type has disadvantages that affect the performance of the vehicle. Therefore, compromises are often made between cost and quality, at the expense of energy efficiency almost all the time. Batteries in combination with supercapacitors are significant improvement and for now this is the system that has the best perspective for future EV.

In the table 1 it is presented the cost per Watt-hour and Specific Energy (Watt-hours per kilogram) for various types of batteries. It is not surprising that the well-known Lead-acid storage batteries head the list. Alkaline cells may be recharged literally dozens of times using the new technology. Recharging alkaline, nickel-cadmium and nickel-metal hydride cells side-by-side in one automatic charger opens up new possibilities for battery selection economy [12].

Battery type	Cost, USD/Wh	Specific Energy, Wh/kg
Lead-acid	0.17	41
Alkaline long-life	0.19	110
NiMH	0.99	95
NiCd	1.50	39
Lithium-ion	0.47	128

Table 1. Batteries cost per Watt-hour and Specific Energy

Costs of lithium-ion batteries are falling rapidly in the race to develop new electric vehicles. The $0.47 price per watt-hour above is for the Nissan Leaf automobile, and they predict a target cost of \$0.37 per watt-hour. Tesla Automobiles uses a smaller battery pack, and they are optimistic about reaching a price of \$0.20 per watt-hour in the near future [12].

There is another type of battery that does not appear in the table above, since it is limited in the relative amount of current it can deliver. However, it has even higher energy storage per kilogram, and its temperature range is extreme, from -55 to +150°C. That type is Lithium Thionyl Chloride. It is used in extremely hazardous or critical applications. The specifications for Lithium Thionyl Chloride are \$1.16 per watt-hour, 700 Watt-hours per kilogram [12].

Several parameters can be considered for selecting the more adequate battery typology: specific energy, specific power, cost, life, reliability, etc. In addition, it is to be considered that batteries for hybrid electric vehicles require higher powers and lower energies than batteries for pure electric vehicles. Among the previously listed typologies, Lead-acid and Nickel-Cadmium andSodium-Nickel Chloride batteries are normally used on board electric vehicles, because of their low specific powers [13].

2.3.5. *Fuel cells*

As far as the fuel cells are concerned, several types are available today, but for vehicle propulsion, Polymer Electrolyte Fuel Cell (PEFC) systems, fed by air and pure hydrogen stored aboard, seem to be highly preferable over other types, mainly because their reduced operating temperature (65-80 degrees depending on the cell design) allow very fast start-up times, and eases the thermal management. A Polymer Electrolyte Fuel Cell is an electrochemical device that converts chemical energy directly into electrical energy, without need of intermediate thermal cycles. It normally consumes H2 and O (typically from Air) as reactants and produce water, electricity and heat. Since cell voltage is so low (less than 1 V), several cells are normally connected in series to form a fuel cell stack with a voltage and power suitable for practical applications.

A fuel cell electric vehicle (FCEV) has higher efficiency and lower emissions compared with the internal combustion engine vehicles. But, the fuel cell has a slow dynamic response. Therefore, a secondary power source is needed during start up and transient conditions. Ultracapacitor can be used as secondary power source. By using ultracapacitor as the secondary power source of the FCEV, the performance and efficiency of the overall system can be improved. In this system, there is a boost converter, which steps up the fuel cell voltage, and a bidirectional DC-DC converter, that couples the ultracapacitor to the DC bus (fig. 8) [13-14].

2.3.6. *New systems*

The priority of the EV future development and its commercial success certainly is optimization of the electric power supply. Besides the usual combinations (batteries and supercapacitors, and supercapacitors), researches are going towards new systems that integrate favorable characteristics of the previously used systems.

Typically, standard ultracapacitors can store only about 5% as much energy as lithium-ion batteries. New hybrid system can store about twice as much as standard ultracapacitors, although this is still much less than standard lithium-ion batteries. However, the advantage of ultracapacitors is that they can capture and release energy in seconds, providing a much faster recharge time compared with lithium-ion batteries. In addition, traditional lithium-ion batteries can be recharged only a few hundred times, which is much less than the 20,000 cycles provided by the hybrid system. In other words, the hybrid lithium-ion ultracapacitors have more power than lithium-ion batteries, but less energy storage. In the future, the hybrid lithium-ion ultracapacitor could also be used for regenerative braking in vehicles, especially if it could be scaled up to provide greater energy storage. Since vehicle braking systems need to be recharged hundreds of thousands of times, the hybrid system's cycle life will also need to be improved [15].

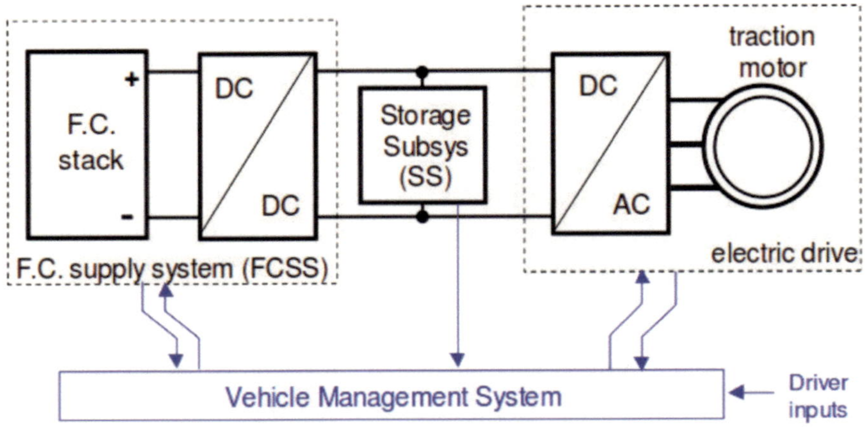

Figure 8. Vehicle with an electrochemical storage system

Using new processes central to nanotechnology, researchers create millions of identical nanostructures with shapes tailored to transport energy as electrons rapidly to and from very large surface areas where they are stored. Materials behave according to physical laws of nature. The Maryland researchers exploit unusual combinations of these behaviors (called self-assembly, self-limiting reaction, and self-alignment) to construct millions -- and ultimately billions -- of tiny, virtually identical nanostructures to receive, store, and deliver electrical energy [16].

2.4. Reduction of losses in the conductors and connectors

From the viewpoint of energy efficiency, choice of supply voltage, as well as quality contacts in the connectors and cable section is very important. The designer is limited by other factors such as the security problem (for battery overvoltage), limited space and cost. Therefore, it is necessary to optimize the supply voltage and the conductor section with given constraints. It is similar to the choice of connectors.

Hybrid and electric vehicles have a high voltage battery pack that consists of individual modules and cells organized in series and parallel. A cell is the smallest, packaged form a battery can take and is generally on the order of one to six volts. A module consists of several cells generally connected in either series or parallel. A battery pack is then assembled by connecting modules together, again either in series or parallel [17]. The pack operates at a nominal 375 volts, stores about 56 kilowatt hours (kWh) of electric energy and delivers up to 200 kilowatts of electric power. These power and energy capabilities of the pack make it essential that safety be considered a primary criterion in the pack's design and architecture [18].

Recent battery fires in electric vehicles have prompted automakers to recommend discharging lithium ion batteries following serious crashes. However, completely discharging a vehicle's battery to ensure safety will permanently damage the battery and render it worthless. Self-discharge effects and the parasitic load of battery management system electronics can also irreversibly drain a battery.

Zero-Volt technology relies on manipulating individual electrode potentials within a lithium ion cell to allow deep discharge without inflicting damage to the cell. Quallion has identified three key potentials affecting the Zero-Volt performance of lithium ion batteries. First, the Zero Crossing Potential (ZCP) is the potential of the negative electrode when the battery voltage is zero. Second, the Substrate Dissolution Potential (SDP) is the potential at which the negative electrode substrate begins to corrode. Finally, the Film Dissolution Potential (FDP) is the potential at which the SEI begins to decompose. The crucial design parameter is to configure the negative electrode potential to reach the ZCP before reaching either the SDP or the FDP at the end of discharge. This design prevents damage to the negative electrode which would result in permanent capacity loss. Figure 9 shows a schematic of the voltage profile during deep discharge of Quallion's Zero-Volt cells [18].

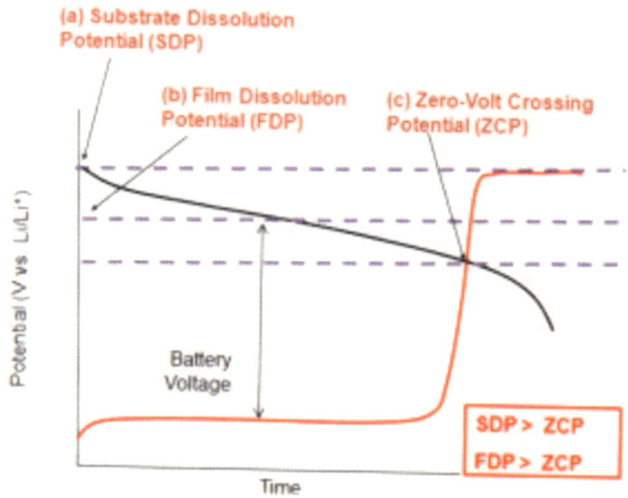

Figure 9. Schematic of key Zero-Volt potentials

Connector contacts are very important, both in terms of energy efficiency (when it comes to high power), and in terms of reliability and security. In recent years, the copper alloy with silver and / or gold is used, but other combinations of metals are to be explored [90.91]. So the compromise between good electrical and mechanical properties, on the one hand, and reasonable prices on the other is required.

Recent literature describes efforts devoted to investigation of copper based alloys in search of improvements in strength and maintenance of strength at high temperatures. The copper-silver alloy is an example of eutectic systems with the eutectic point at 779 °C when the alloy contains 72 % silver and 28 % copper. On both sides of the phase diagram there is a small solubility of the mentioned metals in each other. The maximum solubility of silver in copper is 4.9 at% and the slope of the solvus line indicates the possibility of age-hardening certain alloy compositions. Similar phenomenon - the strengthening of cold worked substitution solid solutions upon annealing up to the re-crystallization temperature is termed anneal hardening. The anneal hardening effect had been observed in Cu-Ag alloys in the annealing temperature range of 140-400 °C, the hardness being increased with the degree of preformation [4,5]. The goal of present work is to investigate corrosion behavior of this alloy obtained by fusion and cast (so called ingot metallurgy - IM) method in different stages of synthesis and thermomechanical treatment. Passivity of copper and its alloys is of interest with respect to basic and applied research due to its wide application in industry. Silver-copper alloys have been investigated elsewhere from the corrosion view point or as an electrode material, but the content of silver in all this alloys overcomes 15 % [19-22].

2.5. Lighting and heating of EV

With the rapid development of high intesive LED technology, it enabled large savings in energy consumption. That fact is crucial for EV. LED and power consumption of exterior vehicle lighting indicated that an all-LED system employing the current generation of LEDs would result in general power savings of about 50% (night time) to about 75% (daytime) over a traditional system. This means that while the long-term fuel cost savings (money) were higher for the gasoline-powered vehicle, long-term distance savings (range) favored the electric vehicle. Now, automotive lighting producer Osram comes to strengthen the idea mentioned above, stating that "micro-hybrids" or mild hybrids, which feature engine stop/start mechanisms to boost the efficiency of conventional vehicles, will benefit greatly from LED lighting by reducing power draw and battery drain, as well as increasing light output during low power mode and startups [23].

Today's roads have very little actual technology incorporated into their design and function. There are many types of technologies which could be incorporated, but we'll begin with what we say is the most important new feature which will soon be applied to actual roads. Since EVs are becoming increasingly popular, while their batteries are still much too weak to assure an anxiety-free drive on the highway, the induction charging (wireless) will begin to be incorporated into one of the lanes, so that these all-electric cars will be able to drive on the highway without using their on-board batteries at all, as they will get their juice straight from underneath the road surface (fig. 10). The idea of inductive charging is simple, and various companies and universities are testing the system now, in view of future mass implementation [23].

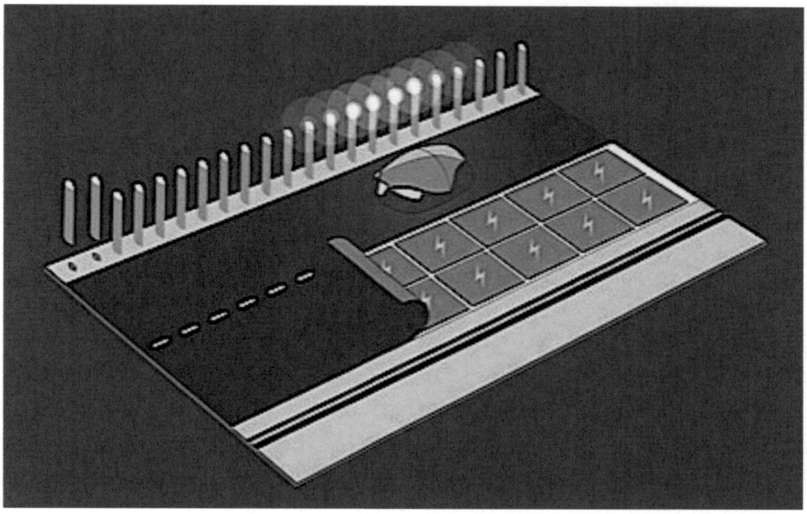

Figure 10. Road surface that charges batteries

Electric vehicles generate very little waste heat and resistance electric heat may have to be used to heat the interior of the vehicle if heat generated from battery charging/discharging can not be used to heat the interior. While heating can be simply provided with an electric resistance heater, higher efficiency and integral cooling can be obtained with a reversible heat pump (this is currently implemented in the hybrid Toyota Prius). Positive Temperature Coefficient (PTC) junction cooling [24] is also attractive for its simplicity — this kind of system is used for example in the Tesla Roadster.

Some electric cars, for example the Citroën Berlingo Electrique, use an auxiliary heating system (for example gasoline-fueled units manufactured by Webasto or Eberspächer) but sacrifice "green" and "Zero emissions" credentials. Cabin cooling can be augmented with solar power, most simply and effectively by inducting outside air to avoid extreme heat buildup when the vehicle is closed and parked in the sunlight (such cooling mechanisms are available as aftermarket kits for conventional vehicles). Two models of the 2010 Toyota Prius include this feature as an option [25].

3. Mechanical losses reduction in EV

3.1. Tyres role in EV

Large impact on the fuel consumption of the cars in general, has tires on its wheels. If the tire optimization is done by the energy efficiency criteria, with acceptable stability, comfort and durability, there is a wide range for development and research. One of the fine examples of the intensive development in this field is racing cars. A modern racing car is a technical masterpiece, but considering the development effort invested in aerodynamics, composite construction and engines it is easy to forget that tyres are still a race car's biggest single performance variable. Average car with good tyres could do well, but it is very known fact that the one with bad tyres, even the very best car did not stand a chance. Despite some genuine technical crossover, race tyres and road tyres are - at best - distant cousins at the moment. An ordinary car tyre is made with heavy steel-belted radial plies and designed for durability - typically a life of 16,000 kilometers or more (10,000 miles). For example, a Formula One tyre is designed to last for, at most, 200 kilometers and it is constructed to be as light and strong as possible. That means an underlying nylon and polyester structure in a complicated weave pattern designed to withstand far larger forces than road car tyres, in [26].

The racing tyre itself is constructed from very soft rubber compounds which offer the best possible grip against the texture of the racetrack, but wear very quickly in the process. All racing tyres work best at relatively high temperatures. For example, the dry 'grooved' tyres used up until very recently were typically designed to function at between 90 degrees Celsius and 110 degrees Celsius [103]. However, electric vehicles can benefit from the years of research and usage of this kind of tyres. The development of the racing tyre came of age with the appearance of 'slick' tyres in the 1960s. Teams and tyre makers realized that, by omitting a tread pattern on dry weather tyres, the surface area of rubber in contact with the road could be maximized. This led to the familiar sight of 'grooved' tyres, the regulations specifying that all tyres had to have four continuous longitudinal grooves at least 2.5 mm deep and spaced 50mm apart. These changes created several new challenges for the tyre manufacturers - most notably ensuring the grooves' integrity, which in turn limited the softness of rubber compounds that could be used, in reference [26].

The 'softness' or 'hardness' of rubber compounds is varied for each road according to the known characteristics of the material that the road was made of. The actual softness of the tyre rubber is varied by changes in the proportions of ingredients added to the rubber, of which the three main ones are carbon, sulfur and oil. Generally speaking, the more oil in a tyre, the softer it will be. Formula One tyres are normally filled with a special, nitrogen-rich air mixture, designed to minimize variations in tyre pressure with temperature. The mixture also retains the pressure longer than normal air would, in [26].

The key characteristics of the new rubber - developed together with the teams in response to the latest aerodynamic regulations - are squarer profiles, increased grip, and softer, more competitive compounds with consistent degradation, optimizing the compounds and profiles to guarantee even better and more stable performance, a longer performance peak, combined with the deliberate degradation that characterized, in [27].

This new measure, which should result in a reduction of aerodynamic down force acting on each tyre, requires a wider and more even contact surface. This objective has been met by having a less rounded shoulder on each tyre and using softer compounds, which produce better grip and more extreme performance, in [27].

Dry weather tyres, known as slicks, are characterized by a tread pattern that is devoid of blocks or channels. Wet weather tyres are characterized by grooves in the tread pattern. The full wet tyres can be easily recognized by the deep grooves in the tread pattern, in reference [28].

At this year's Geneva International Motor Show, one of the tyre manufacturers Goodyear unveils its latest innovation in tyre technology: an extremely low rolling resistance version of its award winning Goodyear EfficientGrip summer tyre with Fuel Saving Technology – specifically developed to fulfill the distinctive requirements of future electric vehicles, in [29]. The look of the tyre inside as well as of the tyre outside is presented in figure 11.

Figure 11. New Goodyear EfficientGrip summer tyre for EV.

The Goodyear EfficientGrip prototype tyre for electric vehicles delivers a range of benefits, including top rated energy efficiency and excellent noise and wet braking performance levels – in combination with Goodyear's latest generation of RunOnFlat Technology for continued mobility after a puncture or complete loss of tire pressure, in [29]. The design of the concept tyre is uniquely suited to complement the performance requirements of electric vehicles. The tyre's narrow shape in combination with its large diameter leads to reduced rolling resistance levels and to a reduced aerodynamic drag and thus reduced energy consumption.

Rolling resistance is mainly caused by the energy loss due to the deformation of the tyre during driving. Less deformation means less energy loss and hence, less rolling resistance. Goodyear engineers used the latest computer simulation technologies to analyze the tyre's potential deformation behavior during driving. The larger rim diameter reduces the overall amount of rubber that is needed, which leads to less rubber deformation during driving. The large tyre diameter requires fewer tire rotations for a certain distance, which in turn results in less heat buildup and tire deformation, which again leads to lower rolling resistance levels and less energy consumption, in [29].

Electric engines often provide a relatively constant torque, even at very low speeds, which increases the acceleration performance of an electric vehicle in comparison to a vehicle with a similar internal combustion engine. This required the development of a modified tread design in combination with a new tread compound to ensure excellent grip especially on dry, and to provide high levels of mileage, in [29].

This EfficientGrip concept tyre showcases our enormous research and development efforts to support the development of electric vehicles with tyres that provide extremely low rolling resistance and noise levels in combination with a very high level of wet performance. Fitted on a standard car this tyre would give 30 percent less rolling resistance which leads to about 6 percent less fuel consumption compared to an average standard summer tyre, in [29].

The effect of tyre pressure on either fuel consumption with regular cars or EV consumption is emphasized. Some researches were done in USA in the last few years. For the control test, the pressure was set at the factory recommended 33 psi in each tire. The subsequent test was done with the pressure set at 45 psi. For each test, the vehicle was driven a total of 550 miles over the course of one week travelling back and forth between the same two cities via the same route. The fuel tank was filled twice per week. Measurement of the quantity of fuel used was taken from the readout on a gas pump at each fill-up. The number of miles travelled was taken from the vehicle's trip odometer, in [190]. Results showed that during the control period, the number of miles travelled per gallon of gasoline consumed was 27. With the tire pressure at 45 psi, the vehicle travelled 30 miles per gallon of gasoline consumed; a difference of 11 percent, in [30].

3.2. Vehicle body

Automotive design and, specifically, the design of electric and hybrid-electric vehicles, involve a variety of challenges that have to be considered by an appropriate design environment. The convergence of more and more electronics with controls and mechanics makes the design process very complex and involves a variety of technical disciplines. With the complex interactions between the individual system parts, a disconnected consideration of each individual domain is not sufficient anymore. Each individual domain requires specific algorithms and modeling languages to achieve optimal performance for the analysis of that specific domain. A single algorithm usually does not perform for all domains equally; therefore the combination of different algorithms via co-simulation expands the design capabilities of the system considerably (Fig. 12) [31].

In recent years simulation programs allow the optimization of vehicle body shapes from the standpoint of energy efficiency. On the other hand, simulations and experiments in the wind tunnel achieve significant energy savings by introducing air turbine, which inevitably airflow into electricity.

3.3. Aerodynamics of EV

Moveable aerodynamic components are nothing new, every time you sit on an airliner you see the wing flaps, ailerons moving around, and often as you come into land you can see the array of hydraulics employed to move them. The systems on a Formula 1 racing car work in essentially the same way. Hydraulic tubes, rods and actuators. But whilst on an Airbus A320 or even a modern UAV or fighter jet there is a huge amount of space to work in, on a grand prix car the opposite is true [32]. EV vehicles could benefit a lot from these technologies.

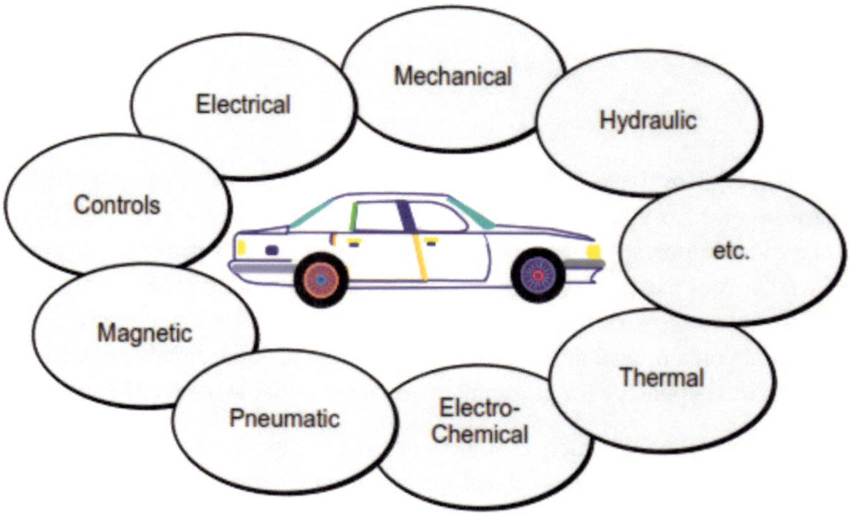

Figure 12. Multi-Domain Design

Racing drivers have a new tool at their disposal, called Drag Reduction System (DRS). It is essentially an adjustable rear wing which can be used to facilitate overtaking. The flap is lifted up at the front and pivots about a point at the trailing edge of the wing, so that in the event of a failure, the flap will drop down into the default, high-down force position. Since the timing loops will be sited after corners, drivers will only be able to deploy the active rear wing as a car goes down a particular nominated straight [32].

The materials used in these systems also require great precision. Today in F1 it is mainly titanium tube, though some of what we do involves peek mainly in the fuel system but primarily titanium. Aluminum and stainless steel are also used. Titanium is favored for its inherent lightness and strength, and it means that it is possible to make the cross section of the material so much thinner than if you were using Aluminum. Over the time, the manufacturers have learned to manipulate titanium tubing in ways, especially in small spaces, and the results of that work will be on cars in the future [32].

This week a row has erupted over the design of two teams' diffusers, after the new Williams and Toyota emerged sporting radically different diffuser designs to the other cars launched so far. Williams came up with a 'double decker' diffuser design, while Toyota initially tested an extension to the middle of their diffuser, and then later added a double decker section of their own. Both these designs raised eyebrows up and down the pit lane, as they appear to stretch the wording of the new rules. [33].

As part of the 2009 package of aerodynamic rule changes designed to reduce down force and increase overtaking, the FIA mandated a smaller diffuser in a more rearward position. With the shock of losing 50 per cent of their down force because of these changes, teams have been working hard to get the bodywork shaped to the new rules to regain the lost down force [33].

4. Additional energy in EV

4.1. Solar cells

Today, world recognizes the synergy between solar panels and electric cars. As the matter a fact there are several car companies that plan to install solar panels in their newer hybrid vehicles. The most important question for most of these manufacturers is: how much extra power will a solar roof panel actually provide? It's very difficult to generate enough power to move a vehicle with energy from the sun's light. So, solar panels at the moment don't have that much of an impact on a hybrid and electric car's efficiency. Solar panels are also made out of silicon, which is too expensive for automakers to use as a viable source [34].

However, there are companies such as Toyota, one of the pioneers in this field, which uses the solar roof panel. Constant technology development will provide better conditions in years that follow for this option. Nowadays, roof panel will power at least part of the hybrid Toyota Prius' air-conditioning unit. Smaller, less power-hungry systems seem to work better with solar power [34].

The most common type of solar panel uses single- or multi-crystalline silicon wafers. Creating the silicon crystal is by far the most energy intensive part of the process, followed by various and sundry manufacturing steps, such as cutting the silicon into wafers, turning the wafers into cells and assembling the cells into modules [35].

The today's electric vehicles consume about 150 watt-hours per kilometer. If the average distance per day is 50 km, then it would be 18,250 kilometers per year. For this calculated consumption, electric vehicle would need to generate 2.75 MWh/year. By this math, mono-crystalline solar panels generate about 263 kWh/m^2 per year in the USA. Therefore, about 10.5 square meters of solar panels to completely offset the energy consumed by today's electric vehicles [35]. The only practical place to put panels on the Roadster is the roof (about 1 square meter). Ideally, this would then generate 263 kWh/year. However, the Roadster won't always be in the sun, and it won't be at its ideal angle. A 60% de-rating would be generous to account for shade and suboptimal angles, so the panel would generate about 150 kWh/year – driving the car an additional 3 kilometers per day [35].

However, there is possibility to put solar cells on the other part of the vehicle's surface. The surface from the vehicle's nose, across the hoods, and all the way to the roof can be used for solar cells as presented in figure 13. Also, technology development will without a doubt make progress in increasing solar energy efficiency.

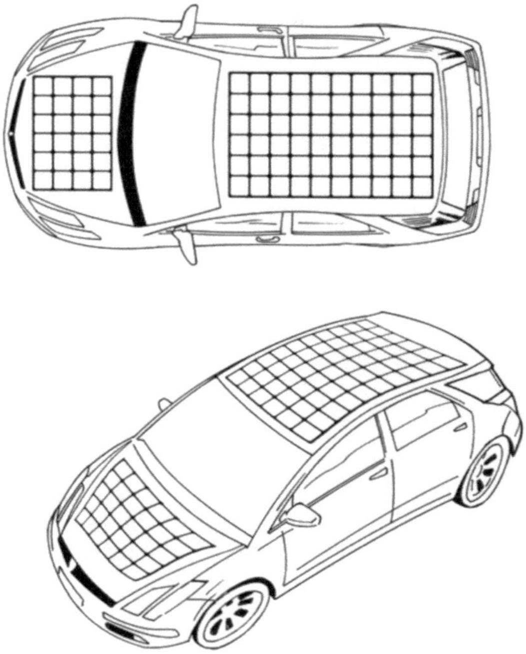

Figure 13. Position of solar cells on the surface of the electric vehicle

4.2. Energy recovery systems

4.2.1. Kinetic energy recovery systems

A kinetic energy recovery system (KERS) is an automotive system for recovering a moving vehicle's kinetic energy under braking. The recovered energy is stored in a reservoir (flywheel or a battery or/and supercapacitor) for later use under acceleration. The device recovers the kinetic energy that is present in the waste heat created by the car's braking process. It stores that energy and converts it into power that can be called upon to boost acceleration, in [36].

The concept of transferring the vehicle's kinetic energy using flywheel energy storage was postulated by physicist Richard Feynman in the 1950s. It is exemplified in complex high end systems such as the Zytek, Flybrid, Torotrak and Xtrac used in F1 and simple, easily manufactured and integrated differential based systems such as the Cambridge Passenger/Commercial Vehicle Kinetic Energy Recovery System (CPC-KERS), in [36].

Xtrac and Flybrid are both licensees of Torotrak's technologies, which employ a small and sophisticated ancillary gearbox incorporating a continously variable transmission (CVT). The CPC-KERS is similar as it also forms part of the driveline assembly. However, the whole mechanism including the flywheel sits entirely in the vehicle's hub (looking like a drum brake). In the CPC-KERS, a differential replaces the CVT and transfers torque between the flywheel, drive wheel and road wheel [36]. KERS Technology is based on a completely new design capable of accumulating power and keeping it in store for the right moment.

KERS Technology works like a turbo charger that provides additional power and acceleration by stiffening the tail of the ski in outturns. The effect: a boost, catapulting the rider into the next turn. Just like when Formula 1 pilots push a button for that extra notch of speed. KERS Technology is an electronic, fully automatic and integrated system. Piezoelectric fibers transform kinetic energy into electrical energy which is stored. Electrical energy is immediately released to areas of the ski, where additional energy is requested. Timing and release are automatically controlled and coordinated. Depending on the flex pattern of different ski models, sensors are programmed beforehand: the more aggressive the ski has to be, the stiffer the tail will become, in reference [37].

The key system features were:

- A flywheel made of steel and carbon fibre that rotated at over 60,000 RPM inside an evacuated chamber
- The flywheel casing featured containment to avoid the escape of any debris in the unlikely event of a flywheel failure
- The flywheel was connected to the transmission of the car on the output side of the gearbox via several fixed ratios, a clutch and the CVT
- 60 kW power transmission in either storage or recovery
- 400 kJ of usable storage (after accounting for internal losses)
- A total system weight of 25 kg
- A total packaging volume of 13 litres, in reference [38]

There are principally two types of system - battery (electrical) and flywheel (mechanical). Electrical systems use a motor-generator incorporated in the car's transmission which converts mechanical energy into electrical energy and vice versa. Once the energy has been harnessed, it is stored in a battery and released when required. Mechanical systems capture braking energy and use it to turn a small flywheel which can spin at up to 80,000 rpm. When extra power is required, the flywheel is connected to the car's rear wheels. In contrast to an electrical KERS, the mechanical energy doesn't change state and is therefore more efficient. There is one other option available - hydraulic KERS, where braking energy is used to accumulate hydraulic pressure which is then sent to the wheels when required, in [36,39].

The first of these systems to be revealed was the Flybrid. This system weighs 24 kg and has an energy capacity of 400 kJ after allowing for internal losses. A maximum power boost of 60 kW (81.6 PS, 80.4 HP) for 6.67 seconds is available. The 240 mm diameter flywheel weighs 5.0 kg and revolves at up to 64,500 rpm. Maximum torque at the flywheel is 18 Nm, and the torque at the gearbox connection is correspondingly higher for the change in speed. The system occupies a volume of 13 liters, in [36].

Nowadays, Formula One has stated that they support responsible solutions to the world's environmental challenges and the FIA allowed the use of 60 kW KERS in the regulations for the 2009 Formula One seasone. Teams began testing systems in 2008: energy can either be stored as mechanical energy (as in a flywheel) or as electrical energy (as in a battery or supercapacitors). As of 2014, in the race cars, the power capacity of the KERS units will increase from 60 kilowatts to 120 kilowatts, in [106].

The aims for introducing KERS technology in the racing world are twofold. Firstly to promote the development of environmentally friendly and road car-relevant technologies in Formula One racing; and secondly to aid overtaking. A chasing driver can use his boost button to help him pass the car in front, while the leading driver can use his boost button to escape. A typical KERS system weighs from 25 to 35 kilograms, in [36,37,39]. For the relevance of the electric vehicles, this energy can be used for supplementing the batteries of electrical engine and thereby adding a few more kilometres to the driving distance at once.

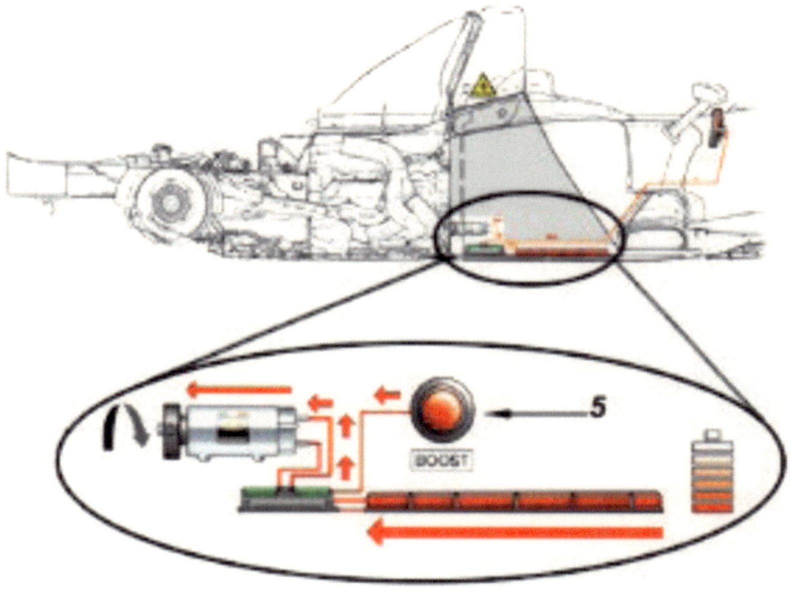

Figure 14. Kinetic Energy Recovery System

Following the current situation, some solutions in KERS packaging has taken a step forwards. Now the energy storage appears to be slightly revised, with the unit inside the gearbox swapped for floor mounted units. The two carbon fiber cases are closed with aluminum tops and are provided with electrical and cooling connections. They sit in the final section of flat floor known as the boat tail, in [40].

Having the units placed on the floor, as opposed to between the gearbox and engine, means they can lower the Centre of Gravity. Also being quite heavy they are placed near the rear axle line to suit the mandatory weight distribution. As mentioned the units are supplied with a common cooling circuit, one pipe routes around the back of the floor to link the devices. There are also a number of electrical connections for both connecting to the KERS Power Control Unit and for monitoring their status. Quickly detachable connectors are used to allow rapid removal of the floor keeping the units in place, in [40].

The future development appear to have found a new mounting position and format for their KERS energy storage with what appear to be floor mounted super capacitors. Super Capacitors (supercaps) are alternative energy storage to Lithium Ion batteries, using very much the same technology as smaller capacitors used in electronics, in [40].

Typically current F1 cars use dozens of Li-ion cells packed into an array forming a 'battery' pack. This KERS Battery Pack is commonly a single part sat under the fuel tank. Although often used as a single battery, the unit can be broken up into a set of batteries in series. In 2011 Red Bull clearly split this part up into several smaller Battery Packs, there being the two aforementioned units either of the gearbox and another in the gearbox. Although interconnecting these parts with cooling pipes, high current cable and sensor cabling ads some weight, this does provide a nicer packaging solution. It's logical to explain these new floor mounted parts as batteries. However they do not look like the battery packs seen in the gearbox last year, or on other cars. Being on the floor of the car they are subject to even more danger from impacts as well as the heat and vibration that caused issues last year, in [40]. The energy stored in a double-layer capacitor, is used to supply power needed by vehicle electrical systems, in [36].

4.2.2. Waste heat energy recovery

In recent years, there has been active research on exhaust gas waste heat energy recovery for automobiles. Meanwhile, the use of solar energy is also proposed to promote on-board renewable energy and hence to improve their fuel economy. New research in thermoelectric-photovoltaic hybrid energy systems are proposed and implemented for automobiles. The key is to newly develop the power conditioning circuit using maximum power point tracking so that the output power of the proposed hybrid energy system can be maximized. This experimental concept can be easily implemented in electric vehicles [41].

According to the recent studies, General Motors is using shape memory alloys that require as little as a 10°C temperature difference to convert low-grade waste heat into mechanical energy. When a stretched wire made of shape memory alloy is heated, it shrinks back to its pre-stretched length. When the wire cools back down, it becomes more pliable and can revert to its original stretched shape. This expansion and contraction can be used directly as mechanical energy output or used to drive an electric generator. Shape memory alloy heat engines have been around for decades, but the few devices that engineers have built were too complex, required fluid baths, and had insufficient cycle life for practical use. Around 60% of all energy in the U.S. is lost as waste heat; 90% of this waste heat is at temperatures less than 200°C and termed low grade because of the inability of most heat-recovery technologies to operate effectively in this range. The capture of low-grade waste heat, which turns excess thermal energy into useable energy, has the potential to provide consumers with enormous energy savings [42].

http://dx.doi.org/10.5772/55237

For practical use, parts of automotive industry nowadays are working to create a prototype that is practical for commercial applications and capable of operating with either air or fluid based heat sources. GM's shape memory alloy based heat engine is also designed for use in a variety of non-vehicle applications. For example, it can be used to harvest non-vehicle heat sources, such as domestic and industrial waste heat and natural geothermal heat, and in HVAC systems and generators [43].

Thermal Energy Recovery Systems for better fuel efficiency proposes solutions for fuel economy and lower CO2-emissions on combustion engines by making use of their exhaust waste heat. This fuel economy is accessible for engines running on gasoline, diesel, bio fuels, hydrogen or any other type of fuel. This solution proposes high power density for mobile applications and rugged solutions for power generation and marine applications, also being recognized by the motorsport world as an important technology for the future in racing and finally a technology that will contribute to the development of electric vehicle [43].

Plug-in hybrid electric vehicles are already noted for their environmental advantages and fuel savings – but now a new breakthrough technology could mean their fuel economy is boosted by a further seven per cent [44]. Most vehicle waste heat recovery systems that are currently being developed utilize a thermoelectric converter to create electricity, as the name implies, directly from heat. These devices depend on a unique property of certain materials which result in the Seeback effect, discovered in 1821, where the application of heat produces an electric current. The devices have no moving parts. You could think of them as similar to photovoltaic cells, except that they respond to heat rather than light [45].

An effective waste recovery system requires three elements:

1. a thermoelectric material package
2. an electric power management system, which directs the electricity injected into the vehicle's electrical system to the place where it will do the most good at any given time
3. a thermal management system, which is essentially a sophisticated heat exchanger [45]

Some other systems in hybrid electric vehicles reduce fuel consumption by replacing a significant portion of the required electric power normally produced by the alternator with electric power produced from exhaust gas waste heat conversion to electricity in a Thermoelectric Generator Module [46].

4.3. Airflow

It was previously mentioned that vehicle body can be designed to reduce downforce and otherwise adverse airflow. Some of the possibilities are presented here.

During forward motion of an electrically-powered vehicle, air is captured at the front of the vehicle and channeled to one or more turbines. The air from the turbines is discharged at low pressure regions on the sides and/or rear of the vehicle. The motive power of the air rotates the turbines, which are rotatable engaged with a generator to produce electrical energy that is used to recharge batteries that power the vehicle. The generator is rotatable engaged with a flywheel for storing mechanical energy while the vehicle is in forward motion. When the vehicle slows or stops, the flywheel releases its stored energy to the generators, thereby enabling the generator to continue recharging the batteries. The flywheel enables the generators to provide a more stable and continuous current flow for recharging the batteries [47].

It is assumed that the vehicle is moving in a calm and steady wind stream with zero wind velocity. If the vehicle is moving at a constant speed of 15 m/s (54 km/h), then we can think a wind stream with15 m/s is flowing around the vehicle. Normally this wind will cause a drag force which is opposite to the direction of the propulsion of the vehicle. At constant speed (zero acceleration) the energy requirements to move the vehicle forward are –To overcome the frictional force (rolling resistance of road) and to overcome wind resistance [48]. At this Condition, if the air stream flowing around the vehicle (which was not interacting with the vehicle previously) is allowed to enter inside and let it flow down to the rear side; then it may be possible to use these air streams to generate power. The vehicle has already interacted with this wind and it d eflects the stream of wind at the two sides of it by stagnation at the front.

This is the energy that had been lost from the vehicle to overcome the aerodynamic resistant. Now if these stream generated by the interaction of the wind and vehicle is captured within the vehicle in such a way that it would not impose an additional drag at the direction of propulsion of the vehicle, some of the energy can be recovered and fed back to the battery by means of conventional energy conversion processes. Placing a wind turbine can serve the purpose. At the same time it will help to increase the pressure at the back side (according to Bernoulli's equation pressure will be increased if velocity is decreased and velocity will be reduced at the back side of the turbine after energy extraction) which will reduce the drag force that existed before with the conventional design of the vehicle. So, vortex shedding will be reduced at the rear side. For this it is necessary to modify the design of a vehicle which gives provision of air flow through the vehicle. On the other hand positioning of the turbines will also be important because they must be placed in such a way that they do not impose or create any additional drag on the vehicle. Symmetrical positioning of the turbine can do t he trick as the thrust acting on the turbines will cancel each other (Fig. 15) [49].

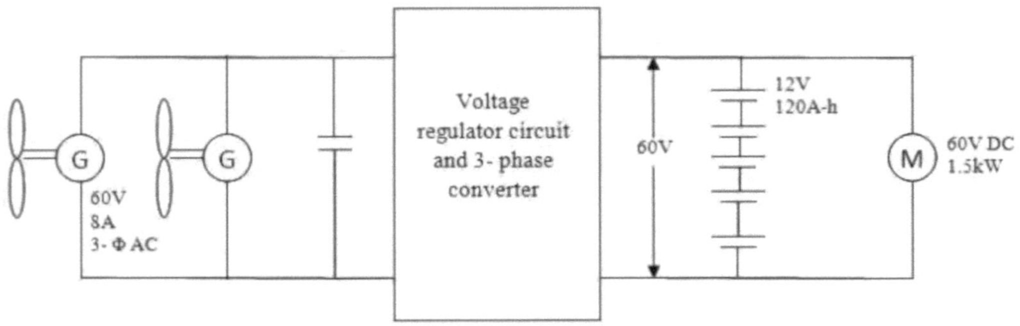

Figure 15. Charging and control circuit of the battery

4.4. Hybrid electric vehicle

Generally, hybrid vehicles could be described as vehicles using combination of technologies for energy production and storage. Two types of the vehicles are in consideration – so called parallel and linear hybrids. Parallel type possesses mechanical connection between power generator and drive wheels, while in linear one such connection does not exist. Serial hybrids have significant advantages in relation to parallel ones because of their mechanical simplicity, design flexibility and possibility for simple incorporation of new technologies [1].

Hybrid electric vehicles (HEVs) combine the internal combustion engine of a conventional vehicle with the high-voltage battery and electric motor of an electric vehicle. As a result, HEVs can achieve twice the fuel economy of conventional vehicles (Fig. 1). In combination, these attributes offer consumers the extended range and rapid refueling they expect from a conventional vehicle, as well as much of the energy and environmental benefits of an electric vehicle. HEVs are inherently flexible, so they can be used in a wide range of applications — from personal transportation to commercial hauling. Hybrid electric vehicles have several advantages over conventional vehicles:

- Greater operating efficiency because HEVs use regenerative braking, which helps to minimize energy loss and recover the energy used to slow down or stop a vehicle;
- Lighter engines because HEV engines can be sized to accommodate average load, not peak load, which reduces the engine's weight;
- Greater fuel efficiency because hybrids consume significantly less fuel than vehicles powered by gasoline alone;
- Cleaner operation because HEVs can run on alternative fuels (which have lower emissions), thereby decreasing our dependency on fossil fuels (which helps ensure our national security); and
- Lighter vehicle weight overall because special lightweight materials are used in their manufacture.

Hybrid electric vehicles are becoming cost-competitive with similar conventional vehicles, and most of the cost premium can be offset by overall fuel savings and tax incentives. Some states even offer incentives to consumers buying HEVs [50].

4.5. Today's high-speed EV

Nowadays, the most powerful high-performance electric vehicle has four electric motors producing a total output of 552 kW and a maximum torque of 1000 Nm. As a result, the gullwing model has become the world's fastest electrically-powered series production vehicle accelerates from zero to 100 km/h in 3.9 seconds [51].

Enormous thrust comes courtesy of four synchronous electric motors providing a combined maximum output of 552 kW and maximum torque of 1000 Nm. The very special gullwing model accelerates from zero to 100 km/h in 3.9 seconds, and can reach a top speed of 250 km/h (electronically limited). The agile response to accelerator pedal input

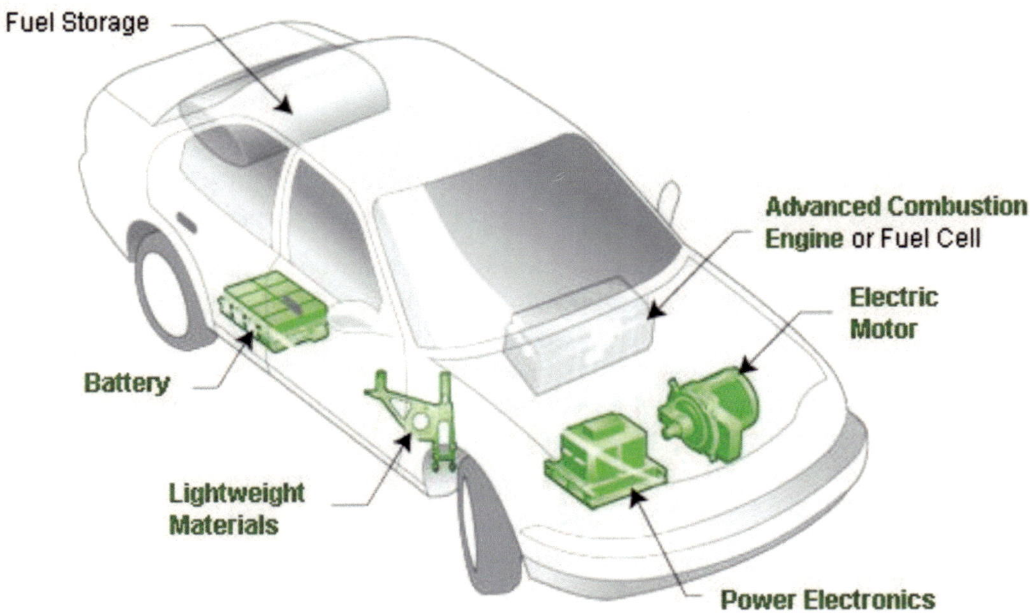

Figure 16. HEV

and the linear power output provide pure excitement: unlike with a combustion engine, the build-up of torque is instantaneous with electric motors – maximum torque is effectively available from a standstill. The spontaneous build-up of torque and the forceful power delivery without any interruption of tractive power are combined with completely vibration-free engine running characteristics [51].

The four compact permanent-magnet synchronous electric motors, each weighing 45 kg, achieve a maximum individual speed of 13,000 rpm and in each case drive the 4 wheels selectively via a axially-arranged transmission design. This enables the unique distribution of torque to individual wheels, which would normally only be possible with wheel hub motors which have the disadvantage of generating considerable unsprung masses [51].

Battery efficiency, performance and weight are by far the most important factors in electric vehicles. The high-voltage battery in the current high-performance electric vehicles boasts an energy content of 60 kWh, an electric load potential of 600 kW and weighs 548 kg – all of which are absolute best values in the automotive sector. The liquid-cooled lithium-ion high-voltage battery features a modular design and a maximum voltage of 400 V. Advanced technology and know-how from the world of Formula 1 have been called on during both the development and production stages [51].

The high-voltage battery consists of 12 modules each comprising 72 lithium-ion cells. This optimized arrangement of a total of 864 cells has benefits not only in terms of best use of the installation space, but also in terms of performance. One technical feature is the intelligent

http://dx.doi.org/10.5772/55237

parallel circuit of the individual battery modules – this helps to maximize the safety, reliability and service life of the battery. As in Formula 1, the battery is charged by means of targeted recuperation during deceleration whilst the car is being driven [51].

A high-performance electronic control system converts the direct current from the high-voltage battery into the three-phase alternating current which is required for the synchronous motors and regulates the energy flow for all operating conditions. Two low-temperature cooling circuits ensure that the four electric motors and the power electronics are maintained at an even operating temperature. A separate low-temperature circuit is responsible for cooling the high-voltage lithium-ion battery. In low external temperatures, the battery is quickly brought up to optimum operating temperature with the aid of an electric heating element. In extremely high external temperatures, the cooling circuit for the battery can be additionally boosted with the aid of the air conditioning. This also helps to preserve the overall service life of the battery system [51].

Ideally the EV is charged with the aid of wall box. As it could be installed in a home garage, this technology provides a 22 kW quick-charge function, which is the same as the charging performance available at a public charging station. A high-voltage power cable is used to connect the vehicle to the wall box, and enables charging to take place in around three hours. Charging takes around 20 hours without the wall box [51].

To ensure maximum safety, the SLS AMG Coupé Electric Drive, one of the most advanced high-performance EV today, makes use of an eight-stage safety design. This comprises the following features:

- All high-voltage cables are color-coded in orange to prevent confusion
- Comprehensive contact protection for the entire high-voltage system
- The lithium-ion battery is liquid-cooled and accommodated in high-strength aluminium housing within the carbon-fibre zero-intrusion cell
- Conductive separation of the high-voltage and low-voltage networks within the vehicle and integration of an interlock switch
- Active and passive discharging of the high-voltage system when the ignition is switched to "off"
- In the event of an accident, the high-voltage system is switched off within fractions of a second
- Continuous monitoring of the high-voltage system for short circuits with potential compensation and insulation monitors
- Redundant monitoring function for the all-wheel drive system with torque control for individual wheels, via several control units using a variety of software

By using this design, EV manufacturers ensures maximum safety during production of the vehicle and also during maintenance and repair work [51].

The intelligent and permanent all-wheel drive concept, with four motors for four wheels guarantees driving dynamics at the highest level, while at the same time providing the best possible active safety. Optimum traction of the four driven wheels is therefore ensured, whatever the weather conditions. According to the developers, the term "Torque Dynamics" refers to individual control of the electric motors, something which enables completely new levels of freedom to be achieved. The AMG Torque Dynamics feature is permanently active and allows for selective distribution of forces for each individual wheel. The intelligent distribution of drive torque greatly benefits driving dynamics, handling, driving safety and ride comfort. Each individual wheel can be both electrically driven and electrically braked, depending on the driving conditions, thus helping to:

- optimize the vehicle's cornering properties
- reduce the tendency to over steer/under steer
- increase the yaw damping of the basic vehicle
- reduce the steering effort and steering angle required
- increase traction

AMG Torque Dynamics system enables optimum use of the adhesion potential between the tires and the road surface in all driving conditions. The technology allows maximum levels of freedom and as such optimum use of the critical limits of the vehicle's driving dynamics [51].

The trailblazing body shell structure of the SLS AMG Coupé Electric Drive is part of the ambitious "AMG Lightweight Performance" design strategy. The battery is located within a carbon-fiber monocoque which forms an integral part of the gullwing model and acts as its "spine". The monocoque housing is firmly bolted and bonded to the aluminum space frame body. The fiber composite materials have their roots in the world of Formula 1, among other areas. The advantages of CFRP (carbon-fiber reinforced plastic) were exploited by the Mercedes-AMG engineers in the design of the monocoque. These include their high strength, which makes it possible to create extremely rigid structures in terms of torsion and bending, excellent crash performance and low weight. Carbon-fiber components are up to 50 percent lighter than comparable steel ones, yet retain the same level of stability. Compared with aluminum, the weight saving is still around 30 percent, while the material is considerably thinner. The weight advantages achieved through the carbon-fiber battery monocoque are reflected in the agility of the electric vehicle and, in conjunction with the wheel-selective four-wheel drive system, ensure true driving enjoyment. The carbon-fiber battery monocoque is, in addition, conceived as a "zero intrusion cell" in order to meet the very highest expectations in terms of crash safety. It protects the battery modules inside the vehicle from deformation or damage in the event of a crash [51].

The basis for CFRP construction is provided by fine carbon fibers, ten times thinner than a human hair. A length of this innovative fiber reaching from here to the moon would weigh a mere 25 grams. Between 1000 and 24,000 of these fibers are used to form individual strands [51].

The purely electric drive system was factored into the equation as early as the concept phase when the super sports car was being developed. It is ideally packaged for the integration of the high-performance, zero-emission technology: by way of example, the four electric motors and the two transmissions can be positioned as close to the four wheels as possible and very low down in the vehicle. The same applies to the modular high-voltage battery. Advantages of this solution include the vehicle's low center of gravity and balanced weight distribution – ideal conditions for optimum handling, which the electrically-powered gullwing model shares with its petrol-driven sister model. Another distinguishing feature is the speed-sensitive power steering with rack-and-pinion steering gear: the power assistance is implemented electro hydraulically rather than just hydraulically [51].

The high-performance ceramic composite brakes are used in the latest electrical vehicles, which boast direct brake response, a precise actuation point and outstanding fade resistance, even in extreme operating conditions. The over-sized discs – measuring 402 x 39 mm at the front and 360 x 32 mm at the rear – are made of carbon fiber-strengthened ceramic, feature an integral design all round and are connected to an aluminum bowl in a radially floating arrangement. The ceramic brake discs are 40 percent lighter in weight than the conventional, grey cast iron brake discs. The reduction in unsprung masses not only improves handling dynamics and agility, but also rides comfort and tire grip. The lower rotating masses at the front axle also ensure a more direct steering response – which is particularly noticeable when taking motorway bends at high speed [51].

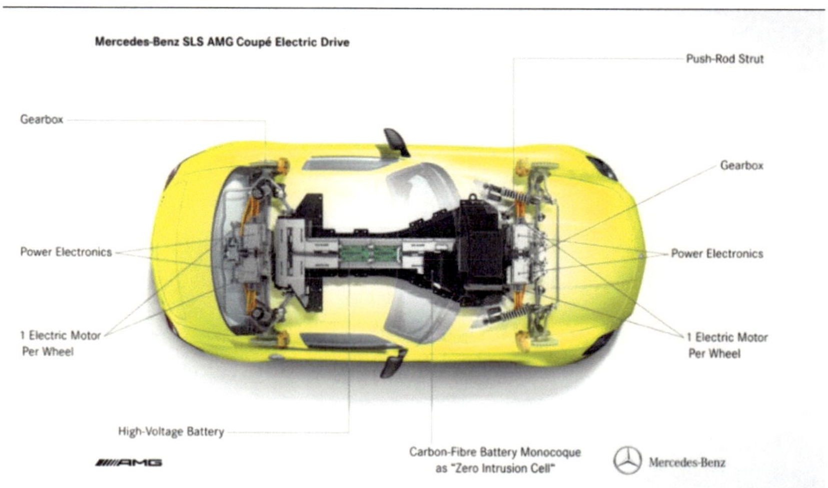

Figure 17. Today's high-speed EV system

5. Driving optimization

5.1. Comfort, information and safety

Minimizing electricity consumption is often in conflict with comfort and even security of vehicles and people. That's why new technologies are being used to increase safety and comfort, and still energy consumption to be on a low level. Some of the current opportunities and trends are presented in here.

5.1.1. Computer control

Nowadays, computers are indispensable part of every vehicle. It monitors and controls virtually all vehicle functions, but also processed and displayed a lot of additional information, which significantly contributes to the comfort and safety. In EV that trend is particularly used. The vehicle is equipped with sensors that provide input data and further processed in a computer. The obtained results act on actuators, or the situation is shown on the display and the decision is left to the man [52-53].

Sensors are elements that receive and convert non-electrical signals into electrical. Temperature shift (translation, rotation, stretching), pressure, brightness, electromagnetic radiation, magnetic fields can be detected and can be converted. The temperature is the most usually measured as the non-electrical input, therefore many types of sensors are developed over the years. There are NTC (Negative Temperature Coefficient) and PTC (Positive Temperature Coefficient) resistors and thermocouples [52-54].

In modern vehicle, for the measurement of ambient temperature, cabin and equipment itself, semiconductor sensors are used. They are the product of modern technology of silicon (Si) integrated circuits, therefore also called Si sensors. Silicon sensors consist of integrated circuits using temperature-active properties of semiconductor compounds. All sensors can be with current or voltage output. In both cases, the output signal is proportional to the absolute temperature. The amplitude of the output signal is relatively high and linear, and the interpretation of the signals can be done without any difficulties. Si sensors temperature range usually is from -50 ° C to +150 ° C. The stability and accuracy of these sensors is good enough to allow readings with ± 0.1 ° C resolution. Thermal imager is used for more complex state visual monitoring used the [55].

For the measurement of other important physical quantities (pressure, force, position, displacement and level), sensors that respond to physical movement and / or movement are used. The most commonly used types are semiconductors and resistant strain gauges, linear voltage displacement transducers (LVDT), resistive potentiometers and capacitive sensors. Although each of these sensors is based on different principles, the output signals of all the sensors are voltage, current and impedance. These signals are directly or indirectly analog voltage expressed, so all the techniques described for the measurement are related to these transducers. Sensors that require external excitation reduce the accuracy of the measurement. Higher excitation levels provide higher levels of the output. However, the higher excitation increases internal power dissipation and measurement error, even with mechanical transducers. Each transducer has its own optimal level of excitation [54].

Flow and velocity quantities are measured using resistive, piezoelectric, thermal, and other transducers. As mentioned earlier, all methods ultimately provide as output an analog voltage, current, or impedance. Types of transducers, such as rotary encoders, turbine, magnetic and optical sensors, have digital or pulse outputs. Speed or number of events can be determined by using digital counters and frequency meter [54,56,57].

Two-way communication between humans and computers is done through the touch screen display. Touch screen allows user to interact with a computer through touching the mark and the image on the screen. It is a visual electronic device that can sense touch and determine its location on the surface. The touch itself means contact between human fingers and the screen. The touch screen can also register contact other passive objects, such as special pens, styluses (used for greater precision and less contaminating the screen). Ability to register touch on the touch screen display depends on the implemented touch technology: ones can register just one touch and its position at a given time (single touch), others are capable of registering two or more simultaneous touch and their position on the screen (multi-touch).

Touch screen displays eliminate constraints on a number of discrete keys that are present in conventional membrane keypad. With a touch screen, combined with digital high-resolution display and integrated software[58-59], now there are virtually millions of switching options available for the user.

In commercial terms, touch screen displays, as devices with touch technology, make computer technology easy to use and accessible to all and also significantly to reduce time and cost of training of its use. They also provide much faster access to information as touch technology simplifies and speeds up the search process, which is crucial to driving. As an assembly that is mounted in front of a video display, touch screen display has an independent XY coordinate system that is calibrated according to the matrix display. To determine the location of the touch in the simplest implementation it requires two measurements, one to determine the coordinates of the X-axis and one to determine the coordinates of the Y axis. These measurements are then converted to the coordinates of the point of contact, which is then sent to the host (PC or microcontroller) via serial communication port [60]. A typical example of the application of TS and microprocessor technology is a GPS navigation system (Fig. 18) [61].

An example of a complete computer in a hermetically closed housing is shown in Fig. 19. Nexcom Company has released transport intended fanless computer - VTC-3300, for vehicles and fleet management [62].

5.1.2. Fire protection in EV

EV and HEV in particular have a lot of critical areas where it can get to the inception of fire. This requires a vehicle equipped elements of fire protection.

Central unit for fire detection and fire alarm, or as it is often called central unit for fire detection and its task is to power supply detectors and detection lines with stable and regulated supply voltage, which should be available in all expected operational situations, able to take a normal signal status, alarm status, interception line signal or removing the detector signal, short circuit signal, to signalize received state at the central unit and to forward signal to the sound and light devices and to ensure that the executive functions of the system that are required. Alarm indication at the control unit can turn on the respective light emitting diodes, or additional information through the display, but also by activating an internal audible alarm, buzzer or

Figure 18. GPS navigation system

Figure 19. Complete computer in a hermetically closed housing

horn. Today, the central fire protection unit connects to the computer, or it is incorporated as software in the computer system.

In addition to the central unit detection system must include detectors, alarms and detection and alarm lines, also the connections to the device that activates the sound and / or light alarms and executive functions. Network that connects the detection system elements is performed mainly by cables and its careful design and selection are essential to the quality, safety and value rationality of the system.

While in conventional systems alarm identification is with group of detectors, central unit and the person that receives information about the group (zone) that alarm is on, however with addressable system each detector gets its code (address) that identifies and tells to the central unit and to the present stuff its state. So the group identification alarm systems, the central unit receive information from a group of detectors (zones) of the alarm or some other event. Event means any change of state of the zone, such as an alarm, a signal failure, signal extraction detectors, fine lines and so on. Whereas in these systems, after the alarm of any zone a person in charge comes to the site, review of the protected object and determines the place where the alarm originated, the addressable system has been known for the receipt of alarms and place of origin, the detector that is in alarm state and the place where it is placed [54].

5.1.3. IR termography

IR thermography is a highly sophisticated measuring technique whose beginnings of extensive using coincide with the beginning of the third millennium. The reasons for this can be found in the fact that the thermography camera as a device that provides a visualization of thermal radiation would have to be consisting of many of the latest developments in science so that they become commercially available and easy to use. Very high breakthroughs in the field of sensors, thin films, optoelectronics, microelectronics and microcomputers are integrated and incorporated into these modern devices adapting them to the requirements of users in almost all areas of human activity.

The word thermography (literally, it would mean see the heat) explains the essence of this concept. Specifically, the point is that the appropriate devices (cameras) translate waves from the infrared region into a selected color of the visible part of the electromagnetic spectrum making them visible to the human eye. Different temperatures at the same time correspond to different colors and shades of colors and it is possible even to choose the color palette in which we want to show the resulting temperature map of the object [55, 63,64]. In modern vehicles is incorporated one or more thermography imagers and monitors the state of driver, equipment, or danger on the road in case of low visibility.

Camera which recorded persons in certain position is fixed, for example in front of mirror. Functionality of system is observed and recorded images were compared with literature data. Especially, it was taken care of record conditions: day-time record, time of taking drugs or active substance (coffee, alcohol, tea), room temperature, personal conditions such as emotions, satiety, hunger and physical activity. It was found that in normal conditions temperature in ocular region of healthy person does not exceed 36.3°C. In case of fever it is significantly higher. Thermograms of healthy person before and after vigorous physical activities show also the temperature changes in means of increase (fig. 20)[65].

The fact that thermography can detect very small differences in temperature gives the ability to detect the presence of persons (fig. 21), or animals (fig. 22) at night or in conditions of dense fog. Thermal detectors can function in the complete absence of any light. This makes them the perfect tool for observation in absolute darkness. Potential danger on the road in such conditions can be detected at a distance of 400 m for some systems, up to several kilometers, depending on the equipment and requirements.

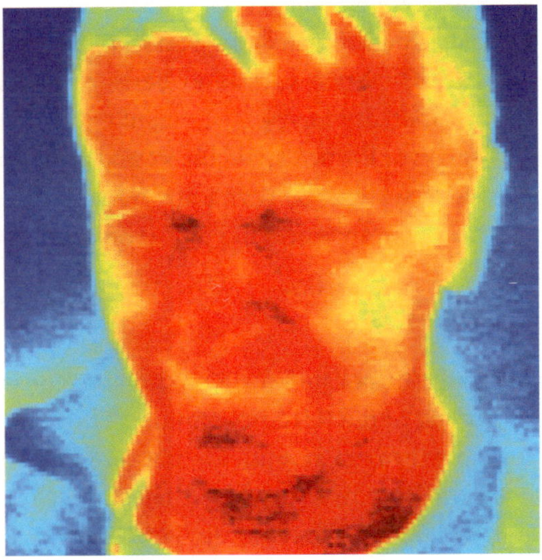

Figure 20. Thermogram of driver

Figure 21. Thermogram of persons on the road

5.2. Route

Route optimization (RO) is an important feature of the Electric Vehicles which is responsible for finding optimized paths between any source and destination nodes in the road network. Recent researches perform the RO for EV using the Multi Constrained Optimal Path (MCOP) problem. The proposed MCOP problem aims to minimize the length of the path and meets constraints on total travelling time, total time delay due to signals, total recharging time, and total recharging cost. The proposed algorithms need to have innovative methods for finding the velocity of the particles and updating their positions with accurate database of the requested roads[66-67].

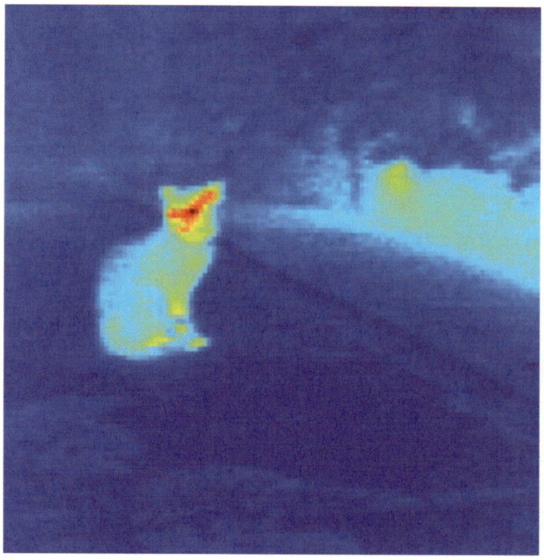

Figure 22. Thermogram of animal on the road

6. Conclusion

Electric drive vehicles are one of the most advanced vehicles at the moment taking into account contamination of environment. Lately there is an increased interest in the world for hybrid vehicles that have smaller fuel consumption and substantially less contamination emission footprint. Hybrid vehicles in most general terms can be described as vehicles comprising combination of energy producing and storing.

In this chapter, possibilities of energy savings in EV and HEV, energy generating in the vehicle itself and measures to improve comfort and safety are presented.

Therefore they must be combined with supercapacitors. Beside the development of standard technologies, development of power supply is crucial for EV. Accumulator batteries and fuel cells still have not reached the level to obscure enough for autonomy and meet the dynamic characteristics of vehicles. Supercapacitors are only available technology today that can provide high power and great cycle numbers at acceptable price. Supercapacitors have other properties that makes them interesting in hybrid vehicles, and it's ability of complete regeneration of energy of braking (so called regenerative braking), which increases energy efficiency, no special maintenance needed, great utilization of electric energy, small toxicity and easy storage after use.

Acknowledgements

This work was financially supported by the Ministry of Science and Technological Development Republic of Serbia (Projects No. 172060 and TR 32043).

Author details

Zoran Stevic[1*] and Ilija Radovanovic[2]

*Address all correspondence to: zstevic@live.com

1 Technical Faculty in Bor, University of Belgrade, Serbia

2 Innovation Center of School of Electrical Engineering, University of Belgrade, Serbia

References

[1] Zoran Stević, Mirjana Rajčić-Vujasinović, Supercapacitors as a Power Source in Electrical Vehicles, Book title: Electric Vehicles – The Benefits and Barriers / Book 1, Edited by: Seref Soylu, Intech, Rijeka (2011)

[2] Miroslav Bjekić, Zoran Stević, Alenka Milovanović, Sanja Antić, Regulacija elektromotornih pogona, Tehnički fakultet, Čačak (2010)

[3] Dragan Milivojević, Zoran Stević and Mirjana Rajčić-Vujasinović, Hardware and Software of a Bipolar Current Source Controlled by PC, Sensors 8 (2008) 1977- 1983

[4] Omar Ellabban, Review of Torque Control Strategies for Switched Reluctance Motor in Automotive Applications, *ieeeexplore.net/search/*

[5] Z. Stević, M. Rajčić-Vujasinović, Chalcocite as a potential material for supercapacitors, Journal of Power Sources 160 (2006) 1511-1517

[6] Zoran Stević, M. Rajčić-Vujasinović, S. Bugarinović and A.Dekanski, Construction and Characterisation of Double Layer Capacitors, Acta Physica Polonica A, Vol. 117 (2010)1, 228-233

[7] Stević, Z. Supercapacitors based on copper sulfides, Ph.D. Thesis, University of Belgrade, 2001.

[8] Arbizzani, C.; Mastragostino, M. & Soavi, F. (2001). New trends in electrochemical supercapacitors. *Journal of Power Sources*, Vol.100, No1-2, (November 2001), pp. 164-170.

[9] M. Rajcic-Vujasinovic, Z.Stankovic, Z. Stevic, The consideration of the electrical circuit analogous to the copper or coppersulfide/electrolyte interfaces based on the time transient analysis, Russian Journal of Electrochemistry 35,3 (1999) 347- 354

[10] Zoran Stević, Mirjana Rajčić Vujasinović, Aleksandar Dekanski, Estimation of Parameters Obtained by Electrochemical Impedance Spectroscopy on Systems Containing High Capacities, Sensors 9 (2009) 7365-7373

[11] Zoran Stević, Zoran Andjelković, Dejan Antić, A New PC and LabVIEW Package Based System for Electrochemical Investigations, Sensors 8 (2008) 1819-1831

[12] http://www.allaboutbatteries.com/Battery-Energy.html

[13] Carmine Miulli, Testing Modeling and Simulation of Electrochemical Systems, Doctoral Thesis, University of Pisa

[14] Samosir, A.S.; Yatim, A.;, "Dynamic evolution control of bidirectional DC-DC converter for interfacing ultracapacitor energy storage to Fuel Cell Electric Vehicle system," *Power Engineering Conference, 2008. AUPEC '08. Australasian Universities*, vol., no., pp.1-6, 14-17 Dec. 2008

[15] http://phys.org/

[16] www.nanocenter.umd.edu

[17] MIT University, USA http://mit.edu/evt/summary_battery_specifications.pdf

[18] http://www.batterypoweronline.com

[19] Grekulović Vesna J., Rajčić-Vujasinović Mirjana M., Stević Zoran M., Electrochemical behavior of Ag-Cu alloy in alkaline media, Chemical Industry, 64 (2) (2010) 105-110

[20] Mirjana Rajčić-Vujasinović, Svetlana Nestorović, Vesna Grekulović, Ivana Marković, Zoran Stević, Electrochemical behavior of cast CuAg4at% alloy, Corrosion, 66(2010)10, 105004-1-5

[21] Mirjana Rajčić-Vujasinović, Svetlana Nestorović, Vesna Grekulović, Ivana Marković, Zoran Stević, Electrochemical behaviour of sintered CuAg4at.% alloy, Metallurgical and Materials Transactions B, 41B(2010)5, 955-961

[22] Vesna Grekulović, Mirjana Rajčić-Vujasinović, Batrić Pešić, Zoran Stević, Influence of BTA on Electrochemical Behavior of AgCu50 Alloy, Int. J. Electrochem. Sci., 7 (2012) 5231 - 5245.

[23] http://www.autoevolution.com

[24] US Patent, US 5889260, Golan, Gad & Yuly Galperin, "Electrical PTC heating device"

[25] "2010 Options and Packages". *Toyota Prius*. Toyota

[26] Formula One official site, Report http://www.formula1.com/inside_f1/understanding_the_sport/5283.html

[27] Formula One official site, Report http://www.formula1.com/news/headlines/2012/1/12953.html

[28] Pirelli official site, Report, http://www.pirelli.com/tyre/ww/en/f1/tyre-range.html

[29] Goodyear official site, Report http://www.goodyear.eu/uk_en/news/80191-goodyear_tire_for_electric_vehicles_delivers

[30] http://thesmartdrive.com/2009/07/the-effect-of-tire-pressure-on-fuel-economy/

[31] Dr. Uwe Knorr, Dr. Ralf Juchem, A complete co-simulation-based design environment for electric and hybrid-electric vehicles, fuel-cell systems, and drive trains, Ansoft Corporation, Pittsburgh, PA

[32] http://www.racecar-engineering.com/articles/f1/drs-the-drag-reduction-system/

[33] http://www.autosport.com/news/report.php/id/73083

[34] http://auto.howstuffworks.com/fuel-efficiency/hybrid-technology/hybrid-cars-utilize-solar-power1.htm

[35] http://www.teslamotors.com/blog/electric-cars-and-photovoltaic-solar-cells

[36] http://en.wikipedia.org/wiki/Kinetic_energy_recovery_system

[37] http://www.head.com/ski/technologies/skis/?region=eu&id=330

[38] http://www.flybridsystems.com/F1System.html

[39] http://www.formula1.com/inside_f1/understanding_the_sport/8763.html

[40] http://scarbsf1.wordpress.com/2012/04/17/red-bull-kers-floor-mounted-super-capacitors/

[41] Xiaodong Zhang; Chau, K.T.; Chan, C.C.; Gao, S.;, "An automotive thermoelectric-photovoltaic hybrid energy system," Vehicle Power and Propulsion Conference (VPPC), 2010 IEEE, vol., no., pp.1-5, 1-3 Sept. 2010

[42] http://www.heat2power.net

[43] http://www.f1technical.net/forum/viewtopic.php?f=4&t=5751

[44] http://www.thegreencarwebsite.co.uk/blog/index.php/2012/10/12/huge-breakthrough-for-plug-in-hybrid-electric-vehicles/

[45] http://www.triplepundit.com/2010/07/vecarious-will-turn-you-car%E2%80%99s-waste-heat-into-energy/

[46] LaGrandeur, J.; Crane, D.; Hung, S.; Mazar, B.; Eder, A.;, "Automotive Waste Heat Conversion to Electric Power using Skutterudite, TAGS, PbTe and BiTe," *Thermoelectrics, 2006. ICT '06. 25th International Conference on*, vol., no., pp.343-348, 6-10 Aug. 2006

[47] US Patent 5,680,032

[48] S.M. Ferdous, Walid Bin Khaled, Benozir Ahmed, Sayedus Salehin, Enaiyat Ghani Ovy, Electric Vehicle with Charging Facility in Motion using Wind Energy, World Renewable Energy Congress 2011, Sweden S.M. Ferdous, Walid Bin Khaled, Benozir Ahmed, Sayedus Salehin, Enaiyat Ghani Ovy, World Renewable Energy Congress 2011-Sweden

[49] http://www1.eere.energy.gov/vehiclesandfuels

[50] http://www.supercars.net/cars/5835.html Story by Daimler AG

[51] I. Radovanovic, I. Popovic, *"Implementation of Smart Transducer Correction Functions"*, YUINFO2012 Conference, Feb 2012

[52] N. Bezanic, I. Popovic, I. Radovanovic, *"Implementation of Service Oriented Architecture in Smart Transducers Network"*, YUINFO2012 Conference, Feb 2012

[53] Zoran Stević, Optoelektronics, University of Belgrade, Technical faculty in Bor (2005)

[54] Zoran Stević, Mirjana Rajčić-Vujasinović, Dejan Antić, Thermovision applications, University of Belgrade, Technical faculty in Bor (2008)

[55] I. Radovanovic, N. Rajovic, V. Rajovic, J. Jovicic, *"Signal Acquisition And Processing in the Magnetic Defectoscopy of Steel Wire Ropes"*, TELFOR Conference, Nov 2011

[56] I. Radovanovic, N. Rajovic, V. Rajovic, J. Jovicic, *"Signal Acquisition And Processing in the Magnetic Defectoscopy of Steel Wire Ropes"*, TELFOR Journal, Vol. 4, No. 2, 2012

[57] Dj. Klisic, M. Zlatanovic,I. Radovanovic, *"Application database of wind measuring stations for testing CFD model of software development tool WindSim"*, 55th ETRAN Conference, Jun 2011

[58] Dj. Klisic, M. Zlatanovic,I. Radovanovic, *"WindSim Computational Flow Dynamics Model Testing Using Databases from Two Wind Measurement Stations"*, ELECTRONICS Vol. 15, Number 2, pp.43-48, Dec 2011

[59] Comparative Study of Various Touchscreen Technologies, M.R. Bhalla, A.V. Bhalla, International Journal of Computer Applications (0975 – 8887) Volume 6– No.8, September 2010

[60] http://gps.toptenreviews.com/navigation

[61] http://www.ipcmax.com

[62] Zoran R. Andjelkovic, Dragan R. Milivojevic, Zoran M. Stevic, Thermovisual camera commands decoding and ISI format encrypting, Journal of Scientific and Industrial Research, Vol. 69, (2010)

[63] www.infraredsolutions.com

[64] Zoran Stevic, Dubravka Nikolovski, Branislava Matic, Computerized Infrared Thermography in Distance Monitoring of Aging People, Proceedings of the XI International scientific-practical conference "Modern information and electronic technologies", Odessa, Ukraine, 24-28 May 2010, p 189

[65] Siddiqi, U.F.; Shiraishi, Y.; Sait, S.M.;, "Multi-constrained route optimization for Electric Vehicles (EVs) using Particle Swarm Optimization (PSO)," *Intelligent Systems Design and Applications (ISDA), 2011 11th International Conference on*, vol., no., pp.391-396, 22-24 Nov. 2011

[66] Siddiqi, U.F.; Shiraishi, Y.; Sait, S.M.;, "Multi constrained Route Optimization for Electric Vehicles using SimE," *Soft Computing and Pattern Recognition (SoCPaR), 2011 International Conference of*, vol., no., pp.376-383, 14-16 Oct. 2011

Chapter 5

Batteries and Supercapacitors for Electric Vehicles

Monzer Al Sakka, Hamid Gualous, Noshin Omar and
Joeri Van Mierlo

Additional information is available at the end of the chapter

http://dx.doi.org/10.5772/53490

1. Introduction

Due to increasing gas prices and environmental concerns, battery propelled electric vehicles (BEVs) and hybrid electric vehicles (HEVs) have recently drawn more attention. In BEV and HEV configurations, the rechargeable energy storage system (RESS) is a key design issue [1–3]. Thus, the system should be able to have good performances in terms of energy density and power capabilities during acceleration and braking phases. However, the thermal stability, charge capabilities, life cycle and cost can be considered also as essential assessment parameters for RESS systems.

Presently batteries are used as energy storage devices in most applications. These batteries should be sized to meet the energy and power requirements of the vehicle. Furthermore, the battery should have good life cycle performances. However, in many BEV applications the required power is the key factor for battery sizing, resulting in an over-dimensioned battery pack [4,5] and less optimal use of energy [4]. These shortcomings could be solved by combination of battery system with supercapacitors [6–8]. In [9], it is documented that such hybridization topologies can result into enhancing the battery performances by increasing its life cycle, rated capacity, reducing the energy losses and limiting the temperature rising inside the battery. Omar et al. concluded that these beneficial properties are due to the averaging of the power provided by the battery system [4,6,9]. However, the implementation of supercapacitors requires a bidirectional DC–DC converter, which is still expensive. Furthermore, such topologies need a well-defined energy flow controller (EFC). Price, volume and low rated voltage (2.5–3 V) hamper the combination of battery with supercapacitors [6,10]. In order to overcome these difficulties, Cooper et al. introduced the Ultra-Battery, which is a combination of lead-acid and supercapacitor in the same cell [11]. The new system encompasses a part asymmetric and part conventional negative plate. The proposed system allows to deliver and to absorb energy at very high current rates. The Ultra-Batteries have been tested successfully in the Honda Insight. However, this technology is still under development. In the last decade, a number of new lithium-ion battery chemistries have been proposed for vehicular applications. In [12–15], it is reported that the most relevant lithium-ion chemistries in vehicle applications are limited to lithium iron phosphate (LFP), lithium nickel manganese cobalt oxide (NMC), lithium nickel cobalt aluminum oxide (NCA), lithium manganese spinel in the positive electrode and lithium titanate oxide (LTO) in the negative electrode. In this chapter, the performance and characteristics of various lithium-ion based batteries and supercapacitor will be evaluated and discussed. The evaluation will be mainly based on the electrical behavior. Then the characteristics of these RESS systems will be investigated based on the electrical and thermal models.

2. Batteries

2.1. Electrical characterization

It is well known that the key consideration in the design of rechargeable energy storage systems in PHEV and BEV applications mainly depend on the power density (kW/kg) and energy density (Wh/kg) due to the design concept. However, the battery technology also should be able to have good performances in the terms of energy efficiency, lifetime, and charging rate [12-15]. In this section all these parameters have been analyzed for 10 lithium-ion battery types as presented in Table 1.

	A	B	C	D	E	F	G	H	I	J
Cathode	LFP	LFP	LFP	NMC	NMC	NCA	LFP	LFP	LFP	LFP
Shape	Cylindrical	Pouch	Pouch	Pouch	Pouch	Cylindrical	Pouch	Cylindrical	Prismatic	Prismatic
Nominal capacity [Ah]	10	10	40	12	70	27	14	2.3	10	40
Nominal voltage [V]	3.3	3.3	3.3	3.7	3.7	3.7	3.3	3.3	3.3	3.3

Table 1. Specifications investigated lithium-ion battery brands [12].

In [16] the main design concepts of PHEV applications are discussed, compared to the three sets of influential technical goals, and explained the trade-offs in PHEV battery design. They mentioned that the energy and power requirements according to the U.S. Advanced Battery Consortium (USABC) should be in the range of 82 Wh/kg and 830 W/kg for PHEV-10 and 140 Wh/kg and 320 W/kg for PHEV-40. Pesaran specified these two battery types as high power/energy ratio battery (PHEV-10) and low power/energy battery (PHEV-40). The first category PHEV-10 is set for a "crossover utility vehicle" weighing 1950 kg and PHEV-40 is set for a midsize sedan weighing 1600kg [16]. In this study, only the battery performance characteristics for PHEV-40 (40 miles All Electric Range) is investigated based on the USABC goals [16].

Figure 1 shows the results of the Dynamic Discharge Performance test (DDP) and the Extended Hybrid Pulse Power Characterization (HPPC) test [12,17-19]. As one can see, the energy density of nickel manganese cobalt oxide (LiNiCoMnO2) based battery types D&E is in the range of 126 – 149Wh/kg while the cells using iron phosphate in the positive electrode show energy density being in the range of 75 – 118Wh/kg. In [20], is reported that the high energy density values for the $LiNiCoMnO_2$ batteries is mainly due to the higher nominal voltage (e.g. 3.7V) and good electrode specific capacities. However, the situation regarding the power density is not clear due the fact that power is varying over a wide range. Figure 1 shows that only cell type D using $LiNiCoMnO_2$ has the highest power density around 2100Wh/kg. This result is mainly due to the good specific impedance [20].

The results indicate also that iron phosphate based battery types B and H have good power performances being in the range of 1580-1650 W/kg. However, based on the USABC goals, all the tested cells can meet the power requirements of 320W/kg with exception of battery F 290W/kg. Although the battery type E has the best energy density, the power capabilities of this battery are limited in comparison to the batteries types B, D and H, which indicates that this battery is more appropriate for BEV applications as reported in [12]. The presented results in Figure 1 are based on the maximum discharge C-rate at 50% state of charge.

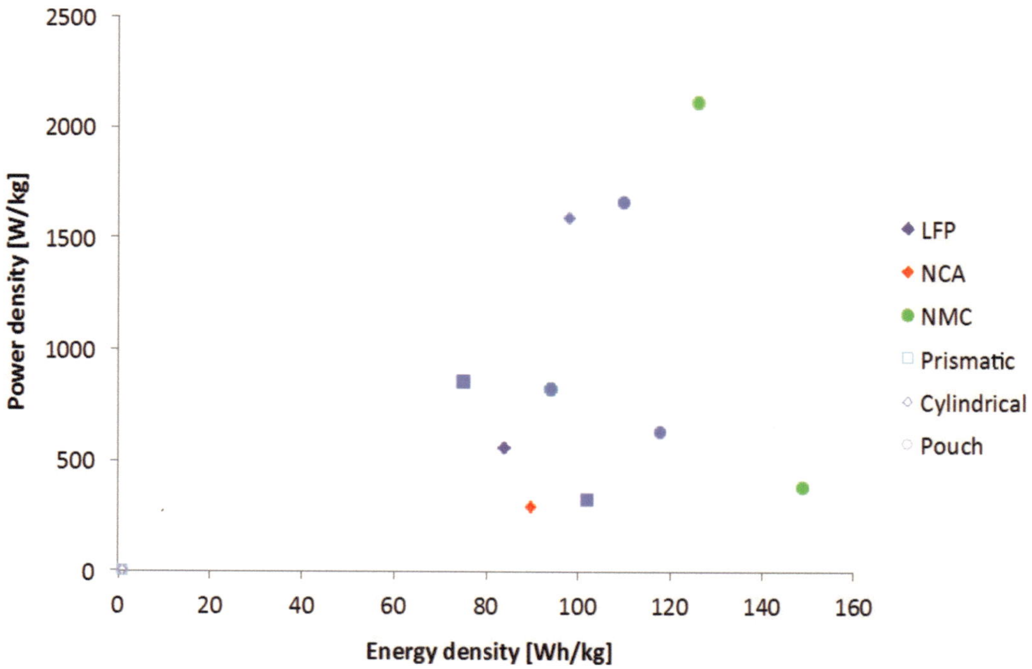

Figure 1. Power density versus energy density at room temperature [12].

2.1.1. Energy efficiency

In PHEV applications, energy efficiency during charge and discharge phases can be considered as one of the key factors. High-energy efficiency is desired to limit the temperature rise inside a battery pack. In this section, the energy efficiency of the proposed battery types has been considered based on the DDP test [19].

It is well pointed out in Figure 2, that the energy efficiency of the nickel manganese cobalt oxide based cells is around 94 – 96%. While the iron phosphate and nickel cobalt aluminum in the positive electrode show generally a lower efficiency in the range of 88 – 93%. The lower energy efficiency for lithium iron phosphate based batteries can be explained due to the relative lower conductivity of cathode material compared to NMC based batteries.

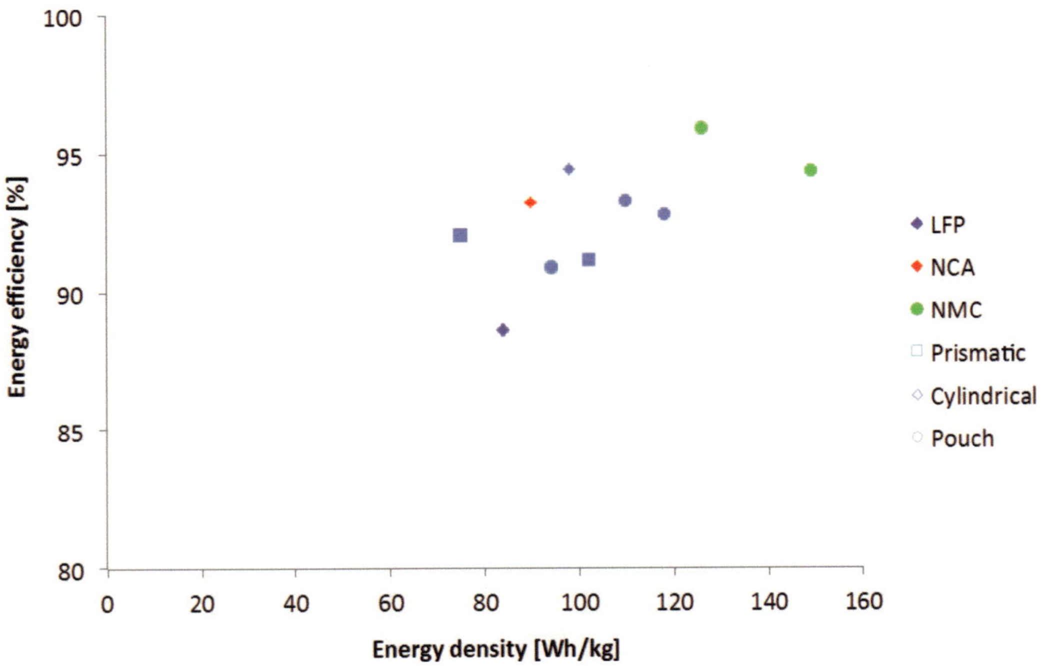

Figure 2. Energy efficiency versus energy density at room temperature [12].

2.1.2. Charge performances

It is generally known that PHEV applications are an important factor for improving the impact of traffic on healthier living environment by emitting a lower amount of CO_2 than the conventional vehicles. However, the advantages of PHEV applications mainly depend of the energy storage device. On the other hand, in order to enhance the suitability of the battery technology in PHEV applications, the battery requires besides good power, energy and energy efficiency performances also acceptable fast charging capabilities. In [21], it is well reported that the charging process of battery typically involves two phases:

http://dx.doi.org/10.5772/53490

- The main charging phase, where the bulk of energy is recharged into the battery (constant current),
- The final charge phase, where the battery is conditioned and balanced (constant voltage).

In this section, the fast charging performances of the different batteries until the main charging phase have been analyzed. In this study the main charging phase has been considered at different charge current rates (0.33 I_t, 1 I_t, 2 I_t and 5 I_t). The reference test current I_t can be expressed as according to the standard IEC 61434 [22]:

$$I_t[A] = \frac{C_n[Ah]}{1h} \tag{1}$$

Figure 3 shows clearly that lithium-ion battery technology have high charge performances. For most lithium-ion batteries, the stored capacity up to V_{max} is above 60% at 5 I_t. Due to the higher charge current rates, the charge time can be reduced with a factor 10. The discharge time is less than 1 hour instead of 8 hours as mentioned in [15]. Here it should be noted that battery cells with high energy density, which are designed for BEVs and PHEVs show high performances between 1 I_t and 2 I_t but indicate less performances at higher current rates (> 2 I_t) [12].

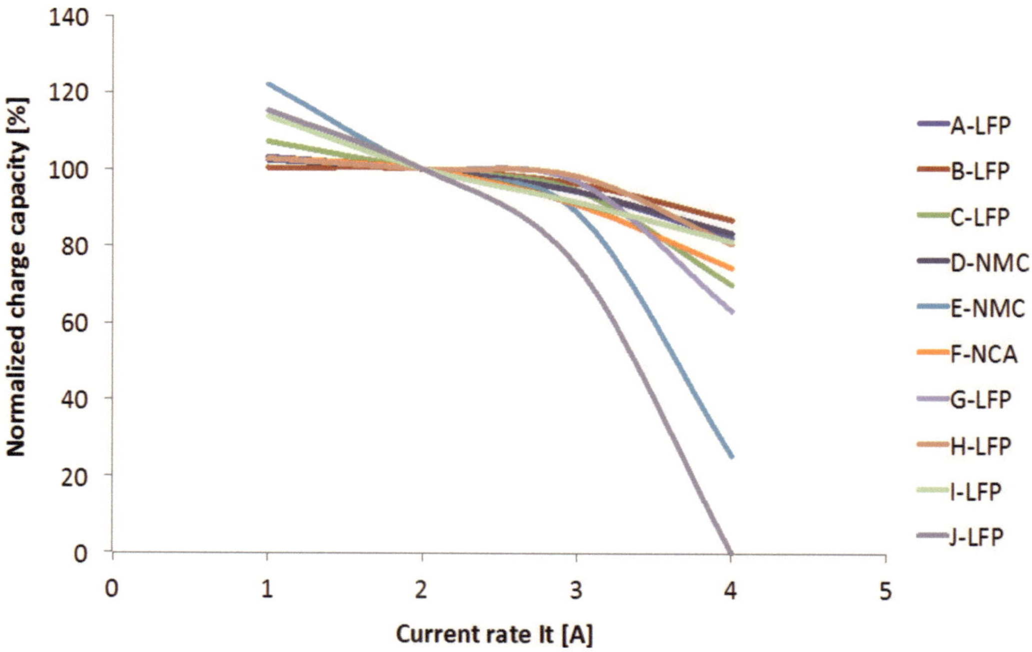

Figure 3. Evolution of stored capacity during main charging phase [12].

2.2. Thermal characterization

According to the United States Advanced Battery Consortium, the battery system in HEVs, PHEVs and BEVs should operate over a wide operating temperate (from -40°C until 60°C) In order to illustrate the battery behavior at different working temperatures, the same dynamic discharge performance test as described above has been performed at -18°C, 0°C, 25°C and 40°C as described in the standard ISO 12405-1/2 and IEC 62660-1 [23-25].

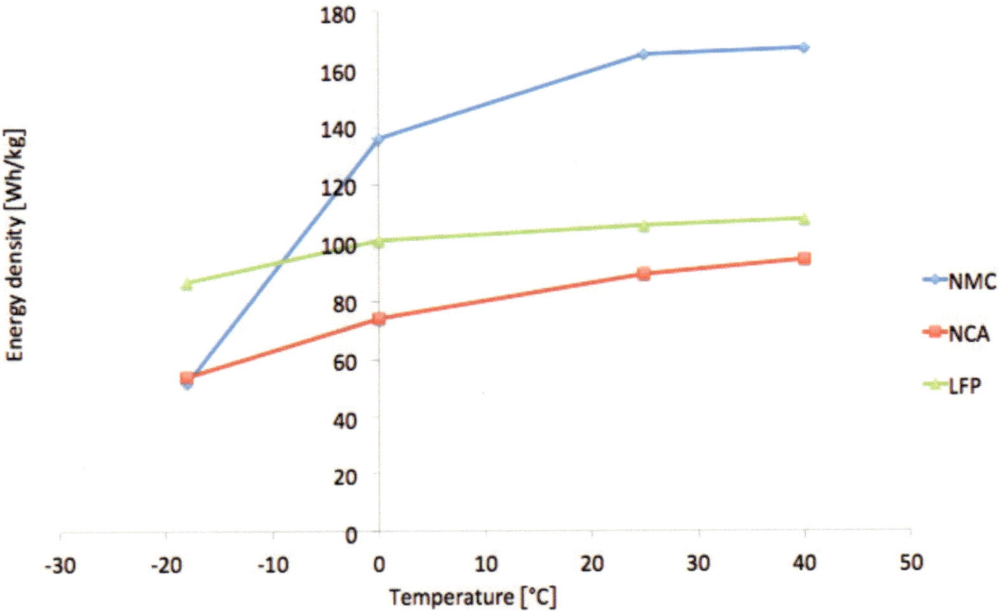

Figure 4. Evolution of energy density as function of working temperature [14].

Figure 4 illustrates that the nickel manganese cobalt oxide based battery (type E) has an energy density of 150-125 Wh/kg in the temperature range of 40°C and 0°C. While the energy density of lithium iron phosphate (type H) and lithium nickel cobalt aluminum oxide in the positive electrode (type F) seem to have less favorable performances 108-101 Wh/kg for LFP and 94 74 Wh/kg, for NCA. However, the performances at -18°C are less beneficial for NMC battery type around 50 Wh/kg against 54 Wh/kg and 86 Wh/kg for NCA and LFP, respectively. These results show that the energy density reduction is 60% for NMC, 40% for NCA and 20% for LFP cells. This means that a heating system will be more than desired for NMC and NCA cells in order to keep the battery cells in the appropriate temperature envelope (40°C-0°C), where the energy performances are relative high. The high energy density in the case of NMC at 40°C and 25°C are due to the good specific capacity and the higher nominal voltage. The obtained energy density for nickel cobalt aluminum in the positive electrode is quite small against what is documented by Burke [9]. The reason is that the investigated cells (see Table 1) are dimensioned for hybrid applications rather than battery propelled electric vehicles. In [26] is reported that the limitation of the energy density at low temperatures is mostly related to the considerable increasing of the internal resistance. However, Figure 5 indicates that the aspect does not apply for LFP based battery. The normalized internal resistance increases in the case of the latter mentioned cell chemistry is 650% compared at the reference temperature (25°C). The internal resistance has been determined at 100% SoC and the applied current was 0.1 I_t and 1 I_t.

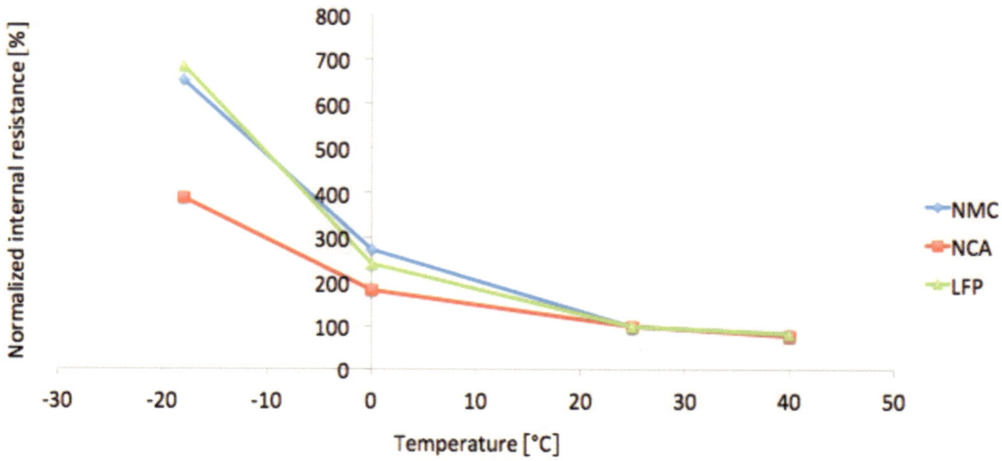

Figure 5. Evolution of the internal resistance as function of the working temperature [14].

In order investigate the behavior of the proposed LFP and NMC based batteries in depth, a number of capacity tests have been carried out at current rates 2 I_t, 5 I_t and 10 I_t at 0°C. Figure 6 and Figure 7 show the favorable performances of the LFP chemistry against the NMC. Especially at 0°C, the LFP battery demonstrates the excellent performances due to the self-heating mechanism that occur at high current rates. In Figure 6, we observe that the voltage at 10 I_t drops fast but remains above the minimal voltage: 2V. Then, the voltage recovers when the battery temperature considerable increases (43°C) due to the higher internal resistance. The battery is able to attain almost the same discharge capacity as at lower current rate and high working temperature as it is illustrated in Figure 8. Here, we can notice that the Peukert number in the temperature range (0°C – 40°C) is close to one as is reported by Omar et al. [7]. However, at low temperatures (-18°C and forward) the Peukert number increases (1.85) due to the reducing of the discharge capacity, which is caused by the significantly high internal resistance. It should be pointed out that in the region 0.33 I_t and 2 I_t, the Peukert number is smaller than 1, which is in contradiction with the Peukert phenomena. The explanation of this behavior is due to the fact that the Peukert relationship has been extracted particularly for lead acid batteries and for relative low current rates and in operating temperatures, which is close to the room temperature. However, for lithium-ion batteries and mainly at low temperatures (-18°C), there are another complex phenomena that occur that only cannot be explained by Peukert.

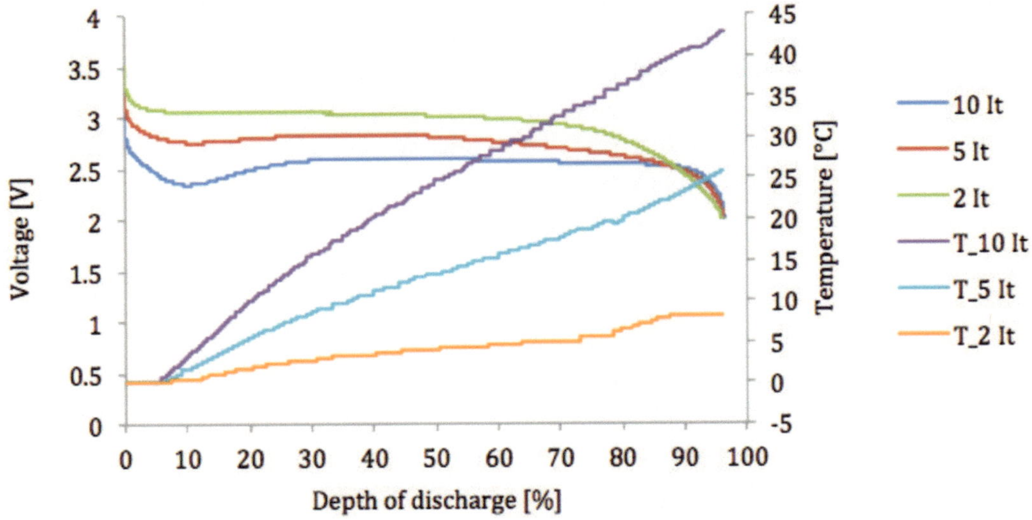

Figure 6. Illustration of the voltage and temperature evolution of LFP based battery versus depth of discharge at different current rates at 0°C [14].

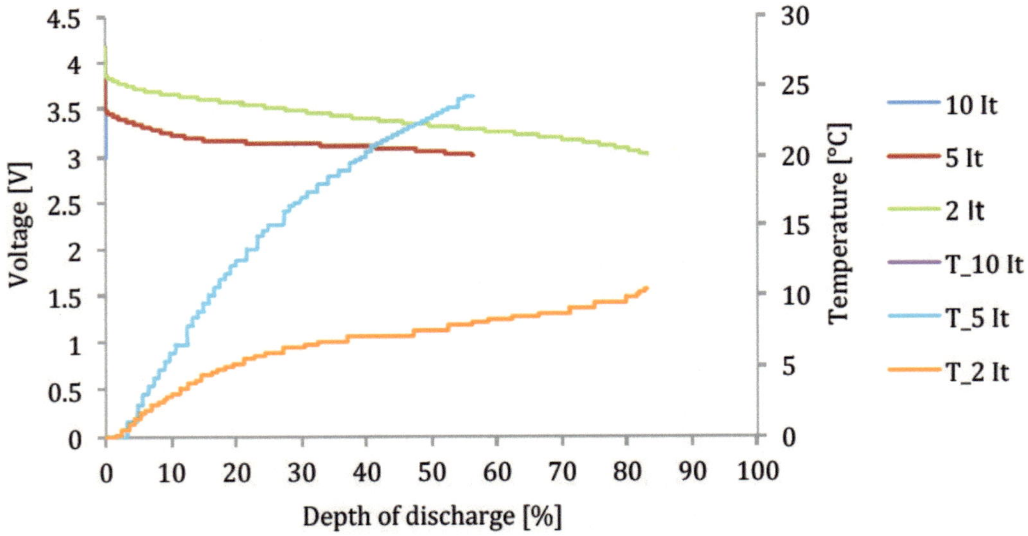

Figure 7. Illustration of the voltage and temperature evolution of NMC based battery versus depth of discharge at different current rates at 0°C [14].

2.3. Electrical and thermal modeling

In development of an appropriate battery pack system, the battery management system can be assumed as a key system [27]. The accuracy and the performances of this system depend on the developed balancing system and an accurate electrical and thermal battery model which can predict the battery cell behavior under all operational conditions. The electrical model is required for prediction of the battery behavior such as energy, power, internal resistance, life cycle and energy efficiency. On the other hand the thermal model is needed to predict the surface temperature of the battery cell for operating of the cooling and heating system when required. Further, the output of the thermal model will be used as an input for the electrical model due to the dependency of the model parameters as a function of the temperature. In this section the performances of the well-known first order FreedomCar battery model will be analyzed by using a dedicated test protocol and a new estimation technique. Then, the analysis is extended with a novel developed thermal model that has been developed at the Vrije Univeriteit Brussel for lithium-ion batteries.

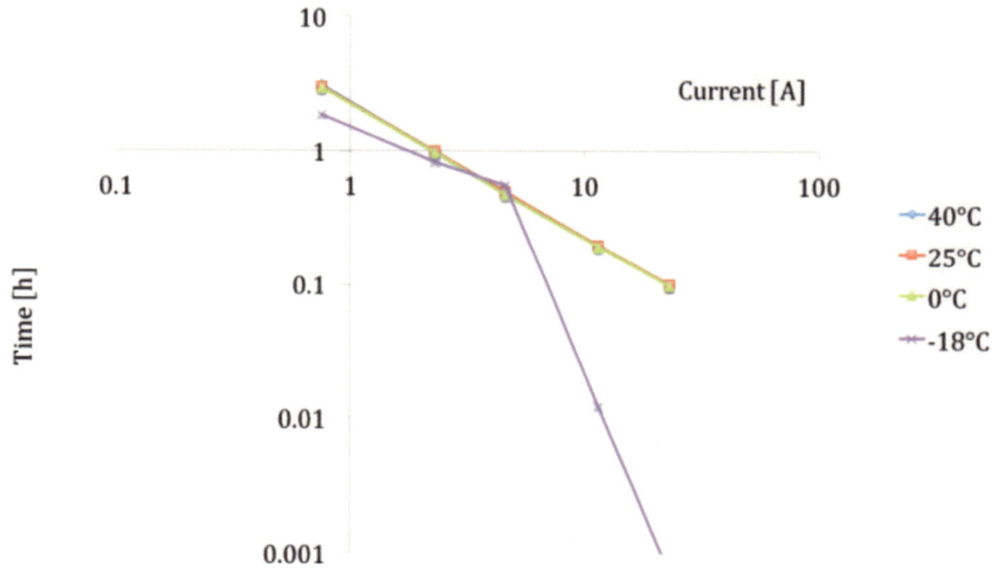

Figure 8. Illustration of the Peukert as function of the operating temperature (LFP) [14].

2.3.1. Electrical model: FreedomCar battery model

As reported above, the BMS requires an accurate electrical battery model for prediction of the battery behavior during the short and long term. Therefore, in the literature, one can find a number of electrical models such as Thévenin, FreedomCar, second order FreedomCar and RC battery model [28, 29]. The Thévenin battery model is a modified model of the FreedomCar battery model as it is presented in Figure 9. The Thévenin model is during steady state operations less accurate than the FreedomCar model due to the absent of the fictive capacitor 1/OCV'. The second order FreedomCar battery model has relatively higher performances than the Thévenin battery model, but this model is also more complicated due to the present of two RC-circuits in the system, which seems in the reality too heavy for BMS in PHEVs and BEVs where 100 battery cells are connected in series. Therefore, the processing unit should be very powerful.

In the framework of this section, only the characteristics of the first order FreedomCar battery model will be addressed and compared with experimental results. As it presented in Figure 9, the FreedomCar model exists mainly of an ohmic resistance (R_o), a fictive capacitor (1/OCV') which represents the variation of the voltage over the time, an open circuit voltage OCV and a RC circuit existing of a polarization resistance R_p and capacitor C. The model assumes that the battery model parameters should be as function of state of charge and temperature. However, the researchers at the Vrije Universiteit Brussel found that the impact of the current rate and cycle life are also important parameters that cannot be avoided [28]. Then, the researchers found also that the ohmic resistance should be divided into two parts: the charge ohmic resistance and the discharge ohmic resistance due to the battery hysteresis [28].

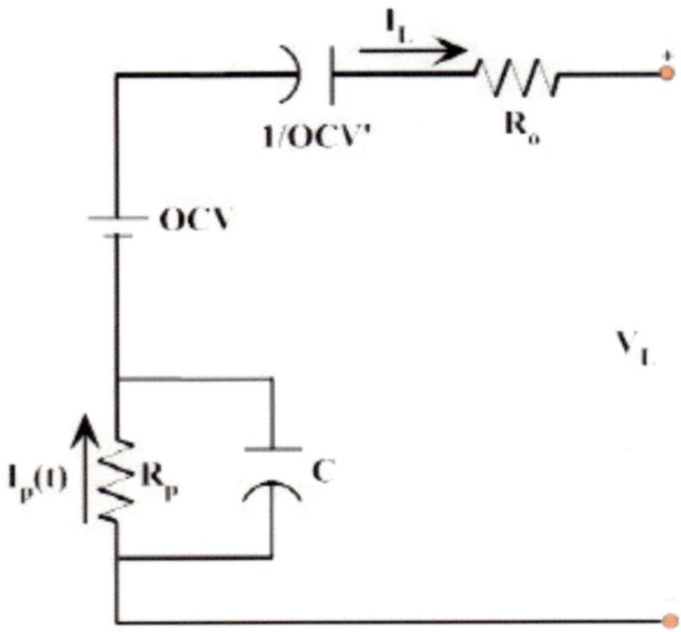

Figure 9. First order FreedomCar battery model [28].

2.3.2. Calibration and validation results

Prior starting with validation of the proposed battery model, the model has been calibrated by performing a new developed test profile at the Vrije Universitiet Brussel as it is presented in Figure 10. As we can observe, there is a good agreement between the simulation and the experimental results. According to these results, the error percentage is not higher than 3.5%. This indicates the powerful performances of the proposed battery model with the developed estimation technique.

http://dx.doi.org/10.5772/53490

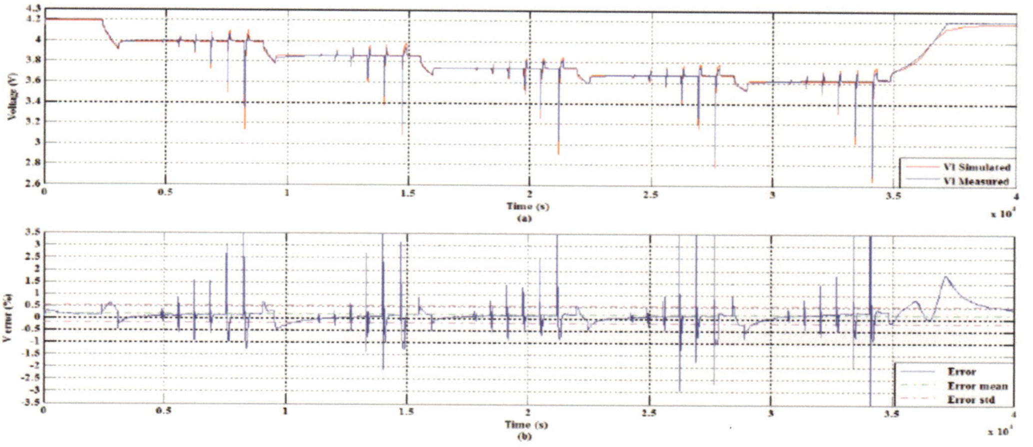

Figure 10. Calibration of the first order FreedomCar battery model at room temperature [28].

2.3.3. Thermal model

Regarding the prediction the thermal behavior of a battery, this can be performed by using high accurate thermal sensors or by dedicated thermal battery models. However, thermal models have many advantages against thermal sensors. The sensors can only measure one specific point. As it is generally know the heat distribution over the surface temperature of the battery is not uniform. In order to have a good sight of the heat development inside the battery, several thermal sensors are needed. This issue will complicate the BMS and the processing time of the BMS will be significantly longer. Therefore, it is more of high interest to issue thermal model which can predict the heat development and distribution over the battery surface. Further, such models allow in advance the battery pack designer to investigate the weakness in the battery pack and to dimension the cooling system more accurately. Finally, the development cost of such battery model is less than the cost of the significant higher number thermal sensors that are needed. In this perspective, a novel thermal model has been developed at the Vrije Universiteit Brussel that can be used for lithium-ion batteries and supercapacitors [30, 31]. In Figure 11 the thermal model is illustrated. As we can observe, the model exists of the following components [30, 31]:

- P_{gen} represents the heat generation (irreversible heat)
- C_{th} stands for the thermal capacitance,
- R_{thi} is the thermal resistance,
- R_{con} represents the convection thermal resistance,

2.3.4. Calibration and validation results

In order to verify the developed thermal battery model, series of comparisons are made based on simulation and experimental results. The first test is presented in Figure 12. As we observe, the model is in good agreement with experimental results. The errors percentage based on this test is in the range of 1°C. In this test, the model has been compared with experimental results based on the load profile as proposed in Figure 13 until the surface temperature has reached the steady stated condition.

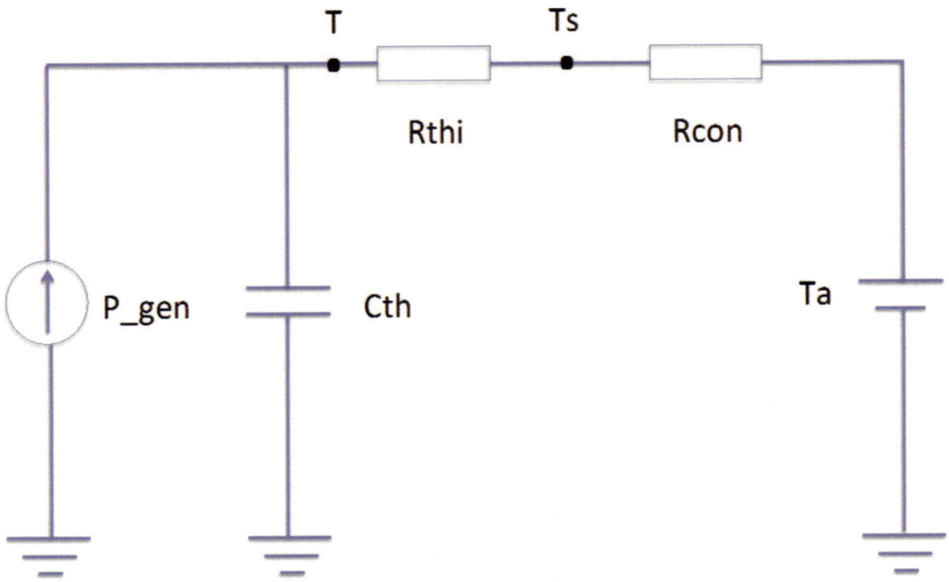

Figure 11. Novel thermal batter model for lithium-ion batteries and electrical double-layer capacitors [30].

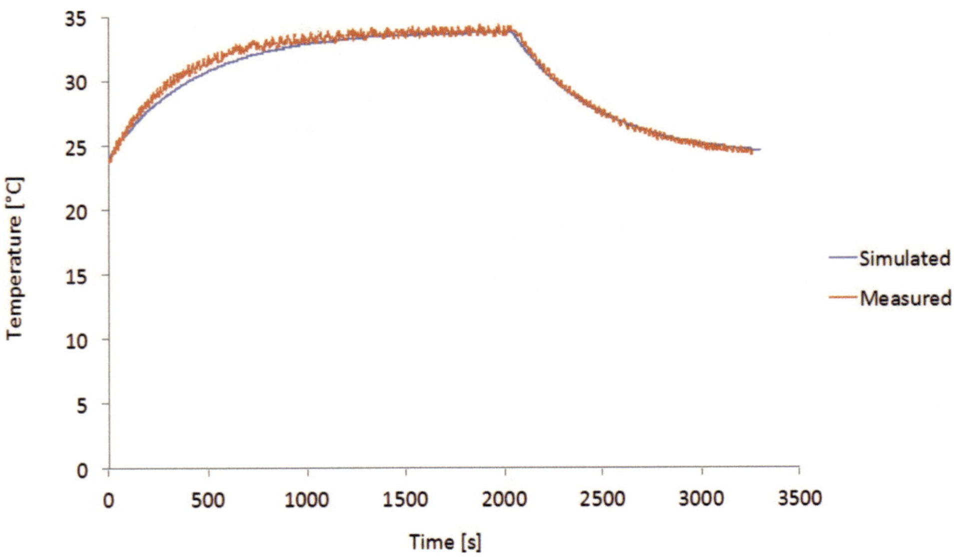

Figure 12. Comparison of simulated and measured at 25°C working temperature [30].

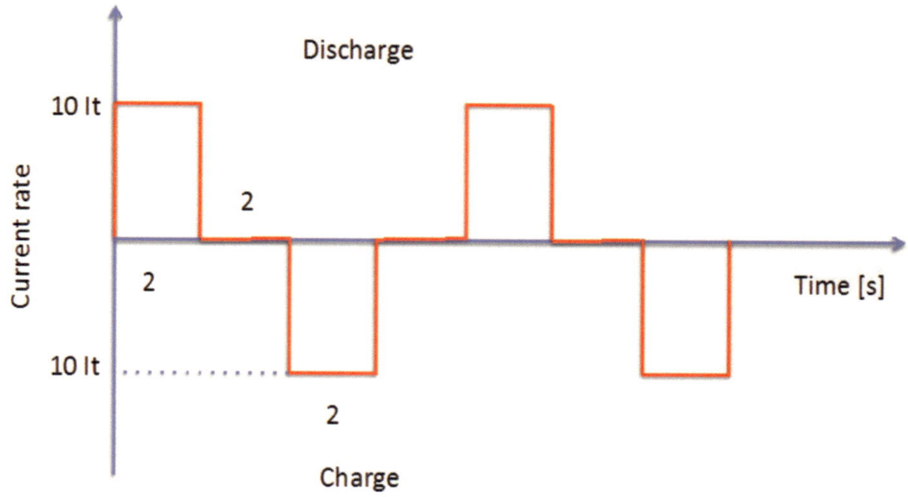

Figure 13. Used load profile for extraction of the thermal model parameters [30].

However, there is a need for validation step to evaluate the performances and accuracy of the developed battery model at other conditions without to perform any calibration in the model. In Figure 14 a validation test has been carried out at room temperature about 24°C. The corresponding simulation and experimental comparison are illustrated in Figure 15. Here again, we recognize that the high accuracy of the battery model against the experimental results. Based on these results, we can conclude that the developed battery model is able to predict the surface temperature of the battery cell with significantly low errors.

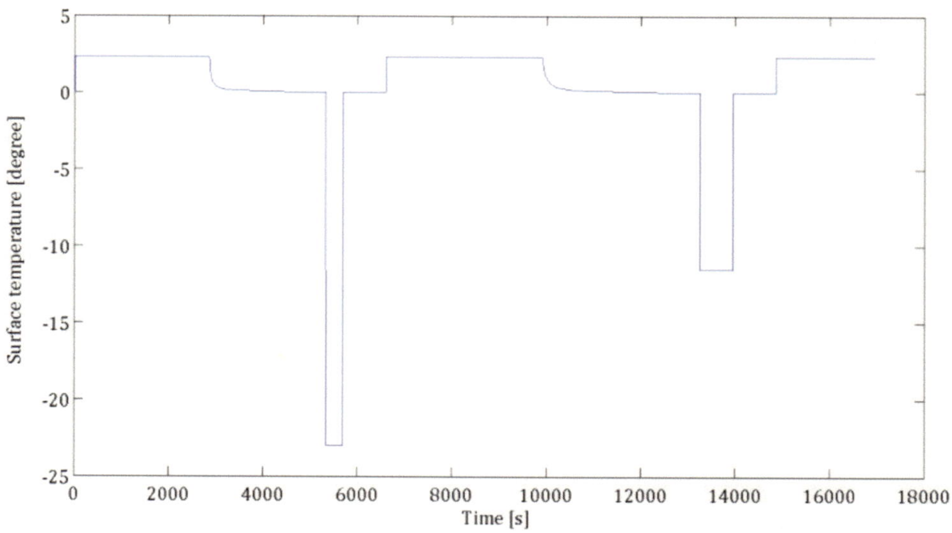

Figure 14. Load profile for validation [30].

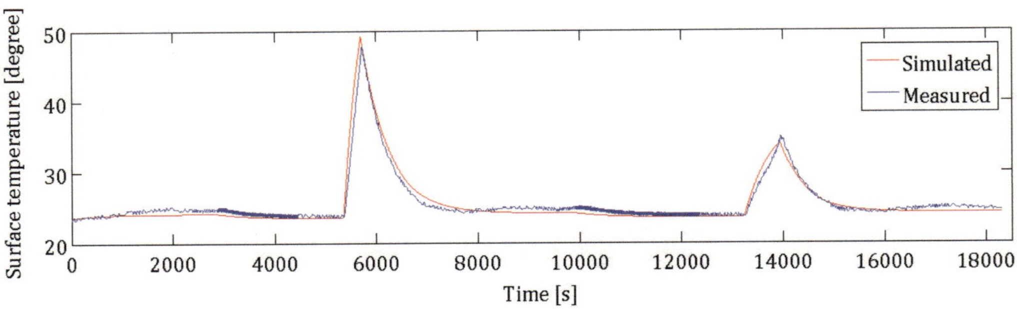

Figure 15. Comparison of experimental and simulation results at room temperature (~24°C) [30].

3. Supercapacitors

Supercapacitors, also known as Electric Double-Layer Capacitors (EDLCs)or ultra capacitors, have a high energy density when compared to conventional capacitors, typically thousands of times greater than a high capacitance electrolytic capacitor. For example, a typical electrolytic capacitor will have a capacitance in the range of tens of milli-farads. The same size supercapacitor would have a capacitance of several farads. Larger supercapacitors have capacitance up to 5000 farads. The highest energy density in production is 30 Wh/kg. Although supercapacitors have very high power density and capacitance values of thousands of Farads are possible, the cell voltage is limited to about 2.7 V to avoid electrolysis of the electrolyte with the consequent emission of gas and deterioration of the supercapacitor cell. The structure of a basic cell is mostly cylindrical. However, there are also now commercial pouch supercapacitors available. The technology achievement is identical to that used for conventional capacitors. The supercapacitors cells used in this study are the BCAP310F and BCAP1500F. Their properties are based on the double layer capacitance at the interface between a solid conductor and an electrolyte. The elementary structure consists of two activated carbon electrodes and a separator impregnated with an electrolyte. The electrodes are made up of a metallic collector, coated on both side with an active material, which has a high surface area part which is required for the double layer. The two electrodes are separated by a membrane (separator), which prevents the electronic conduction by physical contact between the electrodes but allows the ionic conduction between them. This composite is subsequently rolled and placed into a cylindrical container. The system is impregnated with an organic electrolyte. The two electrodes are metalized and connected to the outside (+) and (-) terminal connections of the supercapacitor.

3.1. Electrical characterization

Equivalent series resistance and capacitance of supercapacitor calculation methods:

3.1.1. *Using an Electrochemical Impedance Spectroscopy (EIS)*

Electrochemical impedance spectroscopy (EIS) is used in the characterization of electrochemical behavior of energy storage devices. Impedance analysis of linear circuits is much easier than analysis of non-linear ones. Electrochemical cells are not linear. Doubling the voltage will not necessarily double the current. However, the electrochemical systems can be pseudo-linear. In normal EIS practice, a small (1 to 10 mV) AC signal is applied to the cell. With such a small potential signal, the system is pseudo-linear.

The supercapacitor is polarized with a dc voltage. A small voltage ripple, typically 10mV, is superimposed on the dc component. The ripple frequency is swept between 1 mHz and 1 kHz. The measurement of the current amplitude and phase with respect to the injected voltage permits the determination of the real and imaginary components of the impedance as a function of the frequency. The measurements were performed in a controlled climatic chamber. The supercapacitor capacitance C and the series resistance (ESR) are deduced from the experimental results, respectively.

$$C = \frac{-1}{2\pi \cdot \mathrm{Im}(z) \cdot f} \tag{2}$$

$$ESR = \mathrm{Re}(z) \tag{3}$$

Where:

- Im(z) is the imaginary component of the supercapacitor impedance,
- Re(z) is the real component of the supercapacitor impedance,
- F is the frequency.

The Maxwell BCAP310F and BCAP1500F supercapacitors used in this study are based on activated carbon technology and organic electrolyte. These devices were characterized using the Electrochemical Impedance Spectroscopy (EIS) [32].

Figure 16 and Figure 17 represent the BCAP310F and the BCAP1500F capacitance and ESR as a function of frequency.

At low frequency, the capacitance is maximum, for example at 10mHz the capacitance value is in order of 1660F for the BCAP1500F and 315F for the BCAP310F. At 50mHz the ESR value is in order of 1mΩ for BCAP1500F and 5.2mΩ for BCAP310F. The BCAP310F ESR is relatively high because this device was fabricated, by Maxwell Technologies, especially for these thermal tests; it is including 4 thermocouples type K inside.

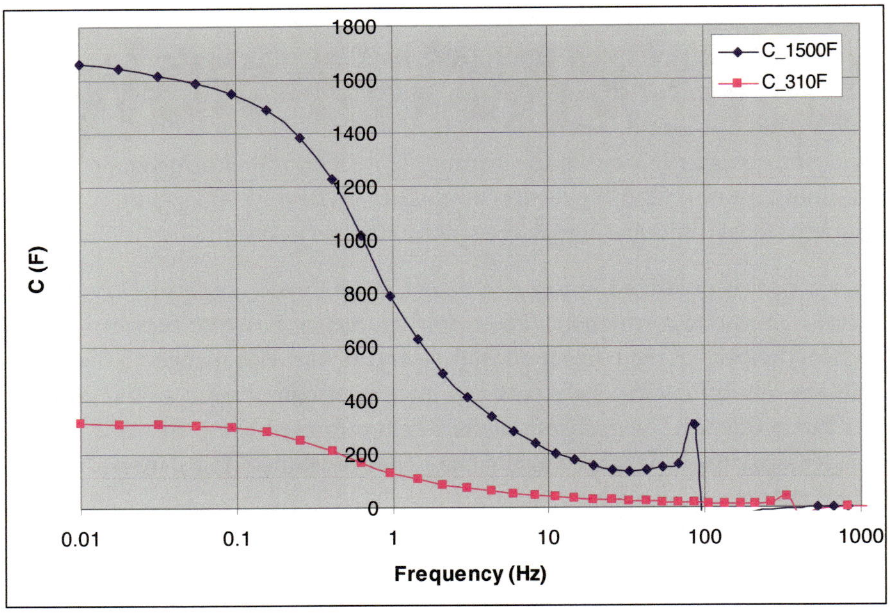

Figure 16. BCAP1500F and BCAP310F capacitance as function of frequency with a bias voltage respectively of 2.7V and 2.5V and a temperature of 20°C.

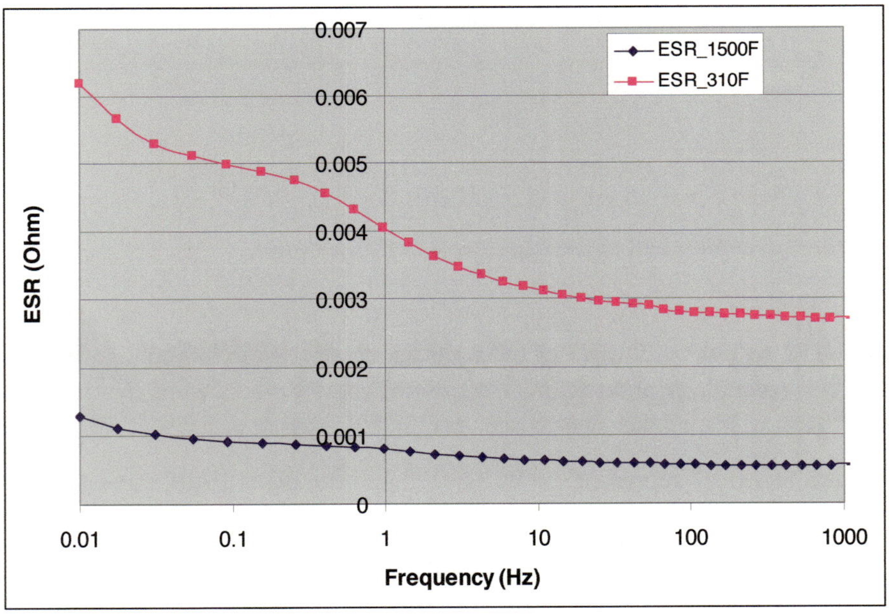

Figure 17. BCAP1500F and BCAP310F series resistance as function of frequency with a bias voltage respectively of 2.7V and 2.5V and a temperature of 20°C.

http://dx.doi.org/10.5772/53490

3.1.2. Based on the IEC 62576 standard

The standard IEC (International Electro-technical Commission) 62576 [33,34] defines the calculation methods of the equivalent series resistance and the capacitance of electric double-layers capacitors.

Figure 18 presents the calculation method of the equivalent series resistance. The supercapacitor is charged at constant current to its nominal voltage, this voltage should be maintained at this value during 30 min. Then, the supercapacitor is discharged at constant current up to 0V. The value of the constant current depends on the applications. The IEC 62576 suggests to choose 10xC, 4xCxUr, 40*C*Ur and 400xCxUr mA for the supercapacitors applied as memory backup (class 1), energy storage (class 2), power unit (class 3) and instantaneous power unit (class 4), respectively [33,34]. Where, C is the capacitance and Ur represents the rated voltage.

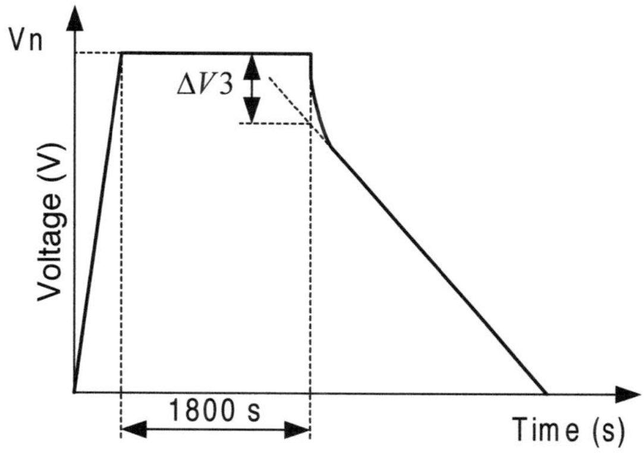

Figure 18. Charge and discharge of the supercapacitor at constant current

The ESR value is calculated based on the following expression:

$$ESR = \frac{\Delta V3}{I} \tag{4}$$

Where ΔV3 is the voltage drop obtained from the intersection of the auxiliary line extended from the straight part and the time base when the discharge starts, and I is the constant discharging current.

Figure 19 presents the calculation method of the capacitance.

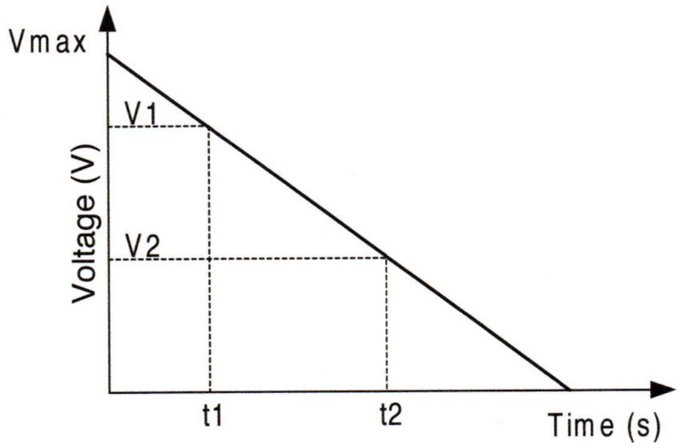

Figure 19. Discharge at constant current of the supercapacitor

The capacitance value is calculated using the following expression:

$$C = \frac{I \cdot \Delta t}{\Delta V} \tag{5}$$

Where I is the constant discharging current, Δt=t2-t1 and ΔV=V1–V2, V1=80%*Vmax, V2=40%*Vmax and Vmax is the maximum voltage of the supercapacitor.

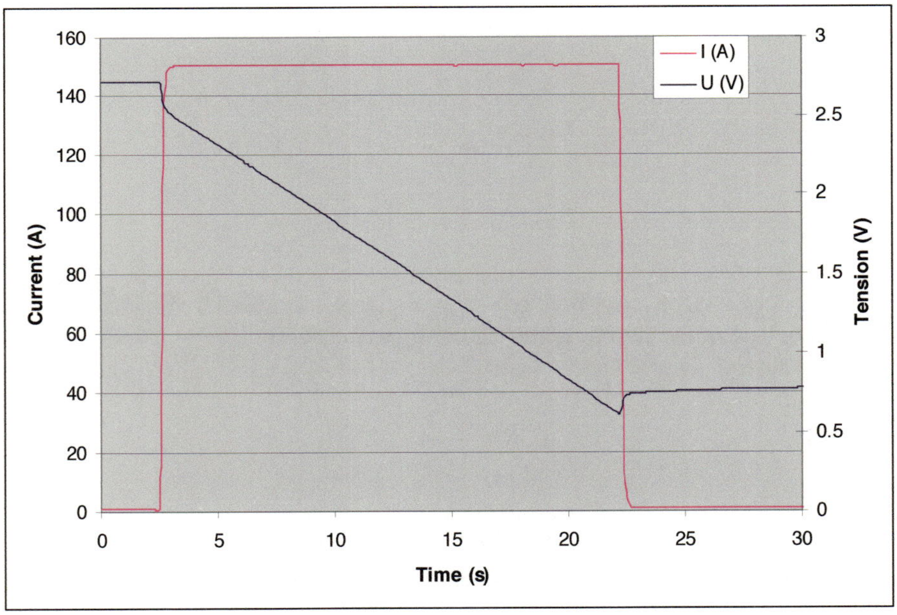

Figure 20. Experimental results of BCAP1500F voltage and current as a function of time.

The BCAP1500F capacitance and ESR were calculated according to the IEC 62576 standard. The supercapacitor is discharged at constant current 100mA/F. Figure 20 represents the BCAP1500F voltage and current versus time during the discharge. ESR and C are 1.07mΩ and 1525F, respectively.

3.2. Thermal characterization

Heat production in supercapacitor is related exclusively to Joule losses. The supercapacitors support currents up to 400A or more depending on cell capacitance and used technology. The repetitive charge and discharge cycles of the supercapacitor cause a significant warming even though the equivalent series resistance value is around the mΩ according to the capacitance. Several authors showed that the supercapacitor ESR varies according to the temperature [35-37]. In [38] the authors have studied the effect of the temperature and the voltage on the supercapacitors ageing. They have established a model which allows analyzing self-accelerating degradation effects caused by elevated voltages and temperatures, this model is a holistic simulation model that combines electrical and thermal simulation of supercapacitor modules with an ageing model.

In the reference [39] the authors have studied and modeled the temperature effect on the supercapacitor self discharge.

This rise in temperature can have the following consequences:

- The deterioration of the supercapacitor characteristics, especially ESR, self discharge and lifetime [39,40], which affect its reliability and its electrical performance.
- The pressure inside the supercapacitor is increased.
- A premature aging of metal contacts, in fact the repetitive heating and significant temperatures can deteriorate rapidly the terminal connections of the supercapacitor.
- The evaporation of the electrolyte and hence the destruction of the supercapacitor if the temperature exceeds 81.6°C which is the boiling point of the electrolyte.

Therefore, it is important to know and understand the heat behavior of supercapacitor cells and modules. This leads to an estimation of the space-time evolution of the temperature.

This study deals with the thermal modeling and heat management of supercapacitor modules for vehicular applications. The thermal model developed is based on thermal-electric analogy and allows the determination of supercapacitor temperature. Relying on this model, heat management in supercapacitor modules was studied for vehicle applications. Thus, the modules were submitted to real life driving cycles and the evolution of temperatures of supercapacitors was estimated according to electrical demands. The simulation results show that the hotspot is located in the middle of supercapacitors module and that a forced airflow cooling system is necessary.

For supercapacitor thermal behavior, the device was characterized by using the EIS for different temperature. Figure 21 presents the Maxwell BCAP0310F ESR variations according to the temperature. The ESR increases at negative temperature values. The ESR variation is higher for negative temperature than for positive one. This is due to the fact that the electrolyte's conductivity is strongly temperature dependent. Above 0°C ESR varies slowly with the temperature. Below 0°C the temperature dependency is stronger. Higher ESR is due to the increase of the electrolyte's viscosity at low temperatures limiting ionic transport speed which increases the resistance of the electrolyte.

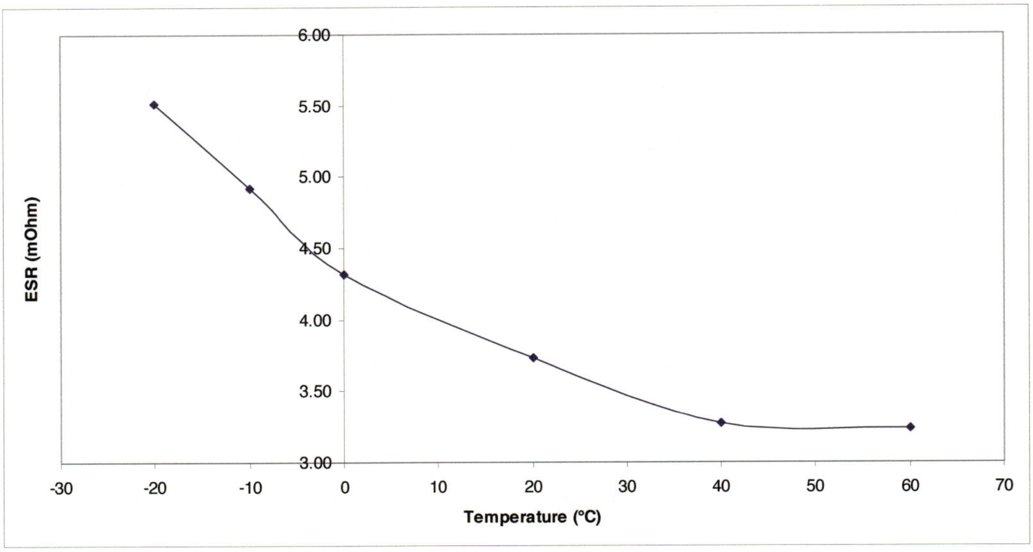

Figure 21. BCAP310F equivalent series resistance as function of temperature.

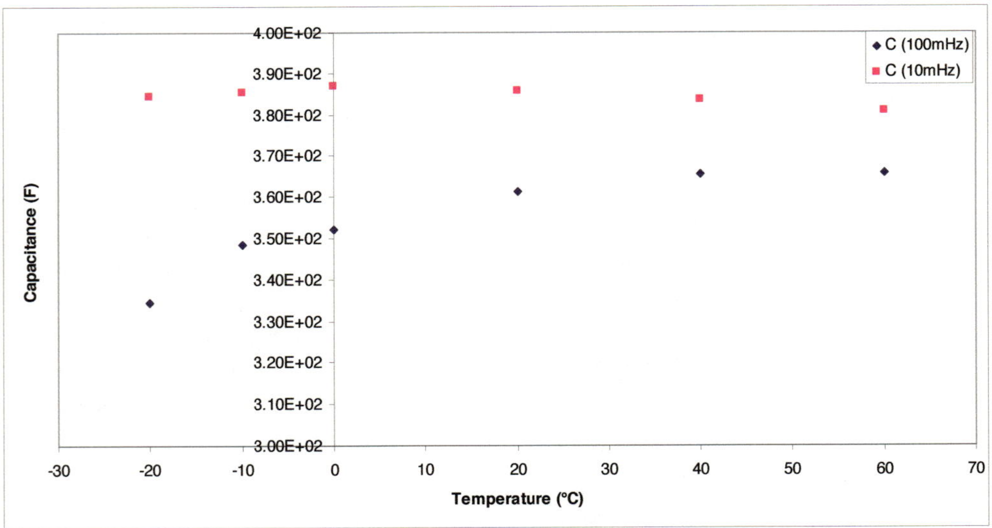

Figure 22. Capacitance evolution according to the temperature for 10mHz and 100mHz

In the case of the capacitance, the experimental results show that the capacitance is lower at negative temperature as shown in Figure 22. For example, at f=10mHz there is no variation of the capacitance with temperature. At 100mHz, C=335F at -20°C whereas C= 361F at 20°C. At negative temperature, the supercapacitor capacitance decreases with temperature.

In conclusion, it is clear that the supercapacitor electric performances and lifetime depend on the temperature.

3.3. Electrical modeling

In literature, several supercapacitors have been developed for different purposes [47-51]. In [47], a model has been proposed by Faranda et al (see Figure 23). The model exists of three branches. The first branch containing R0 represents the fast response of the supercapacitor in term of few seconds. The second branch contains a resistance and a large capacitor. Then the second branch demonstrates the long-term behavior in term of few minutes. However, the analysis that has been carried out by Chalmers University showed that the error between the simulated and experimental results for such model is in the range of 10%, which is statistically high.

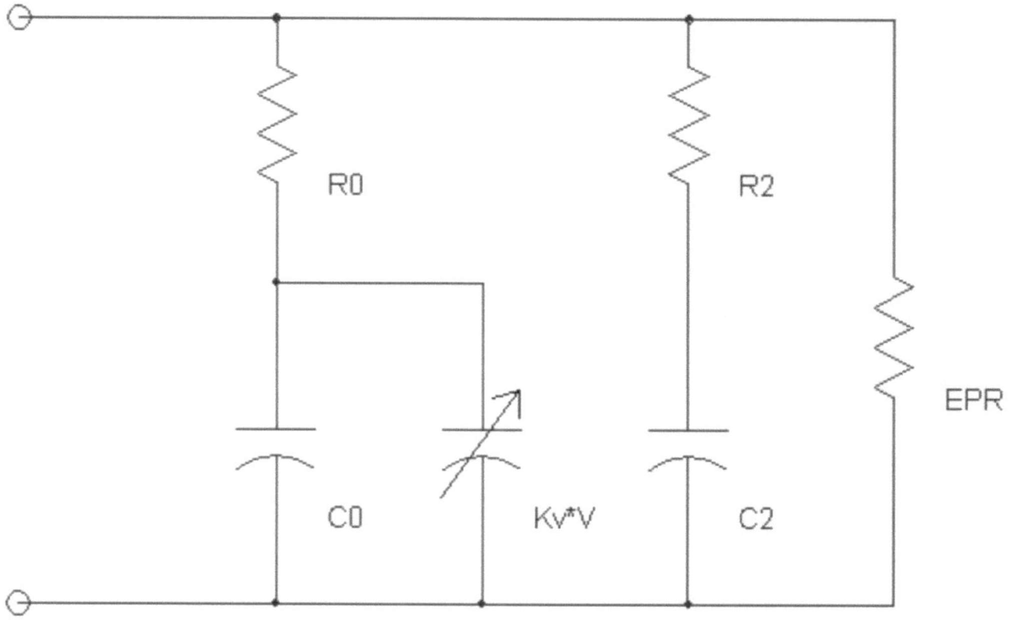

Figure 23. Three branches model [47]

In [48] a second order model has been proposed to demonstrate the supercapacitors behavior. The proposed model is strong similar to the second order Thévenin battery model. The model has significantly higher accuracy (error between the simulated and experimental results <5%) than the previous supercapacitor model due to the non-linear behavior of the model.

In [49-51] a new model has been developed based on electrochemical characterization of the supercapacitors on electrode and electrolyte level. Therefore, the model as presented in Figure 24 below has been proposed.

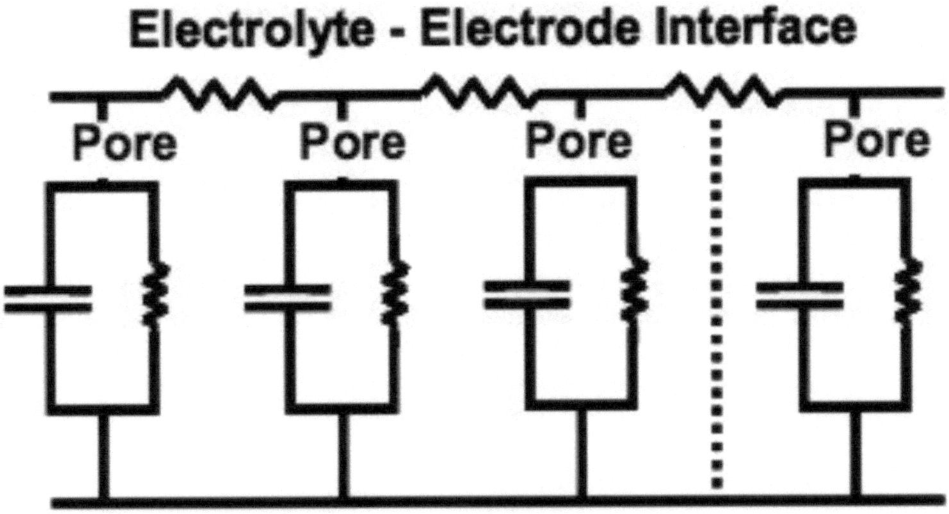

Figure 24. Electrochemical model [49-51]

Here it should be underlined that this model needs dedicated test procedures for determining the model parameters, which only can be carried out by chemists. Therefore, the use of the model in the vehicular applications is useless.

Then for the first two models, the model parameters can be extracted from the electrical approach. However, the simulation time and the complexity of such models is an obstacle in HEV applications. Therefore, in this section the model as presented in Figure 25 seems the most interesting model in real applications.

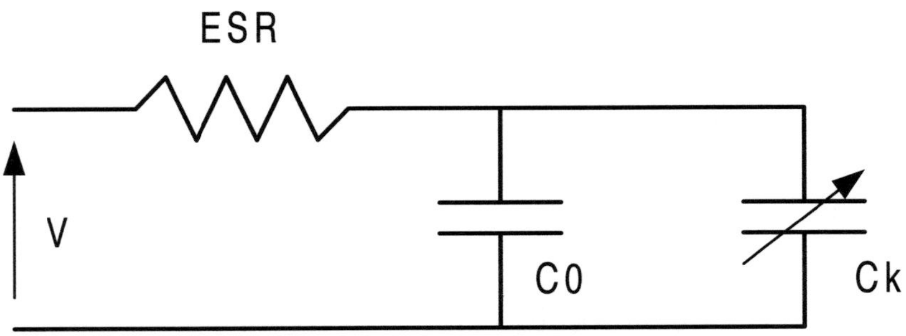

Figure 25. RCC model of the supercapacitor

http://dx.doi.org/10.5772/53490

A supercapacitor cell can be modeled by an equivalent RCC circuit as shown in Figure 25, where ESR is the series equivalent resistance, C0 is a constant capacitor and Ck=k*V varies according to the supercapacitor voltage. These parameters are identified by charging and discharging at constant current [40-46] and the obtained values for the BCAP310F were ESR=4.25mΩ, C0 = 282 F and Ck=46*V. This model is suitable for applications where the energy stored in the capacitor is of primary importance and the transient response can be neglected.

3.4. Thermal modeling

The thermal model developed is based on thermal-electric analogy and allows the determination of supercapacitor temperature inside and at the surface. The developed model can be easily implemented in different simulation programs. It can be used in the modeling of supercapacitors in order to study the heat management of a supercapacitors module. This model makes it possible to size the supercapacitors module cooling system when necessary. This is in order to maintain the temperature of the module within the operating temperature range given by the manufacturer. A Matlab/Simulink® simulation model was developed in order to calculate the R_{th} and C_{th} of a supercapacitor cell. Calculated values were compared to experimental values and the simulation model was validated. Thus a supercapacitor can be modeled as succession of RC and current source circuits. This application permits to calculate the evolution of the temperature in each layer of the supercapacitor cell. It can be used to perform detailed analysis of the temperature variation within a supercapacitor. When using supercapacitor modules which are composed of several cells in series and /or in parallel, it is necessary to study the thermal management of these modules [31]. The aim is to calculate and locate the maximum temperature in order to size the cooling system if needed. In this case, to reduce simulation time, the thermal model can be simplified as shown in Figure 26.

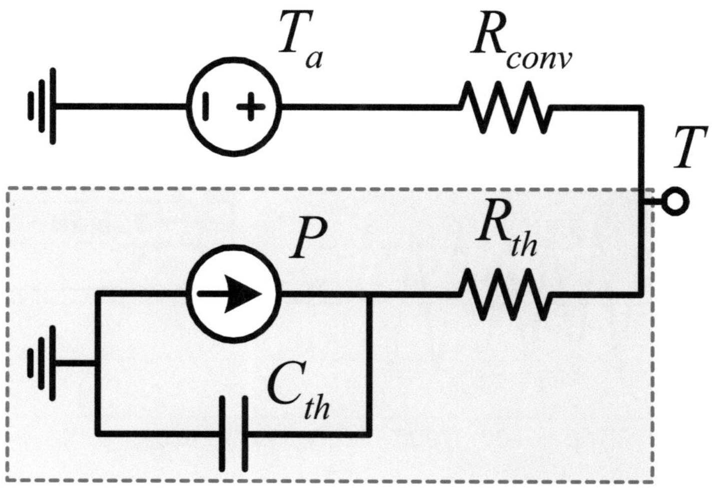

Figure 26. Thermal-electric model of the supercapacitor.

The thermal model gives the evaluation of the temperature on the external surface of the supercapacitor depending on the electrical power, the ambient temperature and the convective heat transfer coefficient. The total power dissipated in the supercapacitor is given by:

$$P(t) = ESR \times I^2(t) \tag{6}$$

Where:

- ESR : the equivalent series resistance of the supercapacitor,
- I(t) : the RMS current value passing through the supercapacitor.

The resistance R_{conv} represents the heat transfer between the surface of the supercapacitor and the ambient air. Its value depends on the convective heat transfer coefficient h and the heat exchange surface of the supercapacitor S_{sc}.

This coefficient can be calculated by using the following expression:

$$R_{conv} = \frac{1}{h \cdot S_{sc}} \tag{7}$$

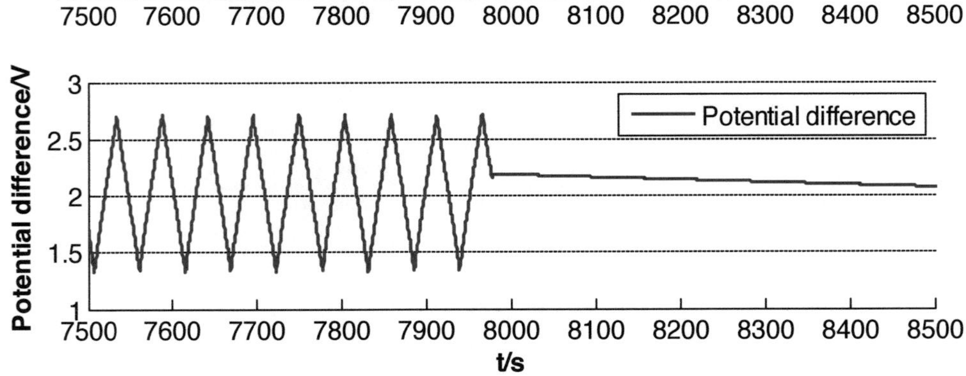

Figure 27. Current and voltage of the 1500F supercapacitor.

http://dx.doi.org/10.5772/53490

In order to validate this model, the parameters were calculated for a 1500F supercapacitor cell. This supercapacitor cell was experimentally tested; it was charged and discharged at 75A with a thermocouple type K placed on the outer surface. Figure 27 shows a zoom of the supercapacitor current and voltage during the receptive cycle which was applied to the 1500F supercapacitor. It shows the warming phase in which the supercapacitor is charged and discharged at 75A constant current then the phase of no cycling where the current is zero.

Figure 28 shows the evolution of the outside surface temperature of the 1500F supercapacitor. The warming phase is about 133 minutes where the supercapacitor is charged and discharged at 75A constant current. The ambient temperature is around 17.5°C.

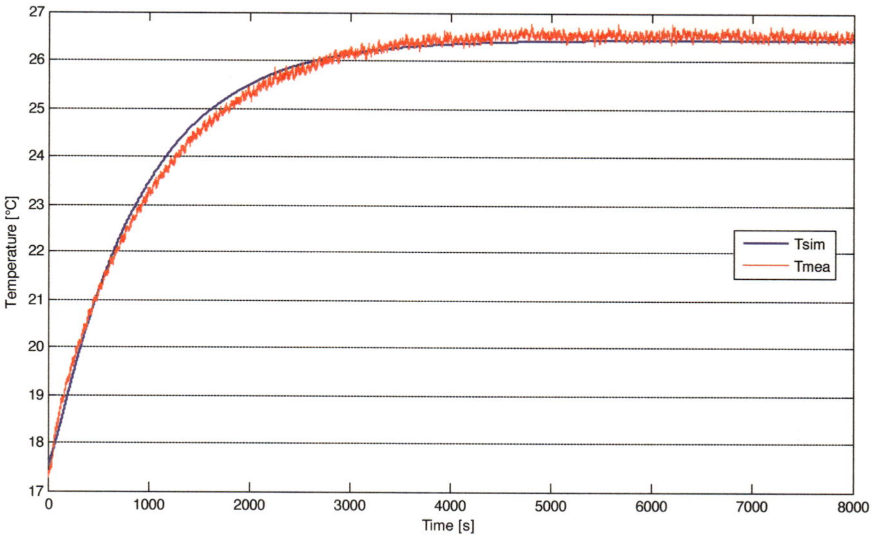

Figure 28. Evolution of measured and simulated temperatures of 1500F cell versus time (75A).

Results presented in Figure 28 show a good correlation between the experimental and simulation. Good agreements were also obtained with 10A and 20A constant currents for charging and discharging cycles.

4. Conclusion

In this chapter, the performance and characteristics of various lithium-ion based batteries are evaluated and discussed taking into account the power and energy densities, the capacity and the current rates. The evaluation is mainly based on the electrical and the thermal behavior. Different types of batteries were characterized at different current rates and different temperatures. The Peukert relationship was evaluated in function of various operating conditions. Electrical and thermal models are developed and presented. The battery electrical model is based on the first order FreedomCar model. The parameters of the electrical model were obtained and calibrated based on a new developed test profile. A battery thermal model is proposed, discussed and validated. Electrical and thermal characterizations of supercapacitors were studied. The different basic calculation methods based on the EIS and the IEC 62576 of the Equivalent Series Resistance (ESR) and the capacitance of a supercapacitor are presented. An electrical model of the supercapacitor based on RCC circuit is presented. A thermal model of the supercapacitor is presented and it is based on the thermal-electric analogy. The model was validated using experimental results of the BCAP1500F supercapacitor cell. The simulation results of the thermal model can be used to find out if a cooling/heating system is necessary for the use of supercapacitor in order to improve its efficiency. The models developed are simple enough to be implemented in different simulation programs and thermal management systems for hybrid electric vehicles.

Author details

Monzer Al Sakka[1], Hamid Gualous[2], Noshin Omar[1] and Joeri Van Mierlo[1]

1 Vrije Universiteit Brussel, Belgium

2 Université de Caen Basse-Normandie, France

References

[1] G. Maggetto, J. Van Mierlo, Annales de Chimie – Science des matériaux, in: Thermatic Issue on "Material for Fuel Cell Systems", vol. 26, 2000, p. 9.

[2] J. Van Mierlo, G. Maggetto, Ph. Lataire, Energy Convers. Manage. 47 (2006) 196.

[3] P. Van den Bossche, F. Vergels, J. Van Mierlo, J. Matheys, W. Van Autenboer, J. Power Sources 26 (2005) 1277.

[4] N. Omar, F. Van Mulders, J. Van Mierlo, P. Van den Bossche, J. Asian Electric Vehicles 7 (2009) 1277.

[5] H. Abderrahmane, B. Emmanuel, Assessment of real behavior of VHE Energy Storage System in heavy vehicles, in: Proceeding of EET-2008 European Ele- Drive Conference, Geneva, March, 2008.

[6] N. Omar, B. Verbrugge, P. Van den Bossche, J. Van Mierlo, Electrochim. Acta 25 (2010) 7534.

[7] J. Cheng, J. VanMierlo, P. Van den Bossche, Ph. Lataire, Super capacitor based energy storage as peak power unit in the applications of hybrid electric vehicles, in: Proceeding of PEMD 2006, Ireland, 2006.

[8] C.R. Akli, X. Roboam, B. Sareni, A. Jeunesse, Energy management and sizing of a hybrid locomotive, in: Proceeding of EPE 2007, Denmark, 2007.

[9] Sh. Lu, K.A. Corzine, M. Ferdowsi, IEEE Trans. Veh. Technol. 56 (2007) 1516.

[10] N. Omar, M. Al Sakka, M. Daowd, Th. Coosemans, J. Van Mierlo, P. Van den Bossche, Assessment of behavior of active EDLC-battery system in heavy hybrid charge depleting vehicles, in: Proceeding of 4th European Symposium on Super Capacitors and Applications, Bordeaux, October, 2010.

[11] A. Cooper, M. Kellaway, Advanced lead-acid – the new battery system for hybrid electric vehicles, in: Proceeding of EET-2008 European Ele-Drive Conference, Geneva, March, 2008.

[12] N. Omar, M. Daowd, B. Verbrugge, G. Mulder, P. Van den Bossche, J. Van Mierlo, M. Dhaens, S. Pauwels, F. Leemans, Assessment of performance characteristics of lithium-ion batteries for PHEV vehicles applications based on a newly test methodology, in: Proceeding of the 25th World Battery, Hybrid and Fuel Cell Electric Vehicle Symposium, Shenzhen, November, 2010.

[13] N. Omar, B. Verbrugge, G. Mulder, P. Van den Bossche, J. Van Mierlo, M. Daowd, M. Dhaens, S. Pauwels, Evaluation of performance characteristics of various lithium-ion batteries for use in BEV application, in: Proceeding of IEEE Vehicle Power and Propulsion Conference, Lille, September, 2010.

[14] N. Omar, M. Daowd, G. Mulder, J.M. Timmermans, J. Van Mierlo, S. Pauwels, Assessment of performance of lithium ion phosphate oxide, nickel manganese cobalt oxide and nickel cobalt aluminum oxide based cells for using in plug-in battery electric vehicle applications, in: Proceeding of IEEE Vehicle Power and Propulsion Conference, Chicago, September, 2011.

[15] A. Burke, M. Miller, Performance characteristics of lithium-ion batteries of various chemistries for plug-in hybrid vehicles, in: Proceeding of the 24th World Battery, Hybrid and Fuel Cell Electric Vehicle Symposium, Stavanger, May, 2009.

[16] J. Axsen, Burke, K. Kurani, Batteries for Plug-in Hybrid Electric Vehicles (PHEVs): Goals and State of the Technology, May, 2008.

[17] N. Omar, M. Daowd, O. Hegazy, G. Mulder, J.M. Timmermans, Th. Coosemans, P. Van Den Bossche, J. Van Mierlo, Standardization work for BEV and HEV Applications: Critical Appraisal of Recent Traction Battery Documents. J. Energies 2012, 5, 138-156

[18] G. Mulder, N. Omar, S. Pauwels, F. Leemans, B. Verbrugge, W. De Nijs, P. Van den Bossche, D. Six, J. Van Mierlo, J, Enhanced test methods to characterise automotive battery cells. J. Power Sources 2011, 196, 100079 – 10087.

[19] IEC 61982-2, Secondary batteries for the propulsion of electric road vehicles - Part 2: Dynamic discharge performance test and dynamic endurance test, August 2002.

[20] A. Amine, C.H. Chen, J. Liu, J. Hammond, A. Jansen, D. Dees, I. Bloom, D. Vissers, G. Hendriksen, Factors responsible for impedance rise in high power lithium ion batteries, Vol. 97, No: 8, 2001, Jul., pp: 684-687.

[21] P. Van den Bossche, B. Verbrugge, N. Omar, J. Van Mierlo, The Electric Vehicle Charged by the Grid: Voltages and Power Levels, PHEV-09, September, 2009.

[22] IEC 61434, Secondary cells and batteries containing alkaline or other non-acid electrolytes - Guide to designation of current in alkaline secondary cell and battery standards, 1996.

[23] IEC 62660-1 Ed. 1: Secondary batteries for the propulsion of electric road vehicles – Part 1: Performance testing for lithium-ion cells, May, 2010.

[24] ISO 12405-1 Electrically propelled road vehicles – Test specification for lithium-ion traction battery packs and systems – Part 1: High- power applications. IEC, 2011.

[25] ISO 12405-2 Electrically propelled road vehicles – Test specification for lithium-ion traction battery packs and systems – Part 1: High- energy applications. IEC, 2011.

[26] K. Sawai, R. Yamato, T. Ohzuku, Impedance measurements on lithium-ion battery consisting of Li(Li1/3Ti5/3)O4 and LiCo1/2Ni1/2O2, Journal of Electrochemistry, No. 51, PP: 1651-1655.

[27] M. Daowd, N. Omar, P. Van den Bossche, J. Van Mierlo, A Review of Passive and Active Battery Balancing based on MATLAB/Simulink, INTERNATIONAL REVIEW OF ELECTRICAL ENGINEERING-IREE, Vol. 6, pp: 2974–2989, 2011

[28] M. Daowd, N. Omar, P. Van den Bossche, J. Van Mierlo, Extended PNGV Battery Model for Electric and Hybrid Vehicles, INTERNATIONAL REVIEW OF ELECTRICAL ENGINEERINGIREE, Vol. 6, pp: 1264–1278, 2010

[29] V.H. Johnson, A. Pesaran, Th. Sack, Temperature-dependent battery models for high-power lithium-ion batteries. Proceedings EVS-17, October 2000, Montréal, Canada.

[30] N. Omar, M. Al Sakka, M. Daowd, O. Hegazy, Th. Coosemans, P. Van den Bossche, J. Van Mierlo, Development of a Thermal Model for Lithium–Ion Batteries for Plug-In Hybrid Electric Vehicles, Proceedings EVS 26, 2012, Los Angeles, USA

[31] M. Al Sakka, Gualous H., Van Mierlo J., and Culcu H. Thermal modeling and heat management of supercapacitor modules for vehicle applications. Journal of Power Sources, 194:581–587, 2009.

[32] Monzer Al Sakka: Supercapacitors and DC/DC Converters for Fuel Cell Electric Vehicle, PhD at Vrije Universiteit Bruseel, Brussels, September 2010, ISBN: 978 90 5487 802 5.

[33] IEC 62576, Electric Double-Layer Capacitors for Use in Hybrid Electric Vehicles - Test Methods for Electrical Characteristics, IEC, 2008.

[34] Yonghua Cheng, "Assessement of Energy Capacity and Energy Losses of supercapacitors in Fast Charging-Discgarging Cycles", Energy Conversion, IEEE Transactions, Volume : 25, Issue. 1, pp. 253 – 261, 2010.

[35] R. Kötz, M. Hahn, R. Gallay, "Temperature behaviour and impedance fundamentals of supercapacitors", Journal of Power Sources, 154 (2006) pp. 550–555, 2006.

[36] H. Gualous, D. Bouquain, A. Berthon, J. M. Kauffmann "Experimental study of supercapacitor serial resistance and capacitance variations with temperature" Journal of Power Sources, Vol. 123, pp.86-93, 2003.

[37] F. Rafik, H. Gualous, R. Gallay, A. Crausaz, A. Berthon "Frequency, thermal and voltage supercapacitor characterization and modelling", Journal of Power Sources, Vol. 165, pp. 928-934, 2007.

[38] Oliver Bohlen, Julia Kowal, Dirk Uwe Sauer "Ageing behaviour of electrochemical double layer capacitors: Part II. Lifetime simulation model for dynamic applications" Journal of Power Sources, Volume 173, Issue 1, Pages 626-632, 2007.

[39] Y. Diab; P. Venet, H. Gualous, G. Rojat, "Self-Discharge Characterization and Modeling of Electrochemical Capacitor Used for Power Electronics Applications" IEEE Transactions On Power Electronics. Vol. 24, Issue 2, pp. 510-517, 2009.

[40] F. Rafik, H. Gualous, R. Gallay, A. Crausaz, A. Berthon, "Supercapacitors characterization for hybrid vehicle applications", Proc. IEEE 5th Power Electronics and Motion Control Conference, Shanghai, China, pp. 1-5, 2006.

[41] A. Hammar, R. Lallemand, P. Venet, G. Coquery, G. Rojat, J. Chabas, "Electrical characterization and modelling of round spiral supercapacitors for high power applications", Proc. 2nd European Symp. on Super Capacitors & Applications, Lausanne, Switzerland, 2006.

[42] L. Zubieta, R. Bonert, "Characterization of double-layer capacitors for power electronics applications", Proc. IEEE 33rd Industrial Applications Society annual meeting, St. Louis, MO, USA, pp. 1149 – 1154, 1998

[43] L. Zubieta, R. Bonert, "Characterization of double-layer capacitors for power electronics applications", IEEE Transactions on Industry Applications Vol. 36 Issue 1, pp. 199-205, 2000.

[44] John M. Miller, Uday Deshpande, Marius Rosu, "CarbonCarbon Ultracapacitor Equivalent Circuit Model, Parameter Extraction, and Application", Maxwell Technologies, Inc Ansoft Corp. San Diego, CA Pittsburg, PA, Ansoft First Pass Workshop, Southfield, 2007.

[45] Bavo Verbrugge, Frederik Van Mulders, Hasan Culcu, Peter Van Den Bossche, Joeri Van Mierlo, "Modelling the RESS: Describing Electrical Parameters of Batteries and Electric Double-Layer Capacitors through Measurements", World Electric Vehicle Journal, Vol. 3, 2009.

[46] Joeri Van Mierlo, Gaston Maggetto, Peter Van Den Bossche Impact, "Models of Energy Sources for EV and HEV: Fuel cells, Batteries, Ultra-Capacitors, Flywheels and Engine-generators", Journal of Power Sources, Vol. 28, N° 128, pp: 76 - 89, 2004.

[47] Faranda, R.; Gallina, M.; Son, D.T.; A new simplified model of Double-Layer Capacitors 2007; ICCEP '07. International Conference on Clean Electrical Power 21-23 May 2007 Page(s):706 – 710; ISBN: 1- 4244-0632-3

[48] Data sheet for supercapacitor from EPCOS with Part No.: B48621-S0203-Q288

[49] Qu, D. Y. and H. Shi (1998). "Studies of activated carbons used in double-layer capacitors." Journal of Power Sources 74(1): 99-107.

[50] Shi, H. (1996). "Activated carbons and double layer capacitance." Electrochimica Acta 41(10): 1633-1639.

[51] Vix-Guterl, C., E. Frackowiak, et al. (2005). "Electrochemical energy storage in ordered porous carbon materials." Carbon 43(6): 1293-1302.

Chapter 6

The Application of Electric Drive Technologies in City Buses

Zlatomir Živanović and Zoran Nikolić

Additional information is available at the end of the chapter

http://dx.doi.org/10.5772/51770

1. Introduction

The environmental concerns and limited reserves of fossil fuels have generated an increased interest for alternative propulsive systems of vehicles. On the other hand, vehicle manufacturers are increasingly facing demands for reducing emissions of harmful gases by the vehicles in accordance with the increasingly stringent legislation.

Buses as means of public transportation could reduce considerably the problems caused by the traffic in the urban areas through the usage, among other things, innovative techniques and technologies of vehicle propulsion systems.

The development of innovative technologies is increasingly oriented towards electrification of vehicle propulsion systems expected to lead to: a reduction of harmful emissions, an increased efficiency of vehicles, improved performances, a reduction of fuel consumption, a reduction of noise, and potentially lower maintenance costs. An electric drive technology implies a technology employing at least one drive device called electric motor. Three key electric drive technologies are: hybrid electric, battery electric, and fuel cell electric technologies.

In this chapter the application of electric technologies in the bus propulsion systems is considered through: an analysis of the state of development of city buses, an analysis of the advantages and shortcomings of electric drive technologies, and identification of the problems standing in the way of their greater commercialization. The presented examples of the developed city buses describe the basic characteristics of the applied propulsion systems and their advantages. The development of hybrid technologies for bus propulsion has grown considerably over the past several years. These technologies have reached massive applications in North America and their expansion to Europe has been initiated during the past several years.

In the part of the chapter dealing with hybrid electric buses a typical bus driving system and its components are reviewed and hybrid systems of the major world manufacturers are presented. Special attention is paid to the comparative analysis the hybrid electric buses and conventional and CNG buses. The available literature data have been critically processed and re-presented. Finally, the characteristics of hybrid city buses of some American and European manufacturers which have found the most widespread applications are reviewed.

The fuel cell powered buses draw special attention of users owing the efficiency of their propulsion system and their ability to cut drastically harmful emissions. Even though they are still not widely used, judging by the number of demonstrated projects the development of fuel cell buses is very intensive throughout the world. Barriers to their wider use are very high costs, lack of an adequate infrastructure, and relatively small radius of movement.

Fuel cell buses are vehicles with zero emissions. CO_2 emissions depend on the type and method of production of fuel for fuel cells. In the part of the chapter dealing with fuel cell buses the typical bus configuration, its subsystems, its ecological characteristics, and costs are reviewed. Some of the development projects and characteristics of a new generation of fuel cell buses are presented.

Battery electric technologies are among technologies which reduce drastically the impact of a vehicle on the environment, however, they are still far from the proven technologies. The reason is the current level of developing technology of the energy storage devices for these vehicles. Influence of the batteries on commercialization of these buses is more pronounced compared to the other electric drive buses. A significant advancement in the area of energy sources has been made over the past several years by the development of lithium-ion batteries which lead to the development of an increased number of prototypes, even to a series production of these vehicles. In the part of the chapter dealing with battery electric buses the characteristics of some of these realized buses are reviewed.

In a separate part of this chapter the characteristics of energy storage device for the electric propulsion systems of the realized buses are presented, and the expectations from further development trends of the energy source devices are outlined.

2. An overview of electric drive technologies

Depending on the degree of electrification of propulsion system, Figure 1, three key electric drive technologies for power the electric vehicles are: hybrid electric, battery electric and fuel cell electric technologies [1].

Hybrid electric technology: A hybrid electric technology uses both an electric motor (EM) and an internal combustion engine (ICE) to propel the vehicle. Vehicles equipped with this technology are called Hybrid Electrics Vehicles (HEVs).

As can be seen from Figure 1, source of energy to power the vehicle with hybrid electric technologies are fuels, including alternative, which can be used in IC engines and electricity stored in the batteries or ultra capacitors. The charge energy storage to electricity is performed via IC engine and/or via the regenerative braking.

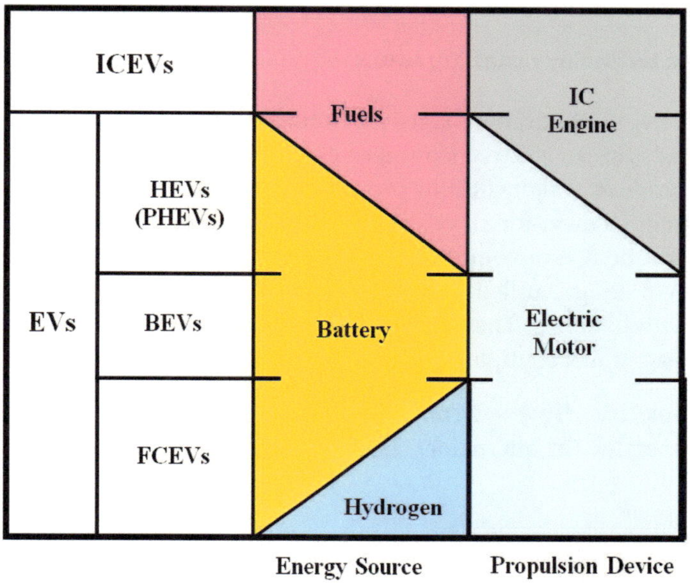

Figure 1. Different degrees of electrification of road vehicles.

Special case of hybrid electric technology is plug-in hybrid electric (PHE) technology. It has a battery that can be charged off board by plugging into the grid and which enables it to travel certain kilometres solely on electricity. Vehicles equipped with this technology are called Plug-in Hybrid Electrics Vehicles (PHEVs).

Battery electric technology: Battery electric technology uses a relatively large on-board battery to propel the vehicle. Battery provides energy for propulsion through an electric traction motor(s) as well as power for all vehicle accessory systems. Some electric vehicles (EVs) can use to drive auxiliary devices like an on board generator, which makes them have characteristics of hybrid solutions. Vehicles equipped with this technology are called Battery Electrics Vehicles (BEVs).

Fuel cell electric technology: Fuel Cells are energy conversion devices set to replace combustion engines and compliment batteries in a number of applications. They convert the chemical energy contained in fuels, into electrical energy (electricity), with heat and water generated as by-products. Fuel cells continue to generate electricity for as long as a fuel is supplied, similar to traditional engines. However unlike engines, where fuels are burnt to convert chemical energy into kinetic energy, fuel cells convert fuels directly into electricity via an electrochemical process that does not require combustion. Vehicles equipped with this technology are called Fuel Cell Electrics Vehicles (FCEVs).

Electric drive technologies also, usually, incorporate other technologies, which reduce energy consumption, for example regenerative braking. That allows the electric motor to re-capture the energy expended during braking that would normally be lost. This improves energy efficiency and reduces wear on the brakes.

2.1. Electric drive technology configurations

Hybrid electric drive configurations consist of a fuel-burning prime power source – generally an ICE–coupled with an electrochemical or electrostatic energy storage device. These two power sources work in conjunction to provide energy for propulsion through an electric traction drive system. Power for all vehicle accessory systems can be provided electrically or mechanically from the ICE or combinations of both. There are currently many different hybrid-electric system designs utilizing ICE, alternative fuels engines, gas turbines or fuel cells in conjunction with batteries. These design options are grouped in three categories: series, parallel and series-parallel configurations.

A series hybrid-electric drive system, Figure 2, consists of an engine directly connected to an electric generator (or alternator). The arrows indicate the mechanical and electrical energy flow.

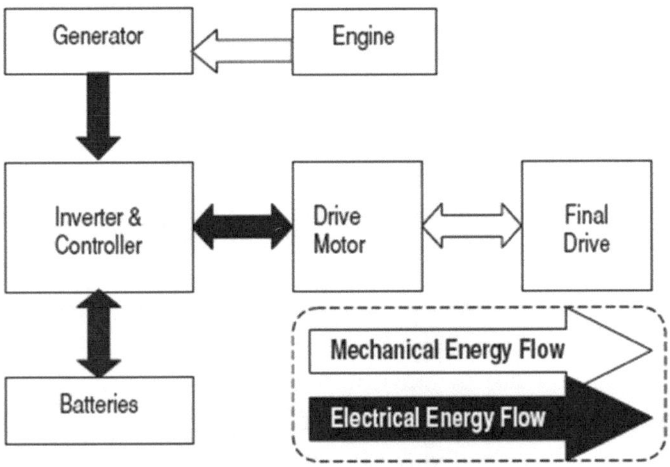

Figure 2. Series hybrid electric drive system.

Power from the generator is sent to the drive motor and/or energy storage batteries according to their needs. There is no mechanical coupling between the engine and drive wheels, so the engine can run at a constant and efficient rate, even as the vehicle changes speed. The serial hybrid technology is the most common hybrid technology.

In a parallel hybrid-electric drive system, Figure 3, both of the power sources (engine and electric motor) are coupled mechanically to the vehicle's wheels. In different configurations, the motor may be coupled to the wheels either through the transmission (pre-transmission

http://dx.doi.org/10.5772/51770

parallel design) or directly to the wheels after the transmission (post-transmission parallel design). Each of these configurations has its advantages.

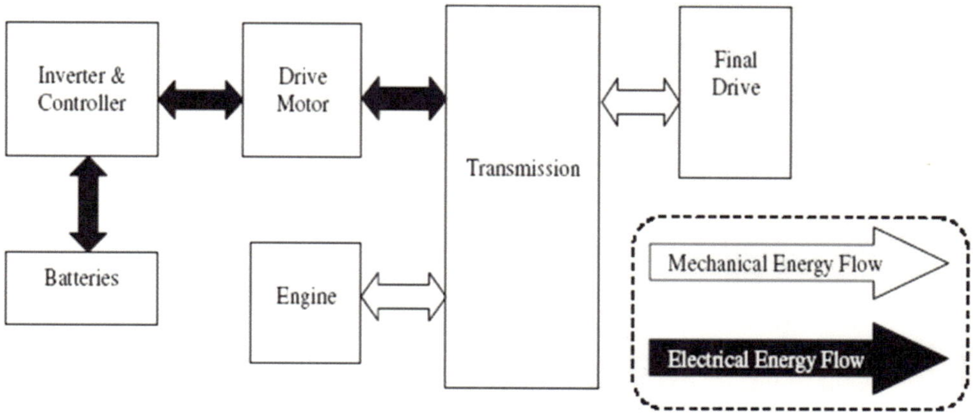

Figure 3. Parallel hybrid electric drive system.

A series-parallel design, also known as power-split or dual mode hybrid system, is interesting because with proper control strategy it can be designed to take advantage of both parallel and series types and avoid their drawbacks.

Battery electric drive system are illustrated in Figure 4. The two types of system configurations are possible depending on the positioning and size of the electric motors.

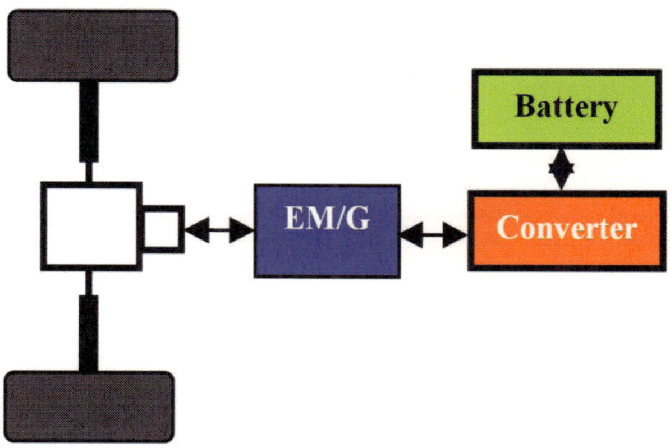

Figure 4. Battery electric drive system.

The central motor type is currently more common. However, the requirement to transfer power from the motor to the wheels does involve some losses in efficiency through friction.

The hub motor type can avoid many of the transmission losses experienced in the central motor type, but are a less regularly used technology.

In fuel cell drive system fuel cell is providing the electric energy needed to run of vehicles. There are two types of fuel cell drive configuration:

- Fuel cell drive without energy storage device (non-hybrid fuel cell vehicles) and
- Fuel cell drive with energy storage device (hybrid fuel cell vehicles), Figure 5.

Fuel cell hybrids operate much like other hybrid electric vehicles but with fuel cells producing electricity that charges the batteries, and a motor that converts electricity from the batteries into mechanical energy that drives the wheels.

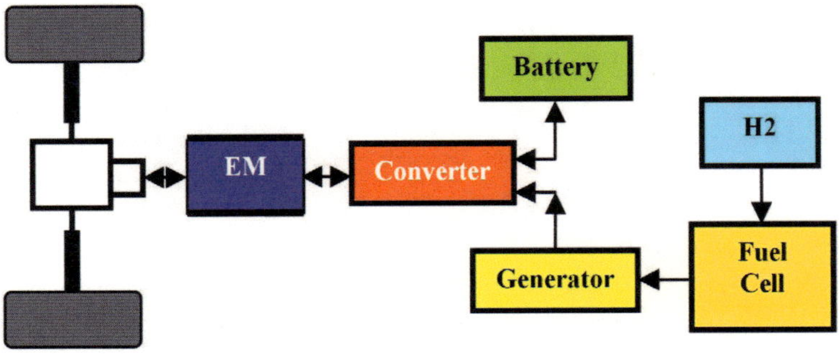

Figure 5. Fuel cell drive system with energy storage device.

2.2. Advantages and disadvantages of electric drive technologies

There are some major advantages of electric drive technologies but there are also some disadvantages. Table 1 summarizes the advantages and disadvantages of a hybrid-electric, plug-in hybrid electric battery and fuel cell drive systems [2]:

Technology	Advantages	Disadvantages
Hybrid electric	Lower fuelling costs; Reduced fuel consumption and tailpipe emissions; Recovered energy from regenerative braking	Higher initial cost; Complexity of two power trains; Component availability
Plug-in hybrid electric	Cleaner electric energy thanks advanced technologies or renewable; Reduced fuel consumption and tailpipe emissions; Optimized fuel efficiency and performance; Recovered energy from regenerative braking; Grid connection potential; Pure zero-emission capability	Higher initial cost; Complexity of two power trains; Component availability-batteries, powertrains, power electronics; Cost of batteries and battery replacement; Added weight
Battery electric	Use of cleaner electric energy; Zero tailpipe emissions; Overnight battery recharging; Recycled energy from regenerative braking; Lower fuel and operational costs; Quiet operation	Mileage range; Battery technology still to be improved; Possible need for public recharging infrastructure

http://dx.doi.org/10.5772/51770

Technology	Advantages	Disadvantages
Fuel cell	Zero tailpipe emissions; Higher energy efficiency than the IC engine; Recovered energy from regenerative braking; Potential of near-zero well-to-wheel emissions when using renewable fuels to produce hydrogen; No dependence on petroleum	Higher initial cost; Increased reliability and durability; Hydrogen generation and onboard storage; Availability and affordability of hydrogen refueling; Codes and standards development; Scalability for mass manufacture;

Table 1. Advantages and disadvantages of electric drive technologies.

3. Hybrid Electric Buses

At the IAA 1969 in Frankfurt, Daimler presented the first electric test bus–an early example of hybrid drive technology. In 1979, Daimler has launched a five-year model trial with a total of 13 Mercedes-Benz OE 305 electric-diesel hybrid buses in regular service. Since 2008 Orion hybrid buses are in regular service on the roads of major U.S. cities, but from 2009 Mercedes-Benz Citaro BlueTec Hybrid is in daily operation [3].

The first sales of serial hybrid city buses in Japan began 1991, when Hino delivered test buses in eight cities.

From 1997 until today leaders by the number of hybrid buses in commercial use are the United States and Canada.

3.1. Hybrid electric bus architectures

A hybrid electric bus (HEB) usually combines an internal combustion engine with the battery and an electric motor. The ICE can be fueled by gasoline, diesel, or other (natural gas, biofuel) and work either in series or in parallel with the electric motor. Regenerative braking capability in HEBs minimizes energy losses by recovering some of the kinetic energy used to slow down or stop a vehicle.

In a series hybrid configuration [4], Figure 6, the ICE drives a generator to feed the electric motor and recharge the battery. Braking energy can be captured and stored in the battery ("regenerative braking"). The engine can be downsized compared to a conventional drivetrain with the same performance, meaning lower ICE weight and higher energy efficiency.

The electric motor powers the drive system, using either energy stored in batteries, or from the engine, or from both as needed. The engine is more efficient at lower speeds and higher load, so the series hybrid is preferred for slow and start-and-stop city driving.

In a parallel hybrid configuration both the engine and the electric motor are linked to the transmission so that either of them, or both at the same time, may provide the power to turn the wheels. Since the parallel hybrid configuration allows the engine to drive the wheels also through a direct mechanical path, it offers better efficiency than a series hybrid configuration, and a more functional and flexible design.

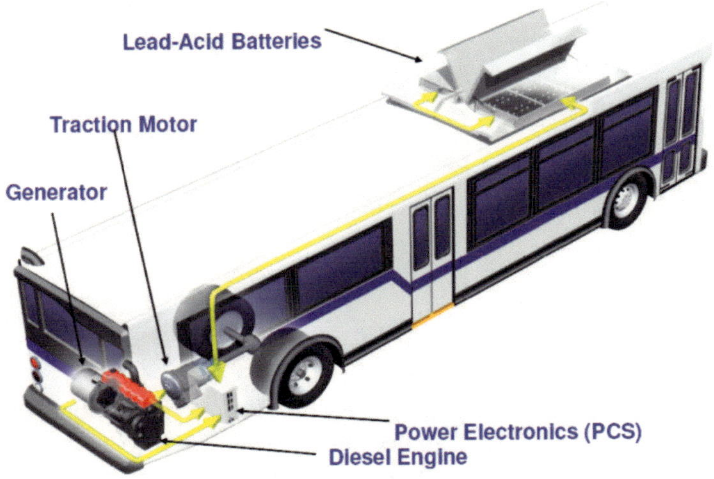

Figure 6. Series hybrid propulsion system.

3.2. Hybrid Electric Bus Components

The principal hybrid-electric bus components include:

1. an Auxiliary Power Unit (APU),
2. a drive motor,
3. a controller and inverter,
4. an energy storage device and
5. other auxiliary systems, such as air conditioning and lighting.

Auxiliary Power Units: Auxiliary Power Units (APUs) used in hybrid-electric buses are available in a number of configurations including IC engines, fuel cells, and with different fuels, such as diesel, gasoline and compressed natural gas (CNG), liquid natural gas (LNG) and propane. The engine is typically sized for the average bus power demand, not peak power demand since the energy storage device provides supplementary power. Engines in hybrid configurations also operate over a narrower range of load and speed combinations compared to engines in conventional buses.

Drive motors: Two primary types of electric motors can be used in electric vehicles, direct current (DC) motors and alternating current (AC) motors. On a power comparative basis, an AC motor generally exhibits higher efficiency, has a favourable power to size/weight ratio, is less expensive and generates regenerative braking energy more efficiently than a DC motor. Electric drive motors are connected to the vehicle wheels either directly, referred to as wheel motors, or through a transmission and differential assembly. Wheel motors are more efficient both in drive cycle and in the regenerative cycle by eliminating the losses in the mechanical transmission and the differential. However, wheel motors are expensive.

Controller and inverter: The electronic controller regulates the amount of energy, (DC power in the case of batteries), that is transferred or converted to AC power by the inverter (in AC motors) for acceleration. It also ensures that voltage is maintained within the specifications required for operating the motor. An electronic controller can also recover electrical energy by switching the motor to a generator in order to capture the vehicle's kinetic energy via regenerative braking. The controller also ensures that the regenerative current does not overcharge a battery.

Energy storage devices: Energy storage devices provide necessary of the energy in hybrid-electric buses to supplement the APU energy when there is a high demand (e.g., acceleration from stop, speed acceleration, climbing an up-hill gradient) and to recover and store the energy generated during deceleration (e.g., braking, down-hill coasting).

3.3. Major manufacturers of hybrid systems

The major manufacturers of hybrid systems are shown in Table 2:

Manufacturer	Propulsion system	Type	Country
BAE	HybriDrive	Series	USA
Allison	$E^{P}40/E^{P}50$	Series-parallel	USA
ISE	ThunderVolt	Series	USA
Siemens	ELFA	Series	Germany
Eaton	EHPS	Parallel	USA
Volvo	I-SAM	Parallel	Canada
Voith	DIWAhybrid	Parallel	Germany

Table 2. Major manufacturers of hybrid systems.

BAE Systems is a major integrator and supplier of integrated hybrid electric propulsion for Orion hybrid electric buses manufactured by Daimler Buses North America. BAE Systems produces the HybriDrive series propulsion system. The HybriDrive is composed of a traction motor and a traction generator to provide power to the vehicle. The liquid cooled, high power-to-weight ratio electric traction motor connects directly to a standard drive shaft and rear axle to provide traction power and regenerative braking. HybriDrive series system powers more than 3.500 buses [5].

Allison has developed EP50 parallel hybrid electric propulsion system with two motors capable of producing 75 kW of continuous power and up to 150 kW of power at full potential. Although designed as a parallel architecture, a power-split or two-mode hybrid electric bus can operate in either a series, or a parallel configuration. There are more than 4,600 buses with the Allison Two Mode Parallel Hybrid Systems in operation across 216 cities and 9 countries. The buses have driven more than 600 million kilometres, saved over 75 million litres of fuel and eliminated more than 197.000 metric tones of CO_2 [6].

ISE Corporation produces the ThunderVolt series drive system who has five key subsystems: Motive Drive Subsystem (electric drive motors, motor controller, gear reduction system, and related components); Auxiliary Power Unit Subsystem (engine, electric generator, and related components); Energy Storage Subsystem (integrated pack of either batteries or ultracapacitors); Vehicle Control and Diagnostics and Electrically-Driven Accessories (electrical power steering and braking systems, air conditioning systems, and related components). ISE components for electric and hybrid drive are manufactured by Siemens (ELFA). ThunderVolt packs have been integrated in over 300 in-service buses, and operated for over 10 million miles (16,09 million km) cumulative [7].

Siemens ELFA hybrid propulsion system was initially developed in the mid 90's for diesel electric buses. To date, Siemens has outfitted more than 700 hybrid busses in the U.S., Italy, Germany and Japan. More than 6 Million operation hours attest to the reliability of the system [8] Rugged liquid-cooled induction motors with power ratings from 50 kW to 180 kW with reduction gearboxes are used as standard for ELFA traction systems. Permanent-magnet generators are used for all of the latest ELFA traction drive generation. The traction converters play a key role in ELFA traction systems. The complete ELFA traction system is controlled using just one standard traction converter software.

Eaton hybrid power system (EHPS) uses a parallel configuration [9]. EHPS consists of an automated clutch, electric motor/generator, motor controller/inverter, energy storage unit, automated manual transmission and an integrated supervisory hybrid control module, takes energy created during braking and regenerates it for later use. An electric motor/generator is located between the output of an automatic clutch and the input to an automated mechanical transmission. The electric motor's peak output is 44 kW.

Volvo has developed a parallel hybrid system [10] with integrated starter alternator motor (ISAM) system that can be used across the complete product range. ISAM is located between the diesel engine and the gearbox and facilitates a compact parallel hybrid package. ISAM integrates the starter motor, electric motor, generator and electronic control unit into one component. The electric motor is used to start and accelerate the bus up to about 20 km/h, while the diesel engine takes over at higher speeds.

Voith presented its parallel DIWAhybrid system at the American Public Transportation Association (APTA) show in 2008. DIWAhybrid system builds on the proven DIWA automatic transmission and is designed for up to 290 kW power input and a maximum input torque of 1.600 Nm. At 150 kW electric traction power, the DIWAhybrid system reduces the load of the diesel engine enough for the latter to be substantially smaller than on conventional diesel buses [11]. Voith has begun production of DIWAhybrid drive system for city buses in the USA at the end of 2011.

3.4. Hybrid electric bus characteristics

An advantage of a hybrid-electric bus over a conventional bus is theoretically better fuel economy and lower exhaust emissions.

Capital, maintenance and operation cost: Hybrid buses can cost up to $500.000, a significant increase over a standard diesel city bus, the cost of which is closer to $300.000 [12]. The major cost associated with hybrid buses is battery replacement, as batteries today are not expected to last the 12-year life of a city bus.

Fuel and maintenance (operating) cost savings over the life of the bus are expected to help recover the higher initial (capital) cost. Specifically, operating cost savings are expected through the following features: increased fuel economy; extended brake life; no transmission to service; less moving parts; less engine wear and less expensive engine.

Emissions and fuel economy: With regard to emissions, clearly hybrids are not providing the zero emission. Nevertheless, some testing on hybrid buses has demonstrated that hybrids offer emissions benefits that are comparable to or better than clean diesel and CNG buses. There are four primary sources of efficiency and emissions reduction found in HEBs [13]: smaller engine size, regenerative braking, power-on-demand, and constant engine speeds and power output.

By adding an electric motor a hybrid electric bus can be equipped with a smaller, more efficient combustion engine.

Regenerative braking recovers energy normally lost as heat during braking, and stores it in the batteries for later use by the electric motor.

Another feature that saves energy and reduces emissions in HEBs is the ability to temporarily shut off the combustion engine during idle or coasting modes.

In a hybrid application, the bus can be designed to use its diesel engine only at the engine's optimum power output and engine speed range.

Barriers to widespread application: The hybrid vehicles are still a "work-in-progress". Many studies noted that hybrid buses still require improvements in three technology areas: energy storage, electrically-driven accessories and on-board diagnostics.

3.5.Comparative analysis between the different buses

An overall comparison of buses with different technologies (diesel, CNG and hybrid electric) have been realized in the COMPRO project and the results are presented in the report "Cost /effectiveness analysis of the selected technologies" [14]. A comparison of buses is based on several parameters, technological, financial, environmental and planning-based, such as reliability, eployment flexibility, fuel price, range, exhaust gas emissions, noise, extra infrastructual needs. The considerations made above are summarized in Table 3. Advantages of each compared technology are featured in green, disadvantages in red.

Diesel Bus	CNG Bus	Hybrid Electric Bus
• Experienced	• New fuel filling station	• High range +
• High range	• Reduced range	• Low emission
• Pollution (PM + NO_x)	• Low emission / no PM	• Reduced consumption
• Noise	• Low fuel costs	• Reduced CO_2
• Increasing fuel costs	• Vehicle costs (+30.000 €)	• Reduced noise
• Dependence on mineral oil	• Dependence on natural gas	• Vehicle costs (+150.000 €)
		• Not yet experienced

Table 3. The comparison between the different technologies of city buses.

Another comparison is based on a West Virginia University's (WVU) study of city bus life cycle cost (LCC) [15, 16]. It covers the folowing bus types: diesel buses using ultra low sulfur diesel (ULSD), compressed natural gas (CNG) buses, and hybrid electric buses. LCC factors included capital costs (bus procurement, infrastructure, and emissions equipment) and operation costs (fuel, propulsion-related system maintenance, facility maintenance, and battery replacement) available from the literature.

A bus 12-year life cycle cost (LCC) analysis for a fleet of size of 100 buses was performed based on information available in the literature, manufacturers' specifications, and fuel economy data gathered by WVU [15, 16]. Only technology-dependent factors relevant to bus propulsion were considered; driver and management cost were excluded. Buses were assumed to operate at a national average speed of 12.7 mph (20.48 km/h), to travel for 32.007 mile (about 51.500 km) per year, and to seat 40 passengers for the purposes of calculation.

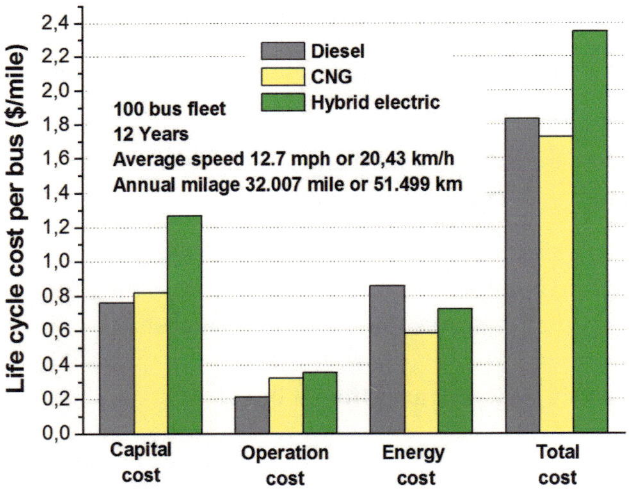

Figure 7. Comparison of life cycle cost between the different bus technologies.

Capital costs for vehicle procurement includes refueling station (CNG bus), depot modification, and emissions reduction equipment (diesel bus). Infrastructure costs for CNG bus technology include two costs: for depot modification and for refueling stations. Operational costs include compression electricity (CNG bus), facility maintenance, propulsion-related system maintenance, battery replacement (hybrid bus), and fuel consumption. Warranty was not considered. Fuel costs were calculated from the product of national annual average mileage, estimated fuel economy, and predicted fuel price. All prices were in 2008 dollars and CNG price data were all converted to the base of diesel (energy) equivalent.

Figure 7 representing total LCC was created for the capital and operation costs (without fuel consumption), and energy cost, per bus per mile.

The capital cost hybrid buses was slightly higher than CNG and diesel buses. However, operation cost analysis was similar for all bus types. Although hybrid buses offered the best fuel economy, this was offset by the battery replacement cost. Generally, the LCC total cost showed that diesel buses are still the most economic technology.

The West Virginia University's report [15] presents estimates for 2007 city bus regulated and greenhouse gas emissions. Tailpipe emissions (particulate matter (PM), nitrogen oxides (NO_x) and greenhouse gas (CO_2)) and fuel economy estimations were based on recent emissions and fuel economy studies, and adjusted with best engineering approach.

For simpler presentation of emissions and fuel economy by the three typical bus fleets (diesel, CNG and hybrid buses), the results given in the WVU study are appropriately processed and presented in Figure 8 [17]. Comparative values in Figure 8 represent average values for three typical driving cycles (MAN, OCTA, and CBD cycles). As can be seen from Figure 8 hybrid buses were attractive in offering emissions advantages. The estimation showed that hybrid buses offered lower tailpipe CO_2, NO_x and PM than the diesel and CNG buses. Figure 8 shows that the diesel hybrid bus fuel economy is better than the diesel bus fuel economy about 19%.

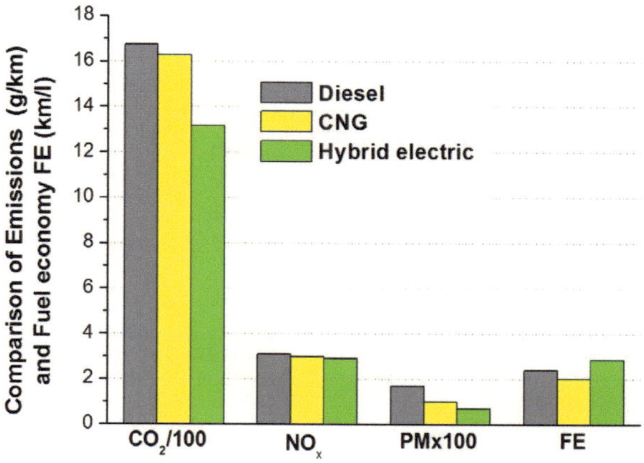

Figure 8. Comparison between the different bus technologies.

3.6. Hybrid electric bus solutions

The use of hybrid technology has become a popular issue in recent years. Hybrid solutions are principally available for all main propulsion types, thus with Diesel, CNG, fuel cells etc.

Table 4 [18, 19] and Table 5 [20] show some typical hybrid electric bus solutions of major bus manufacturers with IC engine.

Hybrid bus	Orion VII Hybrid electric bus	New Flyer Hybrid-Electric bus
Image of bus		
Type of hybrid drive	BAE Systems HybriDrive™ Serial hybrid	ISE ThunderVolt® TB40-HD Serial hybrid
Engine	Cummins ISB, ULSD, 5.9-litre, 194 kW, with a 120 kW traction generator	
Electric motor/generator	AC induction motor, Rated Power 250 hp continuous (320 hp peak)	Dual AC Induction motors, Rated Power 170 kW, Peak Power 300 kW
Energy Storage	Lithium-ion battery	Ultra capacitors
Bus Characteristics	Improve fuel economy 30% and reduce emissions: 90% PM, 40% NO_x, 30% CO_2	Reduce emissions: 25% PM, 32% NO_x and lower fuel consumption and CO2 emission

Table 4. Hybrid electric buses manufactured in North America.

Hybrid bus	MAN Lion's City Hybrid	MERCEDES-BENZ Citaro G BlueTec Hybrid
Image of bus		
Type of hybrid drive	Siemens, Serial hybrid	Serial hybrid
Engine	MAN D0836 LOH CR, EEV, 6.9 litre, 191 kW /260 hp	OM 924 LA, 4.8 litre 160 kW/218 hp
Electric motor/ generator	2x75 kW Siemens asynchronous electric motors. Synchronized generator with an output power of 150 kW	Four electric wheel hub motors on the central and rear axles of the bus, output of 320 kW (4x80 kW). Generator output 160 kW
Energy Storage	Ultra capacitors	Lithium-ion battery

http://dx.doi.org/10.5772/51770

Bus Characteristics	Up to 30 percent lower fuel consumption and CO_2 emission	Diesel consumption and CO_2 emissions reduced by up to 30 %

Hybrid bus	VOLVO 7700 Hybrid	SOLARIS Urbino 18 Hybrid
Image of bus		
Type of hybrid drive	Integrated Starter Alternator Motor (I-SAM) , Parallel hybrid	Voith DIWAhybrid, Parallel hybrid
Engine	New Volvo D5E, 5,0 litre capacity, rated at 210 hp	Cummins ISB6.7 250H engine, 181 kW (246 hp), 6.7 litre capacity
Electric motor/ generator	AC permanent magnet motor, power rating of 160 hp and a continuous rating of 90 hp	The motor provides 85 kW of power continually with a maximum output of 150 kW
Energy Storage	Nickel- Metal-Hydride battery	Ultra capacitors
Bus Characteristics	Fuel savings up to 35 % Lower exhaust emissions	Average fuel saving up to 16%

Table 5. Hybrid electric buses manufactured in Europe.

Besides the mentioned, other manufacturers of hybrid buses are: Gillig, ISE Corporation (North America); Scania, Irisbus Iveco, Van Hool, VDL Bus & Coach, Hess AG (Europe); Tata Motors, Toyota-Hino, Hyundai Motor Company, Mitsubishi Fuso (Asia).

4. Fuel cell buses

Fuel cells for bus applications have generated an enormous amount of attention over the last several years, as they offer the promise of a clean, efficient transportation system no longer dependent on fossil fuels. The fuel cell has a life expectancy about one-half that of an internal combustion engine. Thus, consumers would have to replace the fuel cell twice in order to achieve a vehicle operating lifetime equivalent to that of a traditional engine.

Fuel cell buses (FCBs) require a substantial new infrastructure, support, and training requirements will depend on what type of fuel is used for the fuel cells. Most demonstrations and available buses use pure hydrogen stored in compressed gas form.

4.1. Fuel cell bus configurations

All fuel cell vehicle concepts use electric motors to power the wheels, typically accomplished through the combination of an electric battery storage system and an on-board hydrogen fuel cell. Depending on the degree of hybridization, the battery may provide pure "plug-in" electricity to drive the vehicle some distance. The battery system would be complemented by a hydrogen storage system and a fuel cell, with the goal of extending the driving range to 300 miles (483 km). Early fuel cell bus designs involved an electric drive train, where a fuel cell generates electricity which is directly supplied to an electric motor. A new generation of FCBs is based on the hybrid concept. Figure 9 shows typical arrangement of the components in FCB [21]. Key system components are: fuel cell system, energy storage system, hydrogen storage system, wheel drive, cooling system, and auxiliaries.

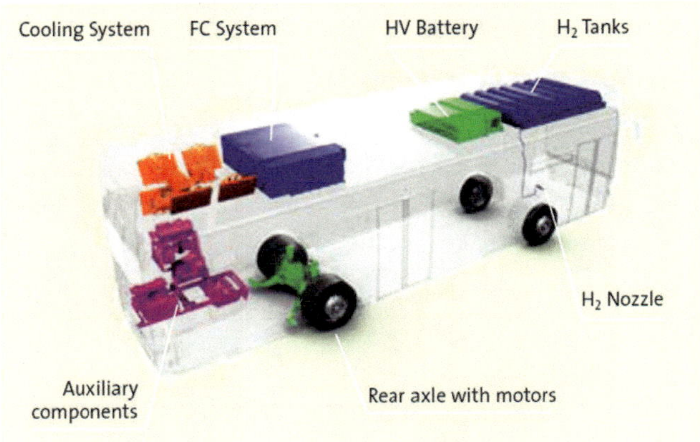

Figure 9. Key components of the fuel cell hybrid bus.

Fuel cell system: A fuel cell is an energy conversion technology that allows the energy stored in hydrogen to be converted back into electrical energy for end use. In a fuel cell vehicle, an electric drive system, which consists of a traction inverter, electric motor and transaxle, converts the electricity generated by the fuel cell system to traction power to move a bus. The fuel cell system and additional aggregates are usually located on top of the roof of the bus.

Fuel cells are classified by their electrolyte and operational characteristics. For application in vehicles mostly used are the Polymer Electrolyte Membrane (PEM) fuel cells. They are lightweight and have a low operating temperature. PEM fuel cells operate on hydrogen and oxygen from air.

Alkaline fuel cells (AFCs) are made by one of the most mature fuel cell technologies. AFCs have a combined electricity and heat efficiency of 60 percent efficient.

A newer cell technology is the Direct Methanol Fuel Cell (DMFC). The DMFC uses pure methanol mixed with steam. Liquid methanol has a higher energy density than hydrogen, and the existing infrastructure for transport and supply can be utilized.

http://dx.doi.org/10.5772/51770

There are some major fuel cell manufacturers supplying fuel cell power plants for heavy-duty applications: Ballard Power Systems and Hydrogenics (Canada), United Technologies Corporation (UTC) Fuel Cells, Enova Systems (USA), Shanghai Shen-Li High Tech Co. Ltd. (China), Siemens and Proton Motor Fuel Cell GmbH, (Germany), Toyota (Japan), Hyundai Motor Co. (South Korea).

Energy storage system: Energy storage systems are generally based on battery packs and/or ultracapacitors (generally up to 100 kW). Maximum power output and storage capacity vary depending on hybrid architecture. Lithium-ion battery technology is the most appropriate of energy storage technology for use in the buses. The batteries are usually located on top of the roof of the bus. FCBs are equipped with regenerative braking.

Hydrogen storage system: Gaseous hydrogen serves as the fuel. It is stored in compressed gas tanks, the number of which is decisive for the maximum range but also confines passenger capacity. The hydrogen storage system has been downsized as a result of the improved efficiency of the drive train. This has led to the reduction in the overall weight of the bus. The cylinders to storage hydrogen on board operate at an increased pressure of 350 bars.

Wheel drive: The electric motor can be either a single main motor or hub–mounted (where the motor is designed within the wheel). The bus may be equipped with a central traction system which will be located at the left hand side in the rear of the bus. The rear axle has 2 wheel hub motors and has been specifically developed to match the required speeds, load capabilities and energy efficiency. It also serves as a generator for energy regeneration during braking.

Cooling system: While running hydrogen through a fuel cell, water is of course being produced. Some of it becomes steam and leaves the system quite easily, as seen at the steam vent at the back of the bus. Yet, because PEM cells are sensitive to high heat the cell stacks must be cooled down. Therefore the byproduct from producing the electricity will always partially turn into liquid water that can accumulate in the stack and slow down the process. This can happen during idle times or at full speed. Therefore all PEM cells need a mechanism that clears the stacks once in a while or else the electricity production will be slowed down. The majority of the stack manufacturers use liquid cooled systems, with radiators to dissipate heat.

Auxiliaries: The auxiliary components in the FCB may be driven electrically. This means that they operate only on demand and are not driven continuously. This solution is typical for FCBs, which are based on the hybrid concept. This will result in a higher efficiency and lower maintenance of the components.

4.2. Fuel cell bus characteristics

Fuel cells offer a number of potential benefits that make them appealing for transport use such as greater efficiency, quiet and smooth operation, and, if pure hydrogen is used on board the bus, zero emissions in operation and extended brake life. Infrastructure, buses, fuel, and maintenance costs associated with hydrogen fuel cells are currently prohibitively expensive. The cost of facilities has ranged from several hundred thousand dollars up to $4,4 million for a maintenance facility, fueling station, and bus wash [22]. Currently, fuel cells for buses are not a commercial product. The existing fuel cell buses are prototypes, manufactured in fairly small numbers. Fuel cell buses can cost $1 to $3 million (or more) since they are hand-built prototypes utilizing a pre-commercial technology.

The hydrogen fuel itself is also currently very expensive. Costs range depending on the method of hydrogen production. One of the major constraints for use of the fuel cell buses is the refueling time for hydrogen buses. Filling over 30kg of hydrogen in less than 5 minutes is not currently feasible without pre-cooling the hydrogen (as the temperature increase at these high fill rates would damage the hydrogen tanks) [23].

Some comparisons with different bus technologies (diesel, hybrid diesel-hybrid and fuel cell), in terms of CO_2 emissions per km traveled, have shown significant benefits of FCBs [23]. CO_2 emissions of fuel cell buses ranging from 0 to 1,8 kg/km. Zero emissions are related to renewable hydrogen and electricity. Emissions of diesel hybrid buses are 0,69 to 1,2 kg/km, but emission of diesel buses are 1,05 to 1,5 kg/km.

4.3. Fuel cell bus projects

The introduction of new types of buses in urban public transport is sometimes a challenging process that includes testing, demonstration and limited production with a tendency to increase the number of vehicles. Fuel cell-powered buses continue to be demonstrated in public transport service at various locations around the world. Many demonstration projects have been launched in the last 10 years in various stages of implementation. Many have been completed, and some of them are still active. An overview of mainly fuel cell city bus development projects is given below [21]:

The HyFLEET:CUTE project has involved the operation of 47 hydrogen powered buses in regular public transport service in 10 cities on three continents. The Project started in 2006 and concluded at the end of 2009. HyFLEET:CUTE was co-funded by the European Commission and 31 Industry partners through the Commission's 6th Framework Programme.

ECTOS (Ecological City Transport System) was an initiative to test three Citaro fuel cell buses in Reykjavik, Iceland. The project was financially supported by the European Commission. The project started at mid-2003 and now the buses are decommissioned.

STEP (Sustainable Transport Energy Project) is a project to explore the use of alternative transport energies in Perth, Australia. It includes the operation three Citaro fuel cell buses. It is funded by the government of Western Australia with support from the Australian government. The project started at mid-2004 and now the buses are decommissioned.

CHIC (Clean Hydrogen In European Cities) is the active project which involves integrating 26 FC buses in daily public transport operations and bus routes in five locations across Europe – Aargau (Switzerland), Bolzano/Bozen (Italy), London (UK), Milan (Italy), and Oslo (Norway). The CHIC project is supported by the European Union and has 25 partners from across Europe, which includes industrial partners for vehicle supply and refueling infrastructure. The project will guide additional regions in Europe in their first fleet application of fuel cell hybrid buses in public transport from 2012 onwards [24]. The buses in the CHIC project will be supplied by three different manufacturers - Mercedes-Benz (Germany), Van Hool (Belgium) and Wrightbus (UK).

"NaBuZ demo" (Sustainable Bus System of the Future – Demonstration) is German-funded project [25] in which the Hamburger Hochbahn AG produced four Mercedes-Benz Citaro FuelCELL Hybrid buses. Since 2011, the first Citaro FuelCELL Hybrid bus is involved into service.

The National Renewable Energy Laboratory (NREL) has recently published a status report documenting progress and accomplishments from demonstrations of fuel cell city buses in the United States. The report describes the status and challenges of fuel cell propulsion for transport and summarizes the introduction of fuel cell city buses in North America. Three major programs are [26]:

1. *Federal Transit Administration's (FTA) National Fuel Cell Bus Program (NFCBP)* includes developing the 11 new buses, expanding the fuel cell manufacturers beyond Ballard and UTC Power to include Hydrogenics and Nuvera, and exploring multiple bus sizes and hybrid propulsion designs.

2. *Zero Emission Bay Area (ZEBA) Group Demonstration* includes 12 next-generation fuel cell city buses with redesigned Van Hool chassis, the newest UTC Power fuel cell power system, and fully integrated hybrid propulsion system.

3. *BC Transit Fuel Cell Bus Demonstration* includes 20 fuel cell buses in Whistler, Canada. The buses are from New Flyer, use Ballard fuel cells, and have hybrid propulsion from ISE/Siemens.

Other projects: In China more than 50 FCBs were used during the Asian Games in 2010 and the Olympics Games in 2008. In Japan, Toyota-Hino has launched several dozen fuel cell buses in the period since 2003. In Korea, a Hyundai fuel cell bus has operated since 2006 [27].

4.4. New generation of fuel cell buses

Mercedes-Benz fuel cell hybrid buses: Mercedes-Benz has promoted the use of fuel-cell hybrid buses around the world over the past nine years. Since 2003, a total of 36 Mercedes-Benz Citaro buses equipped with second-generation fuel cells were driven a combined total of more than 2,2 million kilometres in a total of approximately 140.000 hours of operation [28].

Production of the second generation of Mercedes fuel-cell hybrid buses started in November 2010 under the CHIC project. Compared with the fuel cell buses which were tested in Hamburg in 2003, the new Citaro FuelCELL Hybrid, Figure 10, provides several significant new features [25]: hybridization with energy recovery and storage in lithium-ion batteries, powerful electric motors with 120 kW of continuous output in the wheel hubs, electrified power take-off units and further enhanced fuel cells. These should achieve an extended service life of at least six years or 12.000 operating hours.

New additions are the lithium-ion batteries which for example store recovered energy. With the power stored there the new Citaro FuelCELL Hybrid can drive several kilometres on battery operation alone. In general, the design of the new FuelCELL buses is largely the same as that of the Mercedes-Benz BlueTec Hybrid buses that run in regular service; these buses also get electrical energy from a diesel generator. Thanks to improved fuel cell components and hybridization with lithium-ion batteries the new Citaro FuelCELL Hybrid saves on almost 50 % in hydrogen usage compared with the preceding generation. Overall fuel cell system efficiency has also been improved. The fuel cell bus has a range of around 250 kilometres.

Figure 10. The new Citaro FuelCELL hybrid bus.

Van Hool fuel cell hybrid buses: VAN HOOL (Belgium) is the largest independent manufacturer of integral buses and coaches in Western Europe. More than 80% of the company's production is exported: two thirds stay in Europe, the remainder goes to America, Africa and Asia. In a joint effort with UTC Power (United Technologies Corporation), a supplier of fuel cell systems, Van Hool developed fuel cell buses for the European and U.S. markets. Siemens supplied the twin AC induction electric motor, 85 kW each, converters, and adapted traction software.

Figure 11. The new Van Hool fuel cell bus.

Within the project ZEBA demonstration includes 12 new generation fuel cell hybrid buses and two new hydrogen fueling stations [29]. The buses are 12 m, low floor buses built by

Van Hool with a hybrid electric propulsion system that includes a UTC Power fuel cell power system (120 kW) and an advanced lithium-ion battery (rated energy: 17.4 kWh and rated power: 76 to 125 kW). Eight Dynetek, type 3 cylinders, 350 bars, are mounted on the roof. The new buses, Figure 11, feature significant improvements over two previous generations of fuel cell buses that were demonstrated in California, Connecticut, and Belgium. Improvements include a redesigned Van Hool chassis that is lighter in weight, shorter in height, and has a lower center of gravity for improved weight distribution. The bus has a top speed of 55 mph (88 km/h). The bus purchase cost is about $2,5 million.

4.5. Operating cost and fuel economy of fuel cell buses

Many transport operators continue to aid the FCB industry in developing and optimizing advanced transportation technologies. These in-service demonstration programs are necessary to validate the performance of the current generation of fuel cell systems and to determine issues that require resolution.

By the end of June 2011, nine of the twelve new Van Hool fuel cell buses had been delivered and seven of those were in-service. The buses have accumulated more than 80.000 miles (128.000 km) and a total of 7.653 hours on the fuel cell systems. The results presented here are early/preliminary information from the first five fuel cell buses that have been placed into service at AC Transit [29].

Table 6. presents the comparative test results for the Fuel cell and Diesel buses during the evaluating period.

	Fuel Cell	**Diesel**
Fuel Cost ($/km)	0.96	0.42
Total Maintenance Cost ($/km)	0.94	0.41
Total Operating Cost ($/km)	**1.87**	**0.83**

Table 6. Operating costs of different buses per kilometre.

During 2011, NREL completed data collection and analysis on new generation FCB demonstrations at three transport operators in the United States: SunLine California, CTTRANSIT Connecticut, and AC Transit California [30]. The current-generation FCBs in service at AC Transit, CTTRANSIT, and SunLine were all of the same basic design: Van Hool (12m) buses with ISE Corp. hybrid-electric drives and UTC Power fuel cell power systems.

Table 7 shows the fuel economy of the buses at each location. Data are given in miles per diesel gallon equivalent and in km/litre (1mile per Gallon =0.425 km/litre).

The FCBs at the three locations showed fuel economy improvement ranging from 48% to 133% when compared to diesel and CNG baseline buses. This table also illustrates the variability of the results from fleet to fleet. The results are affected by several factors, including duty-cycle characteristics (average number of stops, average speed, and idle time). Average speed for AC Transit –9,8 mph (15,8 km/h); CTTRANSIT –6,5 mph (10,4 km/h), SunLine –13 mph (20,9 km/h) [30].

	FC bus		Diesel bus		CNG bus	
	mile/gallon	km/litre	mile/gallon	km/litre	mile/gallon	km/litre
AC Transit	6,8	2,9	4,2	1,8	-	
CTTRANSIT	5,5	2,3	3,7	1,6	-	
SunLine	8,0	3,4	-	-	3,5	1,5

Table 7. Fuel economy of the fuel cell buses at different locations.

Data about NO_x and PM emissions (per year) presented by Van Hool in its promotional materials for three different fleets of buses, are shown in Table 8 [31]. The data are calculated for 50.000 km per year, average speed 20 km/h, and power consumption 50 kW/h.

Equivalent emissions reduction potential of 100 hybrid fuel cell buses gives a CO_2 reduction equal to the uptake of 3.100 acres of forest and a NO_x reduction equal to 10 km of 4 lanes of cars [31]. The presented results show all the environmental advantages of the buses with fuel cell technologies.

	NOx (per year)	PM (per year)
100 Diesel Euro III buses	62,5 tons	1,25 tons
100 CNG buses	25 tons	0,25 tons
100 Hybrid fuel cell buses	zero	zero

Table 8. Comparative characteristics of NOx and PM emissions for different buses.

5. Battery electric buses

Electric vehicles are a promising technology drastically reducing the environmental impact of road transport. At the same time, EVs are still far from proven technology. Reality is such that battery technology is simply not the whole answer. This is because (especially for large buses) batteries do not carry enough energy to power the bus for a full day.

According to IDTechEx report [32], "Electric Vehicles 2010-2020" it is estimated that worldwide there are about 480.000 electric buses, mostly small ones - with about 135.000 being bought each year as the fleets grow. Although only 12% of these new buses are electric, it is expected that by 2020 investment in the purchase of these buses will be measured in million of dollars. The main increase in electric buses will be running the free versions to be used without appropriate infrastructure along the route.

5.1. Full- size battery electric buses

Proterra's Battery Electric Bus: Proterra is the pioneering innovator and manufacturer of clean commercial transport solutions for city buses. Three Proterra's zero emission fast-charge battery electric buses, EcoRide BE35, Figure 12, have been put in service in 2010 [33].

The buses feature Proterra's revolutionary clean-transport technology, including the Proterra TerraVolt Energy Storage System, which allows a full battery recharge in less than 10 minutes. Additional bus features include:

Flexible ProDrive and vehicle control system that can operate in battery electric mode or with any small APU to extend vehicle range when needed;

Regenerative braking system, allowing the recapture over 90% of the vehicle's kinetic energy available during braking;

Light-weight composite body resulting in 25% reduction in weight, significantly lower maintenance costs and 40% longer life than traditional diesel buses.

Figure 12. Proterra's battery electric bus.

Proterra's EcoRide BE35 battery powered buses can operate on standard routes for up to three hours—a range of 30-40 miles (48 to 64 km)—and after that, require just 10 minutes of charging to get back on the road. The buses can accommodate as many as 68 passengers and according to Proterra, will provide $300.000 in savings over the course of their lifetime thanks to lower fuel and transportation costs.

BYD electric bus: China's Build Your-Dream (BYD) hybrid buses, Figure 13, were showcased during the 2008 Beijing Olympics. BYD has developed and is currently marketing a lithium-ion battery - powered hybrid-electric bus and all electric buses.

BYD, manufacturer of the first long-range (>300 km) all-electric bus (eBUS-12), has been selected as the sole eBUS provider for the 2011 International Universiade Games held in Shenzhen, China. At the core of the eBUS technology is BYD's in-wheel motor drive system and the Iron Phosphate battery technology. The eBUS also integrates BYD solar panels on the bus roof, converting solar energy to electricity which is stored in the batteries and can completely offset the eBUS air-conditioning load (extending the range on sunny days) [34].

Figure 13. BYD eBUS-12 electric bus.

BYD has signed a Letter of Intent with the city of Frankfurt, Germany to introduce BYD's all-electric, long-range, eBUS. BYD will supply three all-electric buses eBUS-12, two DC charging stations and technical support in the first quarter of 2012.

BYD is set to trial a full-size, all-electric bus in the Danish capital Copenhagen. Two K9 buses will initially be deployed on ordinary passenger routes from the second half of 2012.

The City of Windsor, Ontario has signed a letter of intent to purchase up to 10 BYD fully electric buses for the community's transport services in 2012. It will become the first City in North America to launch long-range, all-electric buses. BYD has delivered over 300 all-electric buses worldwide and claims orders for over 1.300 more in 2012, making it the largest electric bus manufacturer in the world.

e-Traction buses: Netherlands' company e-Traction (European integrator of low-floor fleet buses), delivered in the year 2010 the first of two e-Busz electric drive buses to Rotterdam's public transportation authority. The e-Busz is a VDL Bus & Coach Citea CLF bus converted with the third generation of the e-Traction system [35].

e-Traction specializes in development of TheWheel as a direct-drive in-wheel motor system with integrated power electronics and fluid cooling. TheWheel is designed to deliver very high torque at low revolution. Since 2001, e-Traction is continuously developing in-wheel direct-drive motors for applications ranging from 400Nm to 10.000Nm per wheel. The vehicles with TheWheel save up to 40% traction energy and are 50% more fuel efficient compared to the standard diesel equipped bus. The e-Buzs is a "battery dominant" hybrid bus. This means that it has the ability to run on battery only, with the diesel generator turned completely off. The diesel unit (with diesel generator) can be replaced and, importantly, the bus returned to revenue service in roughly one hour.

In cooperation with e-Traction, Hybricon, a Swedish company converted two Volvo 7700 city buses to fully electric city bus with rapid charge technology. The buses with the name Arctic Whisper, Figure 14, are from November 2012 in public service in the city of Umeå, Sweden [36]. e-Traction and Hybricon removed the whole diesel driveline and replaced it with two e-Traction SM/500-3 wheel motors mounted on a rear axle construction. A 100 kWh Valence Li-Fe battery pack (for the purposes of the prototype) and pantographs for the charging station form the basic configuration, with a 50 kW diesel generator as the back-up system.

Figure 14. Arctic Whisper plug in hybrid bus.

The Arctic Whisper's bus is fast charged for 10 minutes at the end of its route to achieve nearly 100% all-electric operation but with the reliability of diesel. Without fast charging, the Arctic Whisper has an all-electric runtime of about 2-3 hours with the 100 kW batteries before the diesel generator needs to turn on.

Future plans include using different battery chemistries capable of faster charging and higher charging rates of over 200 kW as well as extending this architecture to 18 meter articulated buses.

SMG battery electric bus: The Seoul Metropolitan Government (SMG) has started commercial operation of full-size battery electric buses since 2010, Figure 15. The government has been working on a project to develop the buses with local technology after reaching an agreement with Hyundai Heavy Industries and Hankuk Fiber back in September, 2009. It now has a goal of putting 120.000 of the vehicles to use in the city by 2020 – this will account for 50% of all public transport vehicles [37].

The SMG electric buses are a low floor design, 11,00 m in length and can travel as far as 83 km on a single charge. Using high-speed battery chargers they can be fully charged in less than 30 minutes and have a maximum speed of 100 km/h. Four battery charges are being provided. They use high-capacity lithium-ion batteries and regenerative braking. To reduce their weight and help maximize the distance they can travel between charges, these buses make extensive use of carbon composite materials, instead of metal.

Figure 15. SMG battery electric bus.

Optare's battery-powered Versa: Optare's battery-powered Versa, Figure 16 - the UK's first commercial full-size battery bus which started on 1 April 2011 [38]. Battery-powered Versa buses are planned to be in service at peak times, providing a capacity increase on the 12-minute interval service. The buses can either be rapid charged, using a special charger, or slow charged. Travel de Courcey intends to re-charge the buses when they get to 30% of full charge. A return journey is around 5 miles (8 km). The Versa's testing is currently underway with the first bus to be delivered, pending the arrival of the other two, and their entry into service. They are not the first electric vehicles to be commercially produced by Optare. It has already delivered Solo EV models, also used on short distance shuttle services [38].

Figure 16. Optare's battery-powered Versa.

Tindo solar electric bus: The Adelaide City Council's electric solar bus, Figure 17, is the first in the world to be recharged using 100% solar energy. Recent advances in battery technology have helped the successful development of pure electric buses with a suitable range between recharges. The Tindo solar electric bus uses 11 Zebra battery modules, giving it unprecedented energy storage capacity and operational range [39]:

Figure 17. Tindo solar electric bus.

Some Tindo bus characteristic are: Length–10,42 m, Motor power peak – 160 kW, Motor power nominal - 36 kW, Speed - 76 km/h, Battery content - 261,8 kWh, Fast Charger Booster Power - 70 kW, Fuel costs - 50% lower than for a diesel bus, Range - 200 km between recharges under typical urban conditions.

5.2. Small battery electric buses

Optare solo EV battery electric bus: Optare offers the Solo EV, Figure 18, fully electric buses available in lengths of 8,1m, 8,8m and 9,5m. Replacing the usual diesel engine is an all-new electric drive, featuring an Enova Systems P120 AC induction motor rated at 120 kW and powered by two banks of Valence Lithium Ion Phosphate batteries. The two packs work in parallel and provide 307 V with a total capacity of 80 kWh. [40].

Around 4.000 Optare Solo EV buses in service worldwide produce zero tailpipe emissions. The model demonstrated in Switzerland is based on a standard 27-seat, 8,8 metre Solo, but the technology can be used on other models in the Optare range with higher passenger capacities. The Solo EV has been designed to perform exactly like a standard diesel powered bus, except that it is smoother, quieter and cleaner. It is completely traffic compatible, with good acceleration and hill climbing capabilities and a top speed of up to 90 km/h. On a full charge it has a range of around 110-130 kilometres depending on load factors and topography.

Solaris Urbino electric bus: Polish bus manufacturer Solaris developed the midi Urbino electric bus, Figure 19. The innovative electric bus is based on the 8.9 m Urbino family midi bus. At the heart of its power system is a 120 kW four-pole asynchronous traction motor supplied by Vossloh Kiepe. Energy is stored in two batteries weighing 700 kg each. These liquid-cooled lithium-ion batteries have a rated voltage of 600 V and the capacity to store 120 kWh [41].

Figure 18. Optare Solo EV battery electric bus.

Figure 19. Solaris Urbino electric bus.

The battery capacity gives the Solaris Urbino electric a range of up to 100 km and a maximum speed of 50 km/h. Batteries are charged with a Walter plug-in connection. A full recharge from the 3x400 V, terminal takes as little as four hours. Even with the 1.400 kg traction batteries, the Solaris Urbino electric is only marginally heavier than its conventional counterparts thanks to the innovative lightweight construction employed.

6. Energy storage systems

Energy storage systems, usually batteries, are essential for electric drive vehicles. Batteries must have a high energy-storage capacity per unit weight and per unit cost. Because the battery is the most expensive component in most electric drive systems, reducing the cost of the battery is crucial to producing affordable electric drive vehicles.

The electrical energy storage units must be sized so that they store sufficient energy (kWh) and provide adequate peak power (kW) for the vehicle to have a specified acceleration performance and the capability to meet appropriate driving cycles. For those vehicle designs intended to have significant all-electric range, the energy storage unit must store sufficient energy to satisfy the range requirement in real-world driving. In addition, the energy storage unit must meet appropriate cycle and lifetime requirements. These requirements will vary significantly depending on the vehicle type (battery or fuel cell powered or hybrid electric).

There are many energy storage technology and battery chemistry and packaging options for electric drive buses.

6.1. Energy storage device used in some electric drive buses

Based on the presentation of realized solutions of electric drive buses, in the previous items of this chapter, it may be concluded that in the energy storage devices the latest technology batteries and ultra capacitors are applied. Summary of main characteristics of energy storage devices, for each of the presented buses is given in Table 9.

As can be seen from Table 9, Li-ion batteries are prevailing in the realized bus solutions. In the latest bus solutions Zebra battery and Iron Phosphate battery technology are used. It is notable that the energy capacity of energy storage devices BEBs are significantly higher than that of HEBs and FCBs.

In a HEBs with an ICE that recharges the battery (where battery operates in charge sustaining mode), a lighter and smaller battery is employed.

HYBRID ELECTRIC BUSES		
BUS	MAN Lion's City Hybrid	MERCEDES-BENZ Citaro G BlueTec Hybrid
Energy storage	Capacitors Energy content: approx. 0,5 kWh, Max. charging/discharging power: 200 kW, Voltage: 400-750 V	Lithium-ion battery Energy content:19,4 kWh, Maximum output of 240 kW and, Located on the roof
BUS	VOLVO 7700 Hybrid	SCANIA Ethanol hybrid bus
Energy storage	Nickel-Metal-Hydride battery Energy content: approx. 4,8 kWh, Weighting approximately 350 kg, Rated at 600 volts, Located on the roof	Supercapacitors Energy available: "/400 Wh, 4x125-Volt Maxwell BOOSTCAP® modules, air-cooled, Design life:10-15 years
BUS	SOLARIS Urbino 18 DIWA Hybrid	ORION VII hybrid electric bus

Energy storage	Super capacitors Energy content: 0,5 kWh, Maxwell, 5x125V, Weight 410 kg	Lithium-Ion battery Energy content: 32 kWh, Weight 364 kg, 6 year design life, Roof-mounted
	FUEL CELL BUSES	
BUS	Mercedes-Benz Citaro FuelCELL Hybrid bus	Van Hool Fuel Cell Hybrid Bus
Energy storage	Li-ion Battery Energy content: 26 kWh, Energy storage power 250 kW	Li-ion Battery Energy content: 17,4 kWh, Rated power: 76 to 125 kW
BUS	New Flyer Fuel Cell Bus	Van Hool Fuel Cell Hybrid Bus (UTC power)
Energy storage	Li-ion Battery Energy content: 47 kWh	NaNiCl (ZEBRA) battery Energy content: 53 kWh
	BATTERY ELECTRIC BUSES	
BUS	Optare Solo EV Battery Electric Bus	Solaris Urbino Electric Bus
Energy storage	Li-ion Battery Energy content: 80 kWh	Li-ion Battery Energy content: 120 kWh, Rated voltage of 600 V
BUS	BYD eBUS-12 Battery Electric Bus	Tindo Solar Electric Bus
Energy storage	Li Iron-Phosphate or "Fe" battery Energy content: 324 kWh Milage: 300 km on a single charge	Li-ion Battery Battery content: 261,8 kWh

Table 9. Characteristics of energy storage devices of several bus solutions.

A BEBs require a larger and heavier battery pack, to provide both high energy density and high energy storage capacity so as to maximize the range between recharges.

In the PHEBs one can expect smaller, intermediate-sized, battery packs capable of either charge-sustaining operation in the blended mode with an active ICE, or charge-depleting operation.

6.2. Current and future development of energy storage devices

A number of different battery technologies exist at present. The lead acid battery has been used to supply vehicle electricity for a number of decades. With the introduction of the first modern EVs in the 1980s, the need for more powerful batteries arose. Nickel-cadmium batteries were originally used, later replaced in hybrid vehicles by nickel-metal hydride batteries. However, none of these battery technologies provide the energy density required for sufficient driving distance in pure electric mode.

Recently, apart from the mentioned, other energy storage devices are in the intensive growth and expansion, such as: Lithium-ion battery (Li-ion), Li-ion polymer battery, Sodium Nickel Chloride battery (NaNiCl), Lithium iron phosphate battery (LiFePO4), Zinc Air battery, and Supercapacitors.

Based on available analysis and current battery data, it appears that the current (2010) battery life should exceed seven years and may be around ten years for 'average' use. The most promising chemistries appear to involve silicon, sulfur and air (oxygen) and another important development is research into nanotechnologies. These trends have been widely recognized and a recent presentation by Limotive researchers showed the following battery technology roadmap, Figure 20 [41].

Silicon is an attractive anode material for lithium-ion batteries because it has about ten times the amount of energy that a conventional graphite-based anode can contain [42]. It also has a specific energy of 1.550 Wh/kg – about four times the energy of a conventional graphite-based anode. Furthermore, silicon is the second most abundant element on the planet and has a well-developed industrial infrastructure, making it a cheap material to commercialize with a cost comparable to graphite per unit of weight.

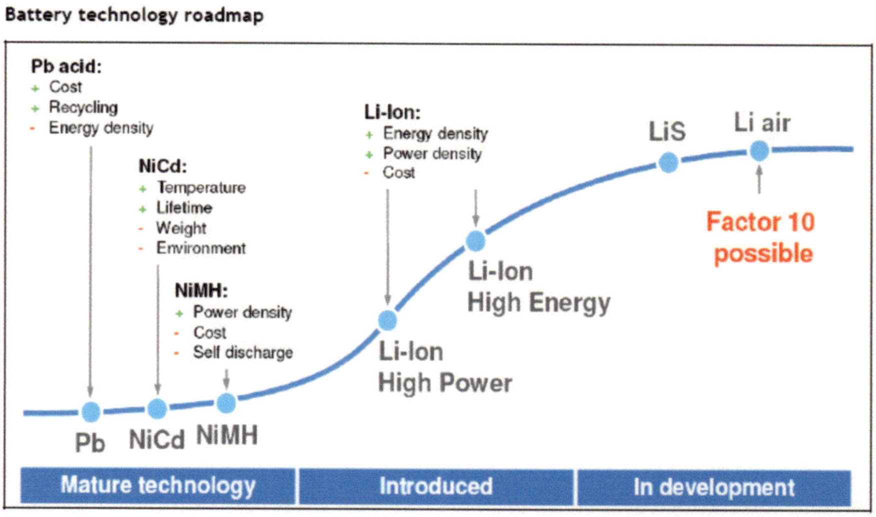

Figure 20. The battery technology roadmap.

The problem with silicon is that it is very brittle and when lithium-ions are transferred during charge and discharge cycles, the volume expands and contracts by 400% which can pulverise the silicon anodes after just the first cycle.

The Li-Ion technology will become more and more the dominant technology for electro mobility. The Li-Ion technology has not yet reached its full potential, further improvements are still possible. Further developments are needed to improve capacity and lifetime, reduce volume and costs (currently around €250-€500/kWh for NiMH and €700-€1.400/kWh for Li-ion), and to be safe and reliable [43].

Although few serious technical hurdles remain to prevent the market introduction of electric powered vehicles, battery technology is an integral part of these vehicles that still needs to be significantly improved. Both current and near-term battery technologies still have a number of issues that need to be addressed in order to improve overall vehicle cost and performance. These issues include[44]:

- *Battery storage capacity* – Batteries for EVs need to be designed to optimize their energy storage capacity, while batteries for PHEVs typically need to have higher power densities.
- *Battery duty (discharge) cycles* – Batteries for various electric powered vehicles have different duty cycles. Batteries may be subject to deep discharge cycles (in all electric mode) in PHEVs or frequent recharge cycles through regenerative braking in conventional HEVs. Batteries for EVs will be subjected to repeated deep discharge cycles without as many intermediate cycles. Current battery deep discharge durability will need to be significantly improved.
- *Durability, life expectancy, and other issues* – Batteries must improve in a number of other respects, including durability, life-expectancy, energy density, power density, temperature sensitivity, reductions in recharge time, and reductions in cost. Battery durability and life-expectancy are perhaps the biggest technical hurdles to commercial application in the near-term.

7. Conclusions

A significant part in the future reduction of consumption of fossil fuels and of the corresponding reduction of emissions of harmful gases will be played by the alternative propulsion systems and alternative fuels. The development of electric drive technologies intended for application in buses is expanding. However, there are many limitations which at this stage slow down these developments. Sustainability of alternative propulsion systems is dependent upon the degree of their technological development and a compromise between the opposed economical, ecological, and social factors [45].

A large number of hybrid buses in North America and Japan and their intense development in Europe over the past several years is a confirmation that their number in the near future will be permanently growing. Although the sale of these vehicles is relatively small, the high cost of fossil fuels and the costs of hybrid vehicles becoming more acceptable will accelerate their further development. The hybrid buses are expected to contribute to further reduction of CO_2 emissions, even though some manufacturers have reached the level of 30%. Further improvements in that direction will be dependent on the degree of hybridization of the propulsion system and electronic control which should contribute to the optimization of operation of ICE and hybrid system as a whole.

The experiences so far acquired, through the development of fuel cell buses and many demonstration projects around the world, are very positive. Some reports indicate that the per-

http://dx.doi.org/10.5772/51770

formances of fuel cell buses in-service are above expectation. One of the many barriers to their wider use is the uncertainty of hydrogen supplies and high production costs of hydrogen. Other barriers are related to the security aspects of usage of these vehicles. Despite the insufficient performances of batteries, the next generation of fuel cell buses will be based on the hybrid concept and Li-ion batteries. Some predictions tell that fuel cells for buses will be commercially available within the next 10-15 years. At present, the high cost of buses is one of the greatest barriers to their commercialization.

Since 2010 increasing the number of battery driven vehicles and buses is evident, thus it can be expected that in the forthcoming times their number will continue increasing. However, barriers to their massive implementation will be the radius of movement, lack of infrastructure for recharging the batteries and, of course, high cost of the batteries and other power equipment (electric motors and control electronics).

Further challenges for electric drive buses will be the development of battery technologies and of other energy sources. Even though a considerable advancement has been made over the past several years by the development of Li-ion batteries, which have achieved energy density of 95-190 Wh/kg, there is still space for further advancements. The only battery chemistries that have a chance of achieving energy densities in the 1,000 Wh/kg range are rechargeable metal-air and other. Other non-chemical energy storage devices include supercapacitors that can reach very high specific power levels for a few seconds, but cannot hold a lot of energy.

The current generation of lithium-ion batteries typically uses a carbon-based anode and a metal oxide cathode. Research on next generation lithium batteries will continue the development of electrode and electrolyte materials and chemistries in order to increase the life and energy density of the battery while reducing size and weight. The most promising chemistries appear to involve silicon, sulphur and air (oxygen) and another important development is research into nanotechnologies.

8. Acronyms and Abbreviations Nomenclature

AC-Alternating Current

AFC-Alkaline Fuel Cell

APTA-American Public Transportation Association

APU-Auxiliary Power Unit

BEB-Battery Electric Bus

BEV-Battery Electrics Vehicle

BYD-Build Your-Dream

CARB-California Air Resources Board

CBD-Central Business District

CHIC-Clean Hydrogen in European Cities

CNG-Compressed Natural Gas

CO-Carbon Monoxide

CO_2-Carbon Dioxide

COMPRO -COMmon PROcurement of collective and public service transport clean vehicles

CUTE -Clean Urban Transport for Europe

DC -Direct Current

DMFC -Direct Methanol Fuel Cell

ECTOS-Ecological City TranspOrt System

EHPS-Eaton Hybrid Power System

EM/G-Electric Motor/Generator

EV-Electric Vehicle

FCB-Fuel Cell Bus

FCEV-Fuel Cell Electrics Vehicle

FTA-Federal Transit Administration

HEB-Hybrid Electric Bus

HEV-Hybrid Electrics Vehicle

ICE-Internal Combustion Engine

ISAM-Integrated Starter Alternator Motor

LCC-Life Cycle Cost

LiFePO4 Lithium Iron Phosphate

Li-ion-Lithium-ion

LNG-Liquid Natural Gas

MAN-Manhattan Cycle

NaNiCl-Sodium Nickel Chloride

NO_x -Nitrogen Oxides

NFCBP-National Fuel Cell Bus Program

NREL-National Renewable Energy Laboratory

OCTA-Orange County Transit Authority

PEM-Polymer Electrolyte Membrane

PHEB-Plug-in Hybrid Electric Bus

PHEV-Plug-In Hybrid Electrics Vehicle

PM-Particulate Matter

SMG-Seoul Metropolitan Government

STEP-Sustainable Transport energy for Perth

ULSD-Ultra Low Sulfur Diesel

UTC-United Technologies Corporation

WVU-West Virginia University

ZEBA-Zero Emission Bay Area

Acknowledgements

Financial support by Ministry of Education and Science Republic of Serbia (Projects TR 35041, TR 35042 and TR 35036) is gratefully acknowledged.

Author details

Zlatomir Živanović[1*] and Zoran Nikolić[2]

*Address all correspondence to: zzivanovic@vin.bg.ac.rs

1 University of Belgrade, Institute of Nuclear Sciences VINCA, Belgrade, Serbia

2 Institute of Technical Sciences of the Serbian Academy of Sciences and Arts, Belgrade, Serbia

References

[1] Chan, C. (2007). The State of the Art of Electric, Hybrid, and Fuel Cell Vehicles. *Proceedings of the IEEE*, 95(4), 704-718.

[2] Battery Hybrid and Fuel Cell Electric Vehicles are the keys to a sustainable mobility, AVERE, European Association for Battery. (2010). *Hybrid and Fuel Cell Electric Vehicles*, http://www.avere.org/www/Images/files/about_ev/Brochure.pdf, accessed 2 November, 2011.

[3] Hybrid Drives. (2012). History of the vehicles with hybrid drive. *DAIMLER*, http://www.daimler.com/dccom/0-5-1200802-1-1400984-1-0-0-1201129-0-0-135-7165-0-0-0-0-0-0-0.html.

[4] New York City Hybrid Bus Overview. (2012). http://ebookbrowse.com/04-compro-ws-hb-labouff-pdf-d296145507, accessed 15 March.

[5] About BAE Systems. (2012). http://www.hybridrive.com/about-bae-systems.asp, accessed 10 May.

[6] European bus fleets reorder Allison hybrid technology to reduce emissions and fuel consumption. (2011). *Allison Transmission*, http://www.propel-technology.com/Article__EuropeanbusfleetsreorderAllisonhybridtechnologytoreduceemissionsandfuel-consumption.aspx, accessed 12 March.

[7] Brecher, A. (2010). Assessment of Needs and Research Roadmaps for Rechargeable Energy Storage System Onboard Electric Drive Buses. U.S. Department Of Transportation, Federal Transit Administration, *Report No. Fta-Tri-Ma-26-7125-2011.1*, accessed 12 May 2012, http://ntl.bts.gov/lib/35000/35700/35796/DOT-VNTSC-FTA-11-01.pdf.

[8] ZF and ISE collaborate on parallel-electric hybrid drive systems, EOBUS. (2009). http://www.eobus.com/news/447.htm, accessed 12 January 2012.

[9] Eaton Powering Business Worldwide. (2012). http://www.eaton.com/Eaton/Products-Services/ProductsbyCategory/HybridPower/SystemsOverview/ElectricHybrid/index.htm, accessed 10 February.

[10] Jobson, E. (2010). Volvo 7700 Hybrid, The worlds most environmentally friendly bus. *Volvo Bus Corporation*, http://www.voev.ch/dcs/users/117/2_7_Volvo_Edward_Jobson.pdf, accessed 20 December.

[11] DIWAhybrid, Voith Turbo GmbH & Co. KG, G 1977 e 2011-10, http://www.voithturbo.com/applications/vt-publications/downloads/1273_e_g_1977_e_diwahybrid_2011-09_screen.pdf

[12] Hybrid Transit Buses, Cost and Maintance. (2010). *The Union of Concerned Scientists' Hybridcenter.org*, http://www.hybridcenter.org/hybrid-transit-buses.html, accessed 10 February 2012.

[13] Hybrid Buses, Costs and Benefits. (2012). *Environmental and Energy Study Institute (EESI)*, Washington, http://www.eesi.org/files/eesi_hybrid_bus_032007.pdf, 21 April 2012.

[14] Zamboni, S, & Normanno, A. (2008). D2.3 Cost/effectiveness analysis of the selected technologies (CNG and HYBRIDS). *COMmon PROcurement of collective and public service transport clean vehicles, Bologna*, http://www.compro-eu.org/doc/COMPRO_deliverable_D2.3_final.pdf, (accessed 13 February 2010).

[15] Nigel, N. C, Feng, Z. W, Scott, W, & Donald, W. L. (2007). Transit Bus Life Cycle Cost and Year 2007 Emissions Estimation. U.S. Department of Transportation Federal

Transit Administration, *FTA-WV-26-7004.2007.1, Final Report,* accessed 27 August 2010, http://www.fta.dot.gov/documents/WVU_FTA_LCC_Final_Report_07-23-2007.pdf.

[16] Nigel, N. C, Feng, Z. W, Scott, W., & Donald, W. L. (2008). Additional Transit Bus Life Cycle Cost Scenarios Based on Current and Future Fuel Prices. U.S. Department of Transportation Federal Transit Administration, *FTA-WV-26-7006.2008.1, Final Report,* accessed 27 August 2010, http://www.fta.dot.gov/documents/WVU_FTA_LCC_Second_Report_11-03-2008.pdf.

[17] Zivanovic, Z., Jovanovic, Z., & Sakota, Z. (2011). A Comparative Analysis of CNG and Hybrid Buses vs Diesel Buses. *10th Anniversary International Conference on Accomplishments in Electrical and Mechanical Engineering and Information Technology, DEMI, Proceedings,* 607-612.

[18] DAIMLER. (2012). Daimler Buses North America. *Orion VII Hybrid Product Overview,* http://www.orionbus.com/Projects/c2c/channel/documents/845706_Orion_Hybrid_Transit_Presentation_Q1_2009.pdf, accessed 21 April.

[19] Diesel Hybrid-Electric Drive System. (2009). *ISE ThunderVolt® Transit Bus Hybrid Diesel (TB-HD),* http://www.isecorp.com/2009/wp-content/uploads/2009/02/diesel-hybrids-09-08-ebook.pdf, (accessed 19 May).

[20] Zivanovic, Z., Diligenski, Dj., & Sakota, Z. (2009). The Application of Hybrid Drive Technology in City Buses. Belgrade. *XXII. JUMV International Automotive Conference Science & Motor Vehicles,* Proceedings on CD, 1-15.

[21] HYDROGEN TRANSPORTS. Bus Technology & Fuel for TODAY and for a Sustainable Future. (2010). *A Report on the Achievements and Learnings from, The HyFLEET:CUTE Project 2006,* http://hyfleetcute.com/data/HyFLEETCUTE_Brochure_Web.pdf, accessed 30 August.

[22] Allison, B., Spencer, C., Hannah, P., Claire, P., & Kayla, R. (2008). The Feasibility of Alternative Fuels and Technologies. *An Assessment of Addison County Transit Resources' Current and Future Options,* Middlebury College, http://www.middlebury.edu/media/view/255373/original/BUSFINALREPORT.pdf, accessed 27 December 2010.

[23] Zaetta, R, & Madden, B. (2011). Hydrogen Fuel Cell Bus Technology State of the Art Review. *Project EC FCH-JU-2008-1 Grant Agreement Number 245133, Report Version 3.1,* http://nexthylights.eu/Publications/Clean-3_D3-1_WP3_EE_State_of_the_Art_23rd-FEB-2011.pdf, accessed 19 May 2012.

[24] CHIC In Brief. (2012). http://chic-project.eu/about/background/chic-in-brief, accessed 10 June.

[25] Hamburg putting its first Citaro FuelCELL Hybrid buses into service, Green Car Congress. (2011). http://www.greencarcongress.com/2011/08/hamburg-putting-its-first-citaro-fuelcell-hybrid-buses-into-service-hamburger-hochbahn-ag-has-acquired-four-mercedes-benz.html, accessed 4 March 2012.

[26] Eudy, L., Chandler, K., & Gikakis, C. (2009). Fuel Cell Buses in U.S. Transit Fleets: Current Status. National Renewable Energy Laboratory, *Technical Report NREL/TP-560-46490.*

[27] Ahluwalia, R, Wang, X, & Kumar, R. (2012). Fuel Cell Transit Buses. *Argonne National Laboratory, Argonne, IL,* http://www.ieafuelcell.com/documents/Fuel_Cells_for_Buses_Jan_2012, accessed 10 June 2012.

[28] New Citaro Fuel Cell Bus Debuts in Hamburg. 50% Lower Fuel Consumption. *Green Car Congress, 2009,* http://www.greencarcongress.com/2009/11/citaro-fc-20091116.html, accessed 24 March 2011.

[29] Kevin, C., & Leslie, E. (2011). Zero Emission Bay Area (ZEBA). *Fuel Cell Bus Demonstration: First Results Report, Technical Report, NREL/TP-5600-52015,* , http://www.nrel.gov/hydrogen/pdfs/52015.pdf, Accessed 10 May 2012.

[30] Leslie, E. (2011). VII.5 Technology Validation: Fuel Cell Bus Evaluations. *FY Annual Progress Report, DOE Hydrogen and Fuel Cells Program,* 1036-1039, http://www.hydrogen.energy.gov/pdfs/progress11/vii_5_eudy_2011.pdf.

[31] Hybrid fuel cell buses. (2012). *Van Hool NV, Belgium,* http://www.vanhool.be/FRA/transport-public/hybride-pile-a-ombustible/Resources/folderFuelCell.pdf, Accessed 2 May.

[32] The future of electric buses. (2011). *Energy Harversting Journal,* http://www.energyharvestingjournal.com/articles/the-future-of-electric-buses-00002792.asp?sessionid=1, Accessed 2 November.

[33] Goldman, J. (2009). Fast Charge Battery Electric Transit Buses, Economically and Environmentally Sustainable, Simple and Safe Advanced Technology. accessed 18 May 2012. *Proterra LLC, ARB ZEB Workshop,* http://www.arb.ca.gov/msprog/bus/zeb/meetings/0509workshops/proterra.pdf.

[34] Largest Electric Bus Fleet in World Launches- BYD Delivers First Orders. (2011). *CHINA4AUTO.com,* http://www.china4auto.com/ennews/html/Commercial_Vehicle/20110513/130526697818626.html, accessed 5 May 2012.

[35] E-Traction- Company Profile. (2012). *E-Traction Europe B.V,* http://www.e-traction.eu/about-e-traction/our-company, accessed 12 June.

[36] Project: Hybricon City Busses. (2012). *Umeå, Sweden, E-Traction Europe B.V,* http://www.e-traction.eu/projects/hybricon, accessed 12 June.

[37] Lucas, P. (2010). Electric buses begin operation. *TheGreenCarWebsite.co.uk,* http://www.thegreencarwebsite.co.uk/blog/index.php/2010/12/28/electric-buses-begin-operation/, accessed 12 June, 2012.

[38] Holley, M. (2012). Battery Versa goes on trial. *Optare, an Ashok Leyland Company,* May 11, http://www.optare.com/images/pdf/De%20Courcey%20Versa%20EV%201.pdf, accessed 12 June 2012.

[39] Tindo. (2012). The World's First Solar Electric Bus. *The Adelaide City Council*, http://www.adelaidecitycouncil.com/assets/acc/Environment/energy/docs/tindo_fact_sheet.pdf, accessed 12 June.

[40] Optare Introduces Battery-Electric Bus. (2009). *Green Car Congress*, http://www.greencarcongress.com/2009/03/optare-introduces-batteryelectric-bus.html, accessed 12 June 2012.

[41] Urbino electric. Characteristic. *Solaris Bus & Coach*, http://www.solarisbus.pl/en/electric.html, accessed 12 May 2012.

[42] Gopalakrishnan, D., Essen, H., Kampman, B., & Grünig, M. (2011). Impacts of Electric Vehicles-Deliverable 2. *Assessment of electric vehicle and battery technology, Delft, CE Delft*.

[43] ETSAP (Energy Technology Systems Analysis Program). (2012). *TECHDS Energy Technology Briefs*, http://www.iea-etsap.org/web/Demand.asp, Accessed 11 April.

[44] Technology Roadmap. (2011). Electric and plug-in hybrid electric vehicles. *International Energy Agency (IEA)*, http://www.iea.org/papers/2011/EV_PHEV_Roadmap.pdf.

[45] Nikolic, Z., Filipovic, Z., & Janjusevic, Lj. (2011). State of development of the electric and hybrid vehicles, energetic and ecological aspect of applications. *Industry*, 34(4), 267-292.

Section 2

Modeling and Design of Electric Vehicles

Chapter 7

Modeling of Full Electric and Hybrid Electric Vehicles

Ferdinando Luigi Mapelli and Davide Tarsitano

Additional information is available at the end of the chapter

http://dx.doi.org/10.5772/53570

1. Introduction

Full Electrical Vehicles (FEVs) and Hybrid Electrical Vehicles (HEVs) are vehicles with many electric components compared to conventional ones. In fact the power train consists of electrical machines, power electronics and electric energy storage system (battery, super capacitors) connected to mechanical components (transmissions, gear boxes and wheels) and, for HEV, to an Internal Combustion Engine (ICE). The approach for a new vehicle design has to be multidisciplinary in order to take into account the dynamic interaction among all the components of the vehicle and the power train itself. The vehicle designers in order to find the correct sizing of components, the best energy control strategy and to minimize the vehicle energy consumption need modeling and simulation since prototyping and testing are expensive and complex operations. Developing a simulation model with a sufficient level of accuracy for all the different components based on different physic domains (electric, mechanical, thermal, power electronic, electrochemical and control) is a challenge. Different commercial simulation tools have been proposed in literature and they are used by the automotive designer [1]. They have different level of detail and are based on different mathematical approaches. In paragraph 2 a general overview on different modeling approaches will be presented. In the following paragraphs the author approach, focused on the modeling of each component constituting a FEV or HEV will be detailed. The authors approach is general and is not based on vehicle oriented simulation tools. It represents a good compromise among model simplicity, flexibility, computational load and components detail representation. The chapter is organized as follows:

- paragraph 2 describes the different approaches that can be find in literature and introduced the proposed one;
- paragraphs 3 to 10 describe all the components modeling details in this order: battery, inverter, electric motor, vehicle mechanics, auxiliary load, ICE, thermal modeling;
- paragraph 11 presents different cases of study with simulation results where all the numerical models has been validated by means of experimental test performed by the authors.

2. FEV and HEV modeling

As shown in Figure 1, the whole vehicle power-train model is composed by many subsystems, connected in according to the energy and information physical exchanges. They represent the driver (pilot), the vehicle control system, the battery, the inverter, the Electrical Motor (EM), the mechanical transmission system, the auxiliary on board electrical loads, the vehicle dynamical model and for, HEVs and Plug-in Hybrid Electrical Vehicles (PHEVs), also an ICE and a fuel tank are considered. To correctly describe them, a multidisciplinary methodology analysis is required. Furthermore the design of a vehicle requires a complete system analysis including the control of the energy given from the on-board source, the optimization of the electric and electronic devices installed on the vehicle and the design of all the mechanical connection between the different power sources to reach the required performances. So, the complete simulation model has to describe the interactions between the system components, correctly representing the power flux exchanges, in order to help the designers during the study. For modeling each component, two different approaches can be used: an "equation-based" or a "map-based" mode [1]. In the first method, each subcomponent is defined by means of its quasi-static characteristic equations that have to be solved in order to obtain the output responses to the inputs. The main drawback is represented by the computational effort needed to resolve the model equations. Vice versa using a "map-based" approach each sub-model is represented by means of a set of look-up tables to numerically represents the set of working conditions. The map has to be defined by means of "off-line" calculation algorithm based on component model equation or collected experimental data. This approach implies a lighter computation load but is not parametric and requires an "off-line" map manipulation if a component parameter has to be changed. For the model developing process, an object-oriented causal approach can be adopted. In fact the complete model can be split into different subsystems. Each subsystem represents a component of the vehicle and contains the equations or the look-up table useful to describe its behavior. Consequently each object can be connected to the other objects by means of input and output variables. In this way, the equations describing each subsystem are not dependent by the external configuration, so every object is independent by the others and can be verified, modified, replaced without modify the equations of the rest of the model. At the same time, it is possible to define a "power flux" among the subsystems: every output variable of an object connected to an input signal of another creates a power flux from the first to the second subsystem ("causality approach"). This method has the advantage to realize a modular approach that allows to obtain different and complex configuration only rearranging the object connection.

A complete model can be composed connecting the objects according two different approaches: the "reverse approach" (also called "quasi-static approach" - see Figure 2) and the "forward approach" (also called "dynamic approach" - see Figure 3). Figure 2 and 3 show simplified models of a HEV, where V is the vehicle model, GB the gear box, PC the power converter, B the battery pack, FT the fuel tank, AL is the auxiliary loads block, v and a are respectively the vehicle's speed and acceleration, f is the vehicle traction force, Ω is the EM angular speed, T_{ICE} and T_{EM} are respectively the ICE and the EM torques, Ω_{ICE} is the ICE angular speed, f_c is the fuel consumption, I and V_s are the electrical motor current and voltage, i_{batt} and V_{batt} are the battery current and voltage, P_{InMot} is the power requested by the EM to the power converter, P_B is the total power requested to the battery that is obtained as a sum of the power requested by the power converter P_{InInv} and the

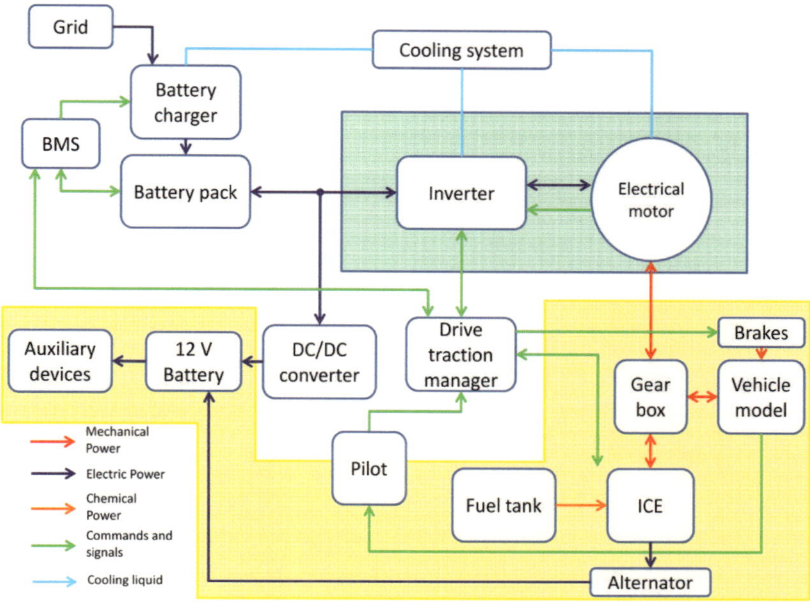

Figure 1. Block diagram of a Plug-In HEV.

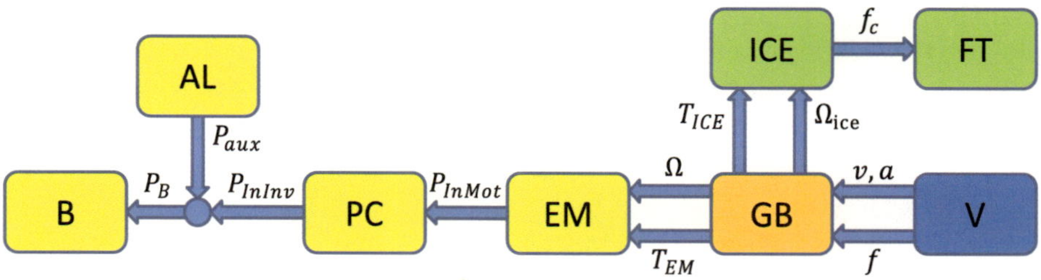

Figure 2. Example of HEV quasi-static modeling approach.

auxiliary loads P_{aux} ($P_B = P_{InInv} + P_{aux}$) and finally i_{aux} is the amount of current requested to the battery for auxiliary electrical loads. Quasi-static method use as input variables the desired speed and acceleration of the vehicle, hence the equations are solved starting from the V model and going back, block by block, to the B model. In the dynamic approach each subcomponent has interconnection variables with the previous and the next blocks. In this way each sub-model is strongly interleaved with the others and its behavior has influence on the total system. The second method requires a higher computational effort but is more accurate and has been applied by the authors in several cases [2–4]. In fact, using the first method, the information flux is unidirectional and the equation set is more simpler often only algebraic. This approach do not take into account the real response and constrain of power train component. On the contrary the dynamic approach produces also a response that runs

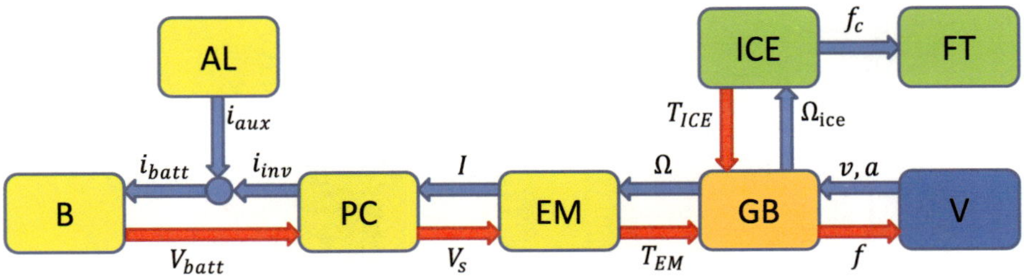

Figure 3. Example of HEV-dynamic modeling approach.

forward the complete model, influencing the output of the following sub-models. In this way, it is possible to study the total behavior including the physical limits of each component and, so, the simulation model is able to describe correctly both the single component and the overall performances of the system. For this method more complex equations (a few number of differential equation) or maps are needed. The following paragraphs describe component by component the proposed method which is based on a simplified dynamic forward approach that could be implemented using both equations or off-line computed look-up tables.

3. Battery modeling

In order to correctly simulate the behavior of a FEV, HEV or PHEV it is important to set up a battery model that evaluate the output voltage considering the State Of Charge (SOC) of the battery itself. Since a battery pack is obtained by a series connection of many cells (n_{cell}), it is quite usual to construct a numerical model considering one single cell. The total battery voltage V_{batt} is obtained using equation (1) assuming that all cells have an uniform behavior and where v_{el} is the voltage of a single cell.

$$V_{batt} = n_{cell} v_{el} \tag{1}$$

The battery model receives as input variables: the current i_{batt} required from the electrical drive model (inverter and electric motor) and the battery temperature ϑ computed by battery thermal model. The model gives as output variables: the battery pack voltage V_{batt}, the SOC and the power losses $P_{LossBatt}$. In order to simulate the battery behavior, instead of a complex electrochemical model, an Equivalent Circuit Model (ECM) can be chosen as a good compromise between accuracy and computational load. For example a first order Randles circuit (represented in Figure 4) can be adopted as dynamic model (see Paragraph 3.2); this model can be easily downgraded imposing $R_1 = 0$ in order to obtain a static model (see Paragraph 3.1). The circuit parameters can be deduced by experimental test or technical literature using the method described in [5].

Furthermore it is fundamental to calculate the battery SOC using equation (2) (where C_n is the rated capacity expressed in Ampere-Hours [Ah] and SOC_0 is the initial state of charge) to evaluate the amount of energy stored into the battery pack.

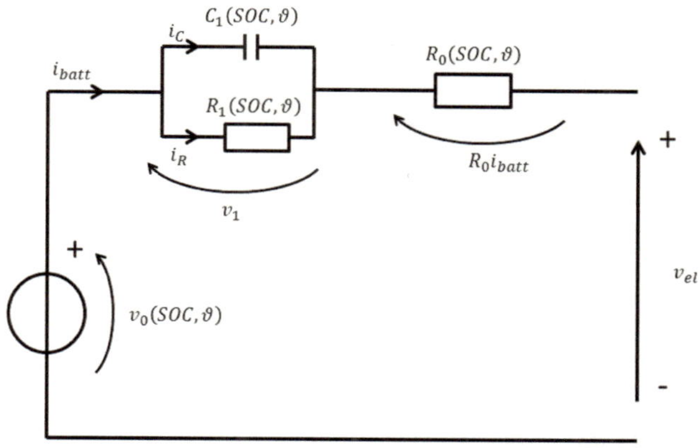

Figure 4. Randles electrodynamical model of a cell.

$$SOC(t) = SOC_0 - \int_0^t \frac{i_{batt}(t)}{3600 \cdot C_n} dt \tag{2}$$

3.1. Static model of battery

Using the manufacturer charge and discharge charts and the data available for different temperature (reported as example in Figures 5-7), it is possible to reconstruct the map of $v_0(SOC,\vartheta)$ and of $R_0(SOC,\vartheta)$ and consequently to calculate $v_{el}(SOC,\vartheta)$ as reported in the static equation (3).

$$v_{el}(SOC,\vartheta) = v_0(SOC,\vartheta) - R_0(SOC,\vartheta) i_{batt} \tag{3}$$

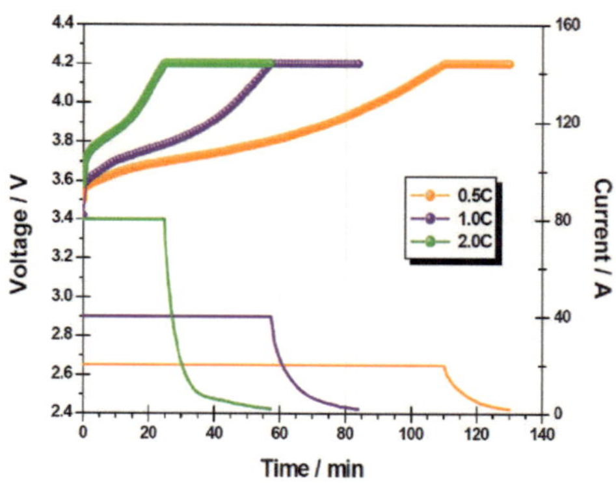

Figure 5. Charging chart for different C-Rates.

A further simplification is to consider the temperature ϑ constant and consequently to calculate and to represent on a map the v_{el} as reported in Figure 8, as a function of the battery SOC and the battery current i_{batt}.

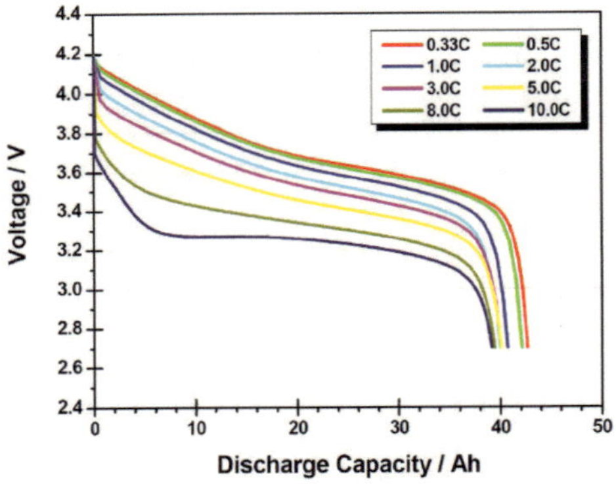

Figure 6. Discharge chart for different C-Rates.

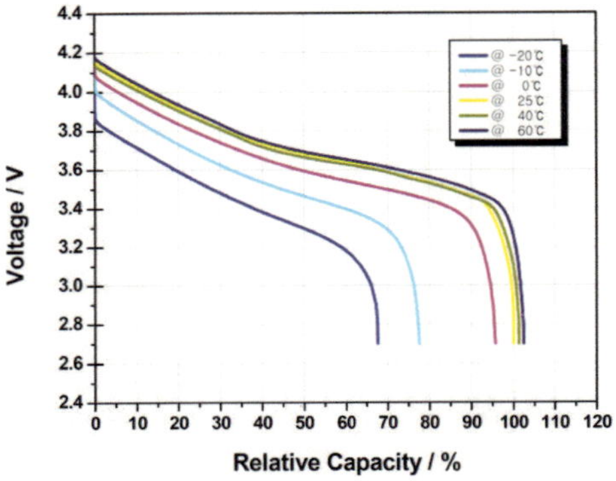

Figure 7. 1C discharge chart for different temperatures.

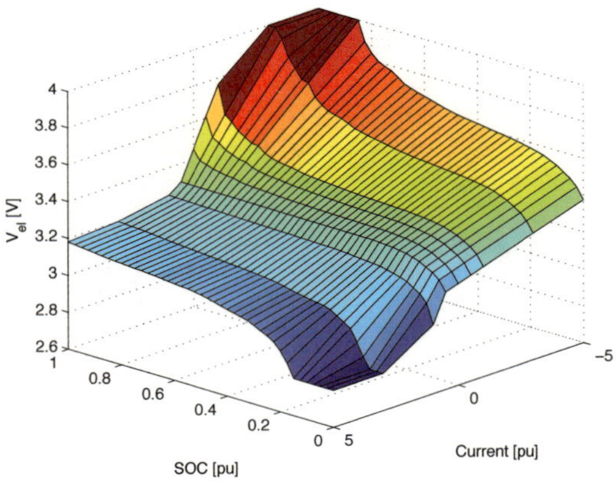

Figure 8. Battery voltage map.

3.2. Dynamical model of battery

Since batteries for traction application are used under heavy dynamic condition with suddenly variation of the supplied current i_{batt}, the static model can not be adopted for all the cases of study where dynamic is fundamental (for example control analysis). Different type of ECM have been developed for simulating battery voltage v_{el} where more that one RC block are used in order to obtain a Ordinary Differential Equation (ODE) of order n and a parasitic parallel branch is added to the ECM to simulate the self discharge phenomenon. Since the main objective is not to simulate all the battery details but the global vehicle behavior a single RC circuit for an enough accurate model can be adopted, as reported in Figure 4.

In order to have good simulation results a fine tuning of the dynamic ECM parameters has to be done. A good procedure for parameter identification, considering also thermal effects, is reported in [5].

It possible to solve the circuit considering the cell voltage v_{el}, as reported in equation (4)[1], where the splitting of the total current i_{batt} into the capacitor C_1 and into the resistor R_1 is considered ad reported in equation (5) and the no load voltage v_0 is SOC dependant.

$$v_{el} = v_0 - R_0 i_{batt} - v_1 \tag{4}$$

$$\begin{cases} i_{batt} = i_c + i_r \\ i_c = C_1 \dfrac{dv_1}{dt} \\ i_r = \dfrac{v_1}{R_1} \end{cases} \tag{5}$$

Finally, substituting i_{batt} obtained from equation (5) in equation (4), is possible to obtain the final dynamic equation of the cell voltage, as reported in equation (6).

$$\frac{dv_1}{dt} = \frac{1}{R_0 C_1}\left(v_0(SOC) - v_{el} - v_0(SOC)\left(1 + \frac{R_0}{R_1}\right)\right) \tag{6}$$

4. Inverter modeling

Different methods are available in the scientific literature in order to evaluate power electronic converter losses [6, 7] and to obtain a consequent energetic model. The most simple approach is to consider the power converter as an equivalent resistive load where the inner power losses are proportional to the square of the flowing current. Since in the most cases the power converter assumes the three phase inverter topology the power losses expression can be formalized as reported in (7), where R_{Inv} is the inverter equivalent resistance and I is the Root Mean Square (RMS) inverter output phase current (that corresponds to the EM phase input RMS current).

$$P_{LossInv} = 3 \cdot R_{Inv} \cdot I^2 \tag{7}$$

[1] In equation (4) (5) (6) where: it has been neglected the dependency of the circuital parameters from battery SOC and temperature ϑ.

The inverter input power can be calculated adding the inverter losses $P_{LossInv}$ to the motor input power P_{InMot} that correspond to the inverter output power P_{OutInv} (equation (8)).

$$P_{InInv} = P_{LossInv} + P_{InMot} = P_{LossInv} + P_{OutInv} \tag{8}$$

A more detailed approach can be described if the simulation model adopted includes the control and inverter modulator details: an instant circuit losses model can be also implemented [6]. The losses are computed considering the basic inverter cell composed of an Insulated Gate Bipolar Transistor (IGBT) and a diode. The inverter is formed by six basic cells divided into 3 arms as reported in Figure 9.

The instantaneous losses of a basic cell p_{cell} can be evaluated using equation (9) where: p_{swT} are transistor switching losses, E_{on} and E_{off} are turn-on and turn off energy, f_s is the inverter switching frequency, E_{recD} and p_{recD} are the recovery diode energy and power losses, v_{ce} and v_{ak} are respectively the transistor and diode forward voltage drop, i_c and i_f are the transistor and diode direct current ad p_{fwT} and p_{fwD} are transistor and diode conduction forward losses. The total inverter instantaneous losses are reported in (10). For the IGBT and diode the typical current Vs voltage curves and the switch on/off energy losses Vs current charts are shown in Figure 11, 12 and 13.

These curves can be simplified as shown in equation (11) where all the parameters (A_{fwT}, B_{fwT}, A_{fwD}, B_{fwD}, B_{onT}, C_{onT}, B_{offT}, C_{offT}, B_{recD}, C_{recD}) can be deduced from the semiconductor device technical data sheet [8, 9]. Equation (12) can be obtained substituting equation (11) into the (10). These equations express the instantaneous losses p_{inv} as a function of semiconductor devices current.

$$\begin{cases} p_{fwT} = v_{ce}(i_c) \cdot i_c \\ p_{fwD} = v_{ak}(i_f) \cdot i_f \\ p_{swT} = [E_{on}(i_c) + E_{off}(i_c)]\, f_s \\ p_{recD} = E_{recD}(i_d) \cdot f_s \\ p_{cell} = p_{swT} + p_{recD} + p_{fwT} + p_{fwD} \end{cases} \tag{9}$$

$$p_{inv} = 6 \cdot p_{cell} \tag{10}$$

$$\begin{cases} v_{ce}(i_c) = A_{fwT} + B_{fwT} i_c \\ v_{ak}(i_f) = A_{fwD} + B_{fwD} i_f \\ E_{onT}(i_c) = B_{onT} i_c + C_{onT} i_c^2 \\ E_{offT}(i_c) = B_{offT} i_c + C_{offT} i_c^2 \\ E_{recD}(i_f) = B_{recD} i_f + C_{recD} i_f^2 \end{cases} \tag{11}$$

http://dx.doi.org/10.5772/53570

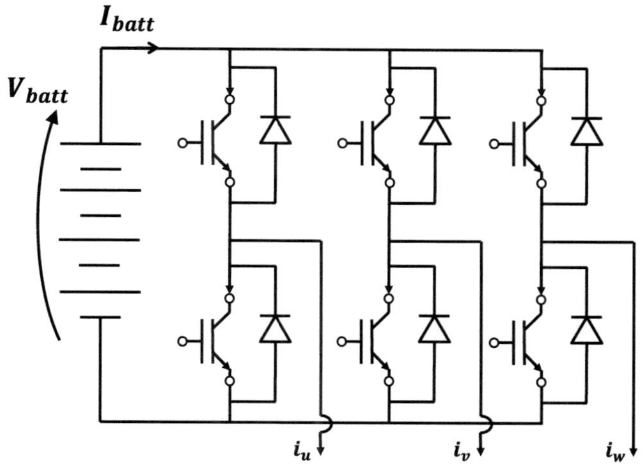

Figure 9. Battery fed three phase inverter

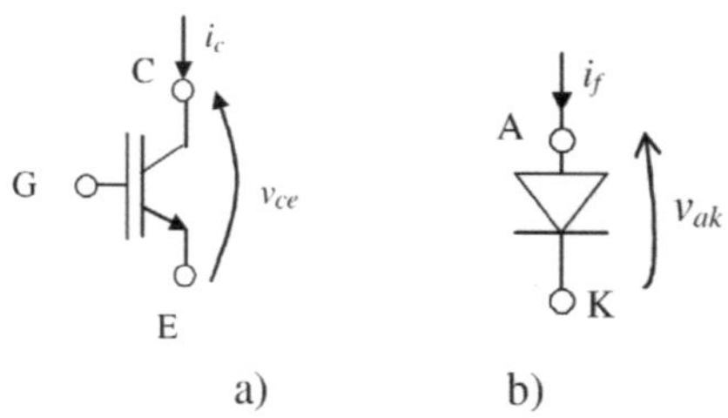

Figure 10. Symbols and definitions for Igbt a) and Diode b).

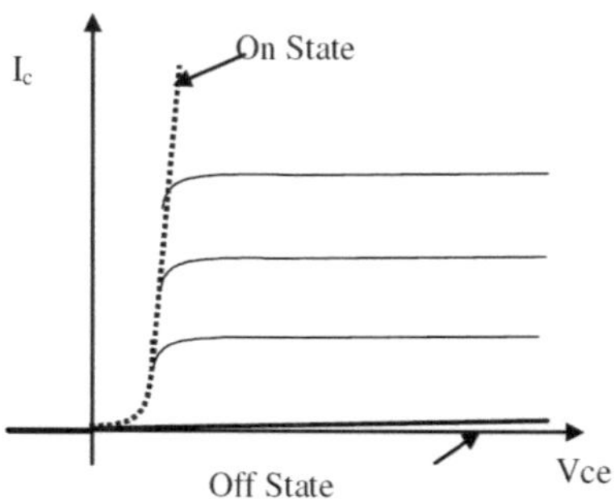

Figure 11. IGBT current Vs voltage diagram.

$$\begin{cases} p_{fwT}(i_c) = A_{fwT} i_c + B_{fwT} i_c^2 \\ p_{fwD}(i_f) = A_{fwD} i_f + B_{fwD} i_f^2 \\ p_{swT}(i_c) = (B_{onT} i_c + C_{onT} i_c^2) f_s + (B_{offT} i_c + C_{offT} i_c^2) f_s \\ p_{recD}(i_f) = (B_{recD} i_f + C_{recD} i_f^2) f_s \end{cases} \tag{12}$$

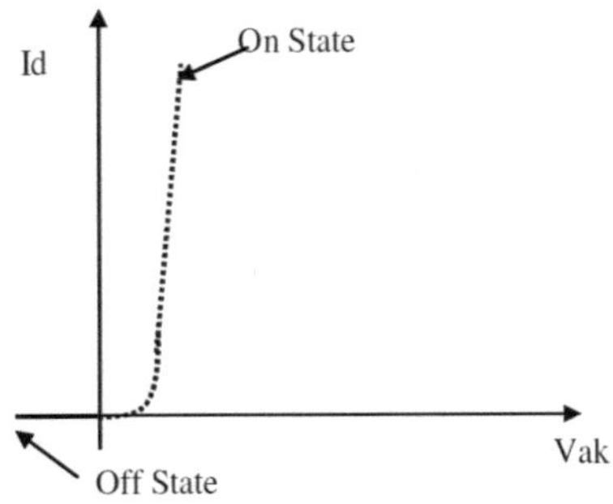

Figure 12. Diode current Vs voltage diagram.

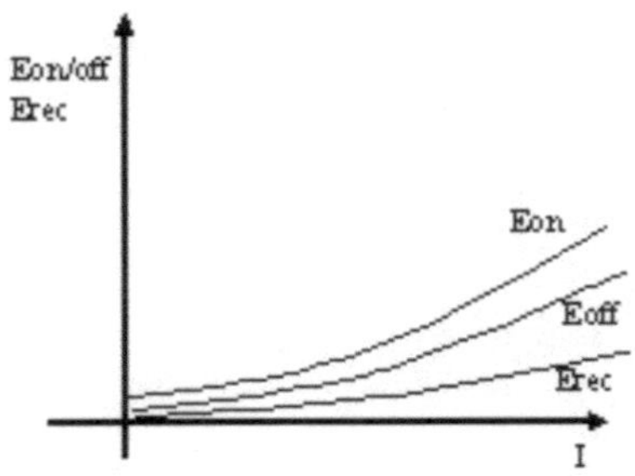

Figure 13. IGBT/Diode switching energy vs current.

The instantaneous inverter losses expressions (10) need, in order to be evaluated, the calculation of the instantaneous alternate three-phase motor current. This fact implies that the simulation model has to be solved with a very short integration step with a consequent high computation load and large simulation time. For FEV and HEV power train modeling purpose such time details and accuracy is not needed but the exact losses calculation is necessary on a larger time scale. An average approach on an alternate quantities period can be adopted. In this way a larger time step is enough and the RMS value of alternate voltage and current can be used. In this method the losses calculation accuracy is assured and very fast phenomena (evolution during a AC current period T) are neglected. This approximation is sufficient for vehicle power train modeling and for energy and power flow analysis. Assuming sinusoidal time dependency for current, as reported in equation (13) where: I_M is the maximum current value, $\omega = 2\pi/T$ is the current angular frequency and φ is the phase angle between motor voltage and current, substituting the (13) into equation (12) and assuming that $i = i_c = i_f$, the instantaneous inverter losses with explicit time dependence can be obtained. Averaging the losses on an Alternating Current (AC) variables period T is possible to obtain the losses mean value [10]. The average relationships are obtained as reported in equation (14) where: T_s is the IGBT switching period, T_{dead} is the dead time between high and low side IGBT switch on operation and $\cos\varphi$ is the motor power factor. The total PWM operation cell average losses P_{PWM} are the all terms sum, while the total averaged inverter losses P_{invPWM} are reported in equation (15).

$$i(t) = I_M cos(\omega t - \varphi) \tag{13}$$

$$\begin{cases} P_{fwT} = \left(\frac{1}{2} - \frac{T_{dead}}{T_s}\right)\left(\frac{A_{fwT}}{\pi}I_M + \frac{B_{fwT}}{4}I_M^2\right) + m\cos\varphi\left(\frac{A_{fwT}}{8}I_M + \frac{B_{fwT}}{3\pi}I_M^2\right) \\ P_{fwD} = \left(\frac{1}{2} - \frac{T_{dead}}{T_s}\right)\left(\frac{A_{fwD}}{\pi}I_M + \frac{B_{fwD}}{4}I_M^2\right) - m\cos\varphi\left(\frac{A_{fwD}}{8}I_M + \frac{B_{fwD}}{3\pi}I_M^2\right) \\ P_{onT} = f_s I_M\left(\frac{B_{onT}}{\pi} + \frac{C_{onT}}{4}I_M\right) \\ P_{offT} = f_s I_M\left(\frac{B_{offT}}{\pi} + \frac{C_{offT}}{4}I_M\right) \\ P_{recD} = f_s I_M\left(\frac{B_{recD}}{\pi} + \frac{C_{recD}}{4}I_M\right) \\ P_{PWM} = P_{onT} + P_{offT} + P_{fwT} + P_{fwD} + P_{recD} \end{cases} \tag{14}$$

$$P_{invPWM} = 6 \cdot P_{PWM} \tag{15}$$

Since the inverter sub-model receives as input V_s, I, $\cos\varphi$ and ω, previously evaluated by the electric motor model, and V_{batt} (the available battery voltage) it can calculate the current required to the battery i_{inv} and the total inverter losses P_{PWM}. The sequence of equations to be solved is reported as follows:

1. total power supplied to the motor calculation: $P_{InMot} = \sqrt{3}V_s I cos\varphi$;
2. inverter AC phase current max. value calculation: $I_M = \sqrt{2}I$;
3. inverter PWM amplitude modulation index calculation: $m = \sqrt{2}V_s/V_{batt}$;
4. total inverter averaged losses P_{invPWM} calculation by means of equation (14) and (15);
5. total inverter input power calculation: $P_{InInv} = P_{InMot} + P_{invPWM}$;
6. inverter input current calculation: $i_{inv} = P_{InInv}/V_{batt}$.

5. Electrical motor modeling

The most adopted motors for FEV and HEV are AC induction motors and AC Permanent Synchronous Magnets Motor (PMSM) regulated by means of a field oriented control or direct torque control. In this section the models of both motors will be presented using a phase vector approach [11, 12] and considering the motor field oriented controlled. For both motor models it is possible to define the input and output variables as follows:

- input: required torque T_{ref}, instantaneous rotating mechanical speed Ω, battery voltage V_{batt};

- output: torque T_{EM}, RMS phase current I, line to line voltage V_s, power factor angle φ, total losses $P_{LossMot}$, Motor input (P_{InMot}) and output power (P_m), electrical frequency f and angular frequency $\omega = 2\pi f$.

For FEV and HEV power train modeling and simulation a complete motor model including the detailed electromechanical dynamic is not required; it is better to use a steady state model that consider the controlled motor including all the energetic phenomena (power losses calculation). The proposed model include also limits and constrains due to the motor power supplier, which is based on batteries and inverter, such as maximum deliverable voltage, power and current.

5.1. Induction motor

For the induction motor the steady state equations [13] are reported in equation (16) where $\bar{V}_s$, $\bar{I}_s$ and $\bar{\psi}_r$ are respectively stator voltage, stator current and rotor flux phasors, R_s, R_r, M, L_k are respectively stator resistance, rotor resistance, mutual inductance and total leakage inductance, n is the pole pairs number, T_{EM} is the torque , $\bar{I}_m$ is the magnetizing current phasor, $\bar{I}_r$ is rotor or torque current phasor, Ω is the mechanical angular speed, x is the relative rotor slip speed, ω is the AC variable angular frequency and j the imaginary unit. Equation (16) can be represented by means of the equivalent circuit reported in Figure 14.

The three phase motor is modeled using a "rational" approach that correspond to have a "single phase'equivalent" model also for energetic relations and torque expression [13]. In fact the amplitude of current phasor $\bar{I}_s$ and the stator voltage phasor $\bar{V}_s$ are related to the RMS phase current I and voltage V by means of equation (17). The induction motor model includes also equation (18) where ψ_{rn} is the induction motor rated flux, ω_n is the rated motor angular frequency, P_{Cu} and P_{Fe} represent respectively the copper and iron losses and Q_{InMot} is the motor reactive input power. Equation (18) allows to calculate all the power terms and stator quantities to be used as inputs for inverter and battery model.

$$\begin{cases} \bar{V}_s = R_s\bar{I}_s + j\omega L_{ks}\bar{I}_s + j\omega M\bar{I}_m \\ 0 = -\dfrac{R_r}{x}\cdot\bar{I}_r + j\omega M\bar{I}_m \\ \bar{I}_s = \bar{I}_r + \bar{I}_m \\ \bar{\psi}_r = M\bar{I}_m \\ x = \dfrac{\omega - n\Omega}{\omega} \\ T = nMI_mI_r = n\psi_rI_r \end{cases} \tag{16}$$

$$\left\{ V_s = V\cdot\sqrt{3} I_s = I\cdot\sqrt{3} \right. \tag{17}$$

http://dx.doi.org/10.5772/53570

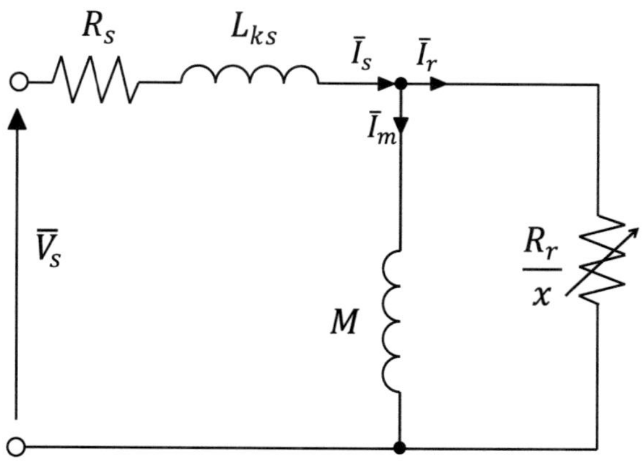

Figure 14. Induction motor steady-state equivalent circuit.

$$\begin{cases} P_{Cu} = R_s I_s^2 + R_r I_r^2 \\ P_{Fe} = P_{Fen} \dfrac{\omega}{\omega_n} \dfrac{\psi_r^2}{\psi_{rn}^2} \\ P_{LossMot} = P_{Cu} + P_{Fe} \\ P_m = T\Omega \\ P_{InMot} = P_{LossMot} + P_m \\ Q_{InMot} = \omega M I_m^2 + \omega L_{ks} I_s^2 \\ \varphi = atan\left(\dfrac{Q_{InMot}}{P_{InMot}}\right) \end{cases} \tag{18}$$

Equations (16) and (18) have to be solved together with equation (19) that define the rotor flux value as function of the rotating speed Ω and of the rated speed Ω_n. Equation (19) represents the field weakening condition for the induction motor. Furthermore it is also necessary to control that the torque request T_{ref} does not exceed the maximum motor torque T_{refMax} and the consequent power request $(T_{ref} \cdot \Omega)$ does not exceed the motor power limit P_{motMax} (see equation (20)).

$$\begin{cases} \psi_r = \psi_{rn} & \text{if} \quad \Omega < \Omega_n \\ \psi_r = \psi_{rn} \dfrac{\Omega_n}{\Omega} & \text{if} \quad \Omega > \Omega_n \end{cases} \tag{19}$$

$$\begin{cases} T_{ref} = T_{refMax} & \text{if} \quad T_{ref} > T_{refMax} \\ T_{ref} = P_{motMax}/\Omega & \text{if} \quad T_{ref} \cdot \Omega > P_{motMax} \end{cases} \tag{20}$$

Moreover the global electrical drive limits verification has to be taken into account in order to avoid that the requested operating point do not correspond to an allowed condition. The three conditions to consider are:

1. maximum RMS input current I_{max} that is related to the inverter current limit (as reported in equation (21));
2. maximum motor voltage limit V_{sMax} that correspond to the maximum deliverable inverter voltage for a given battery voltage (as reported in equation (22));
3. the maximum motor input power limit P_{inMax} that is related to the maximum battery deliverable power (as reported in equation (23)).

These conditions have to be verified and imposed after the calculation of equations (20), (19), (16) and (18).

$$I = \frac{I_s}{\sqrt{3}} < I_{max} \tag{21}$$

$$V_s < V_{sMax} \quad \text{then} \quad V_{sMax} = \frac{V_{batt}}{\sqrt{2}} \tag{22}$$

$$P_{InMot} < P_{inMax} \tag{23}$$

The proposed model can be used for off-line map calculation, that can be included in the simulation model, or calculated directly on-line during the numerical simulation process. The calculus procedure for induction motor can be summarized as follows:

1. verify if the torque request T_{ref} is compliant with absolute motor torque and power limit otherwise saturate T_{ref} using the (20);
2. solve the field weakening conditions (19);
3. solve the (16), (18) using as input variables $T_{EM} = T_{ref}$ and Ω;
4. verify the (21), (22) and (23), in order to impose the motor, inverter and battery limitations;
5. if the condition (21) is not respected reduce T_{ref}, go back to step 3 and iterate;
6. if the condition (22) is not respected reduce ψ_r, go back to step 3 and iterate;
7. if the condition (23) is not respected reduce T_{ref}, go back to step 3 and iterate.

http://dx.doi.org/10.5772/53570

5.2. Permanent magnets synchronous brushless motor

For the Permanent Synchronous Magnets Motor the steady state equation [11] are reported in equation (24) where: V_d and V_q are the stator voltage phasor $\bar{V}_s$ components ($\bar{V}_s = V_d + jV_q$), I_d and I_q are the stator current phasor $\bar{I}_s$ components ($\bar{I}_s = I_d + jI_q$), R_s is the stator resistance, L_s is the stator synchronous inductance, ψ_m is the permanent magnet flux phasor. The other symbols, T_{EM}, Ω, ω and n assume the same meaning that ones indicated in the induction motor description.

Equation (24) has to be solved, also in this case, together with equations (25) and (26). Similarly to the induction motor a pre-process operation on torque request T_{ref} has to be implemented in order to impose the respect of torque and power motor limit. Furthermore the field weakening condition have to be imposed to the motor. It consists in setting the correct value of I_d current [12] by means of equation (27). In fact the current I_d can be maintained equal to zero in the constant torque/flux region and has to be imposed negative in the field weakening zone. Finally also the limit input conditions have to be taken into account using the same equations of the induction motor ((21), (22) and (23)).

$$\begin{cases} V_d = R_s I_d - \omega L_s I_q \\ V_q = R_s I_q + \omega L_s I_d + \psi_m \omega \\ T_{EM} = n \psi_m I_q \\ \Omega = \dfrac{\omega}{n} \end{cases} \tag{24}$$

$$\begin{cases} P_{Cu} = R_s I_d^2 + R_s I_q^2 \\ P_{Fe} = P_{Fen} \dfrac{\omega}{\omega_n} \\ P_m = T_{EM} \Omega \\ P_{LossMot} = P_{Cu} + P_{Fe} \\ P_{InMot} = P_m + P_{LossMot} \end{cases} \tag{25}$$

$$\begin{cases} I_s = \sqrt{I_d^2 + I_q^2} \\ V_s = \sqrt{V_d^2 + V_q^2} \\ Q_{InMot} = V_q Id - V_d I_q \\ \varphi = atan \left(\dfrac{Q_{InMot}}{P_{InMot}} \right) \end{cases} \tag{26}$$

$$\begin{cases} \psi_s = \psi_m & \text{if} \quad \Omega < \Omega_n \\ \psi_s = \psi_m \dfrac{\Omega_n}{\Omega} & \text{if} \quad \Omega > \Omega_n \\ Id = \dfrac{\psi_s - \psi_m}{L_s} \end{cases} \tag{27}$$

Also in this case the model can be used both for off-line map calculation and on-line numerical simulation process.

The calculus procedure for PMSM can be summarized as follows:

1. verify if the torque request T_{ref} is compliant with absolute motor torque and power limit otherwise saturate T_{ref} using equation (20);
2. solve the field weakening conditions (equation (27));
3. solve equations (24), (25) and (26) using as input $T_{EM} = T_{ref}$ and Ω;
4. verify equations (21),(22) and (23), in order to impose the motor, inverter and battery limitation;
5. if the condition (21) is not respected reduce T_{ref}, go back to step 3 and iterate;
6. if the condition (22) is not respected reduce ψ_s, go back to step 2 and iterate;
7. if the condition (23) is not respected reduce T_{ref}, go back to step 3 and iterate.

In Figure 15 is reported, as example, an efficiency map of a 65kW peak power PMSM obtained by means of the proposed model, for a 2500 kg mass FEV. The per unit efficiency η_{EM} can be calculated using equation (28).

$$\eta_{EM} = \frac{P_m}{P_{InMot}} \tag{28}$$

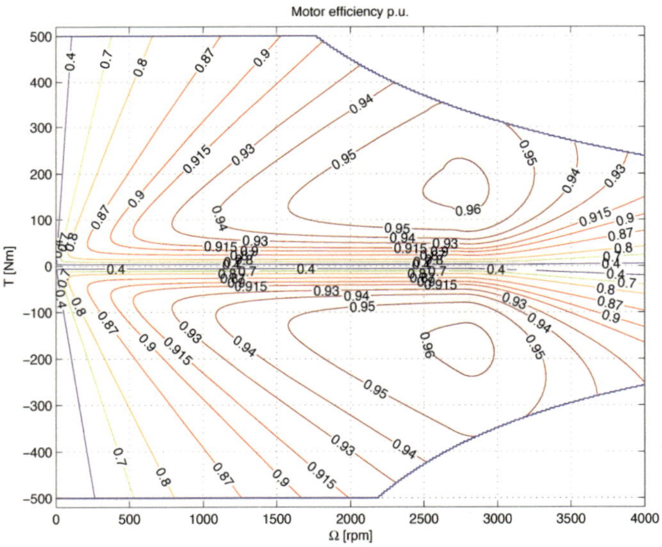

Figure 15. Efficiency map for a PMSM as function of torque and speed.

http://dx.doi.org/10.5772/53570

6. Vehicle longitudinal dynamic modeling

In order to reconstruct the energetic power flow between FEV and HEV components a simple vehcile longitudinal dynamic model has to be considered. In this paragraph the model will be described considering the most general case constituted by an HEV; the model of a FEV can be simply deducted neglecting all the ICE contributions. This model receives as input the torque given by the ICE T_{ICE} and by the EM T_{EM} coming from the respective simulation models and the gear ratio of the mechanical gearbox coming from the pilot model and calculate the vehicle speed $v(t)$ and distance covered $s(t)$.

As first it is necessary to evaluate the total torque at the wheels T_w as sum of the EM torque reported at the wheel T_{EMw} with the ICE torque reported at the wheel T_{ICEw}. For this all the reduction ratios and the efficiencies of the transmission chain have to be considered, as reported in equations (29) and (30), which are specialized for traction condition (29) and for braking condition (30). In these equations τ_{EM} and $\eta_{\tau EM}$ are respectively the reduction ratio of the EM and its efficiency, τ_{ICE} and $\eta_{\tau ICE}$ are respectively the reduction ratio of the ICE and its efficiency, τ_{diff} and η_{diff} are respectively the differential reduction ratio and its efficiency.

$$\begin{cases} T_{EMw} = T_{EM} \cdot \tau_{EM} \cdot \tau_{diff} \cdot \eta_{\tau EM} \cdot \eta_{diff} \\ T_{ICEw} = T_{ICE} \cdot \tau_{ICE} \cdot \tau_{diff} \cdot \eta_{\tau ICE} \cdot \eta_{diff} \end{cases} \tag{29}$$

$$\begin{cases} T_{EMw} = \dfrac{T_{EM} \cdot \tau_{EM} \cdot \tau_{diff}}{\eta_{diff} \cdot \eta_{\tau EM}} \\ T_{ICEw} = \dfrac{T_{ICE} \cdot \tau_{ICE} \cdot \tau_{diff}}{\eta_{\tau ICE} \cdot \eta_{diff}} \end{cases} \tag{30}$$

Usually for an HEV the ICE has a mechanical gearbox with $5 \div 7$ fixed reduction ratios and the EM has an unique fixed reduction ration. For this reason the longitudinal dynamic model receive as input from the driver model the correct gear that has to be considered.

In order to define the longitudinal equivalent dynamic equation it is also necessary to introduce all the resistance forces acting on the vehicle, as reported in equation (31), where: m is the total mass of the vehicle, g is the gravitational acceleration, f_v is the rolling resistance coefficient, ρ is the air density, C_x is the aerodynamic penetration coefficient, S is the total frontal area of the vehicle , α is the slope of the road.

$$F_{res} = m \cdot g \cdot f_v + \frac{1}{2}\rho C_x S v(t)^2 + m \cdot g \cdot \sin\alpha \tag{31}$$

Finally it is possible to evaluate the vehicle acceleration a, as reported in equation (32), where r_w is the radius of the vehicle wheels, m^* represents the equivalent mass of the rotating part of the vehicle (wheels, rotor, shaft)[2].

$$\begin{cases} f = \dfrac{T_w}{r_w} \\ a = \dfrac{T_w/r_w - F_{res}}{(m + m^*)} \end{cases} \tag{32}$$

Using vehicle longitudinal acceleration a from equation (32), it is possible to obtain vehicle speed and position.

$$\begin{cases} v(t) = \int_0^t a(t)dt \\ s(t) = \int_0^t v(t)dt \end{cases} \tag{33}$$

Finally the EM and the ICE speed are obtained as described in equation (34).

$$\begin{cases} \Omega = \dfrac{v(t)\tau_{EM}\tau_{diff}}{r_w} \\ \Omega_{ICE} = \dfrac{v(t)\tau_{ICE}\tau_{diff}}{r_w} \end{cases} \tag{34}$$

7. Auxiliary loads model

7.1. Auxiliary electrical loads

In order to correctly estimate the energy consumption on a FEV it is important to consider all the auxiliary electrical loads that the traction battery has to fed.

Particularly the low voltage loads (12 or $24V_{dc}$), represented for example by light, circulating pump, fan and control units, have to be estimated considering an adequate average value of power consumption during the trip. The energy for these loads is usually delivered by the traction battery through a DC/DC converter. The battery current i_{aux} can be calculated with equation (35) using the power consumption P_{aux} of electrical auxiliary loads, the battery voltage V_{batt} from the battery model and the efficiency of the DC/DC converter $\eta_{DC/DC}$.

[2] As example the equivalent mass representing the EM inertia referred to the vehicle can be evaluated considering the following equation.

$$m^*_{EM} = \frac{J_{EM}\tau^2_{EM}\tau^2_{diff}}{r^2_w}$$

http://dx.doi.org/10.5772/53570

$$i_{aux} = \frac{P_{aux}}{V_{batt}\eta_{DC/DC}} \tag{35}$$

7.2. Pumps

On HEVs and FEVs are usually installed liquid cooled electrical traction devices, in particular motor and inverter. For this reason auxiliary circulation pumps are needed in order to guarantee an adequate heat exchange between the components and the cooling fluid.

It is possible to estimate the hydraulic power P_{hy} required for the pump using equation (36), where ρ is the fluid density, Q the volumetric flow rate, g the gravity constant, h is the total head of the hydraulic circuit and h_l is an equivalent of hydraulic losses expressed in meter of water column. Usually the term h_l, that is responsible of a pressure drop Δp_l, is preponderant with respect to h and strictly depends from the design of the cooling circuit into the component.

$$P_{hy} = \rho Q g (h + h_l) = \rho Q g \left(h + \frac{\Delta p_l}{\rho g} \right) \tag{36}$$

At last, using a pump efficiency (η_{pump}) given by the manufacturer, it is possible to evaluate the electrical power requirement on the auxiliary load using equation (37).

$$P_{el} = \frac{P_{hy}}{\eta_{pump}} \tag{37}$$

8. ICE modeling

Since an accurate model of thermal combustion process require a wide knowledge of ICE design (i.e. intake and exhaust geometry, geometry of cylinder, spark position and timing, ...) a map based model is sufficient in order to estimate the engine fuel consumption and efficiency on drive cycle with a time scale of hundred of seconds.

The structure of the ICE model receive as input the torque request from the energy management control and the ICE speed from the longitudinal dynamic model and gives as output the effective torque T_{ICE}, the instantaneous volumetric fuel consumption f_c and the amount of CO_2 produced. A global structure of the model is represented in Figure 16.

The maps inserted into the ICE block can be obtained directly from the engine manufacturer; otherwise they can be obtained through experimentally tests using an engine test bench or directly on the vehicle using the Controller Area Network (CAN) information. An example of torque and fuel consumption map referred to the vehicle reported in paragraph 11.1 is reported in Figures 17 and 18.

For the volume L of fuel present in the tank equation (38) can be used, where L_0 represents the initial volume condition.

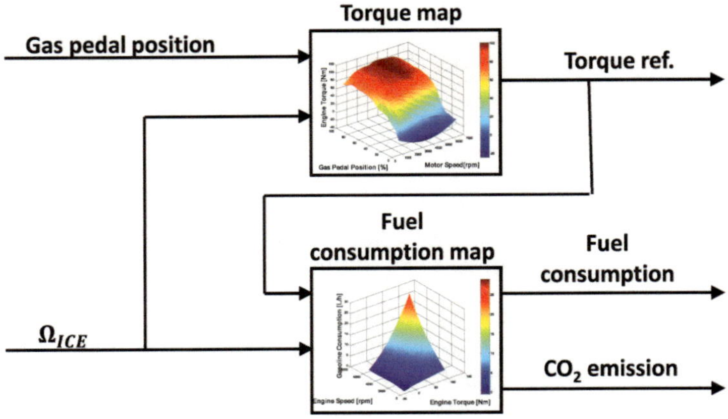

Figure 16. Block scheme for ICE

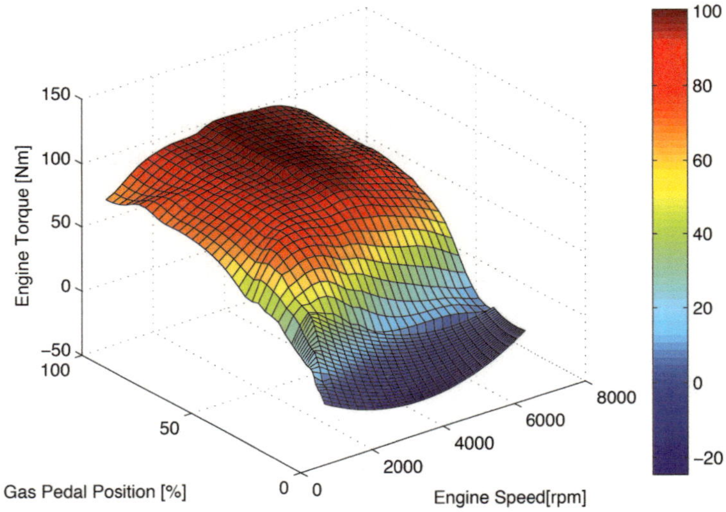

Figure 17. Engine torque map.

$$L = L_0 - \int_0^t f_c dt \tag{38}$$

Other approach for ICE modeling can be settled up using theoretical approaches as reported in [14].

Finally a rough estimation of the CO_2 emission can be established using equation (39), in which ρ_C is the average content of carbon in gasoline, M_{mCO_2} is the molar mass of CO_2, M_{mC} is the carbon molar mass and φ is a coefficient for incomplete combustion.

http://dx.doi.org/10.5772/53570

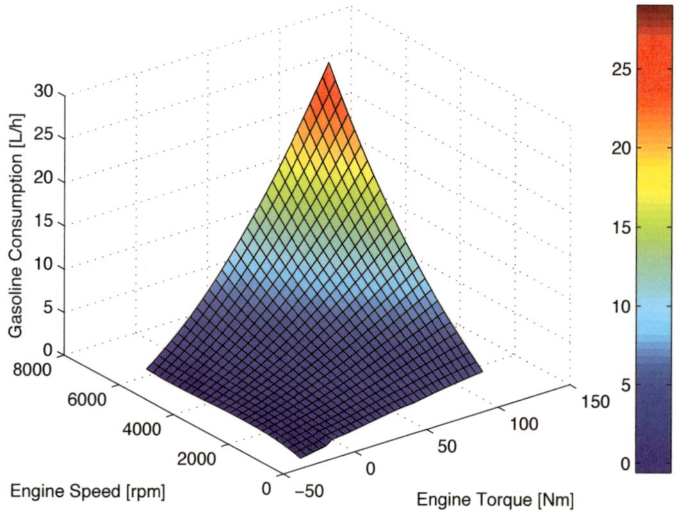

Figure 18. Engine fuel consumption map.

$$CO_2 = f_c \cdot \rho_C \cdot \frac{M_{mCO_2}}{M_{mC}} \cdot \varphi \tag{39}$$

9. Thermal modeling

The different FEV and HEV components and subsystems can be modeled including a simple thermal equivalent network where each component is considered as an homogeneous body. The chosen model is a first order lumped parameters thermal network [15] where: P_{lc} are the total component power losses, C_c is the total thermal capacity, R_c is the total thermal resistance that represent all the transfer heating phenomena (conduction, convention and radiation heat transfer), $\Delta\vartheta_c = \vartheta_c - \vartheta_{mean}$ is the temperature difference between the component inner temperature ϑ_c and the reference temperature ϑ_{mean}. The first order ODE is reported in equation (40) and the equivalent network is reported in Figure 19.

$$\begin{cases} P_{lc} = \dfrac{\Delta\vartheta_c}{R_c} + C_c \cdot \dfrac{d\Delta\vartheta_c}{dt} \\ \vartheta_c = \vartheta_{mean} + \Delta\vartheta_c \end{cases} \tag{40}$$

If the component is natural-air cooled the reference temperature ϑ_{mean} is equal to the ambient temperature ϑ_{amb}. Otherwise, if a forced-air cooling system is adopted, the equivalent thermal resistance R_c assumes different values as a function of the cooling fan status. Therefore if the cooling fan is running the $R_c = R_{cON}$ that corresponds to a lower value than $R_c = R_{cOFF}$ when the fan is stopped. A more sophisticated model can relate the R_c parameter as a function of the fan speed.

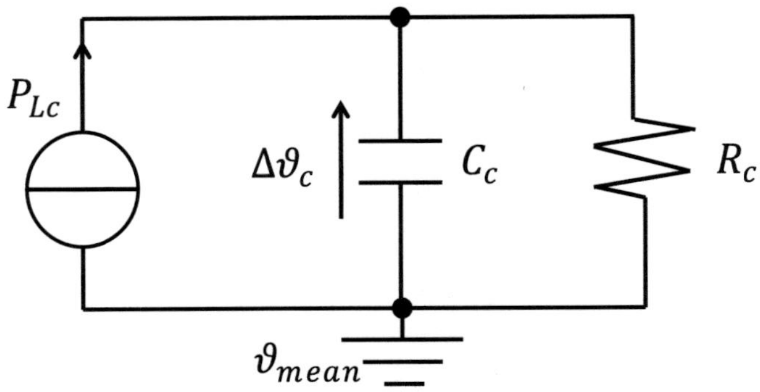

Figure 19. General component thermal model.

A FEV and HEV liquid-cooling system is often adopted especially for ICE, EM and inverter. The cooling system is based on an hydraulic circuit where a cooling fluid (usually a 50 % mix of water and glicole) is pumped into the components to be cooled and in a liquid-air heat exchanger, wich is usually forced air cooled by means of cooling fans. For these situations the thermal model of the liquid-based cooling system is to be considered too. Also in this case a first order ODE reported in equation (41) can be used. The equivalent circuit is reported in Figure 20 where: P_{ltot} are the sum of the total losses of the components that are liquid-cooled, R_{liq} is the equivalent variable thermal resistance of the liquid-air heat exchanger, C_{liq} is the liquid cooling system equivalent thermal capacity, ϑ_{liq} is the average liquid temperature in the cooling liquid circuit and $\Delta\vartheta_{liq}$ is the temperature difference between liquid and ambient.

In this case the reference temperature ϑ_{mean} for the component thermal model of Figure 19 has to be taken equal to the liquid average temperature ($\vartheta_{mean} = \vartheta_{liq}$). The equivalent liquid cooling system thermal resistance R_{liq} is a time-variant parameter since it depends on the air-liquid heat exchanger cooling fan status. For example can "switch" between two values if the fan is ON/OFF controlled (R_{liqON} when fan is on and R_{liqOFF} when is off).

$$\begin{cases} P_{ltot} = \dfrac{\Delta\vartheta_{liq}}{R_{liq}} + C_{liq} \cdot \dfrac{d\Delta\vartheta_{liq}}{dt} \\ \vartheta_{liq} = \vartheta_{amb} + \Delta\vartheta_{liq} \end{cases} \tag{41}$$

10. Driver and energy management control

The model receives as input the drive cycle that the vehicle has to execute; this reference is given to a pilot model that gives as output a signal representative of driver torque request; the pilot model acts as a speed closed loop that compares the required speed to the instantaneous one coming from the vehicle longitudinal dynamic model. Considering the vehicle structure (hybrid or full electric) and the hybrid control logic, the traction manager control splits the pilot request of torque between the ICE, the EM and the mechanical brakes, as reported in Figure 21. In this block, through torque vs speed curves, the required torques, both for the electrical and for the ICE motor, is saturated to the limit values.

http://dx.doi.org/10.5772/53570

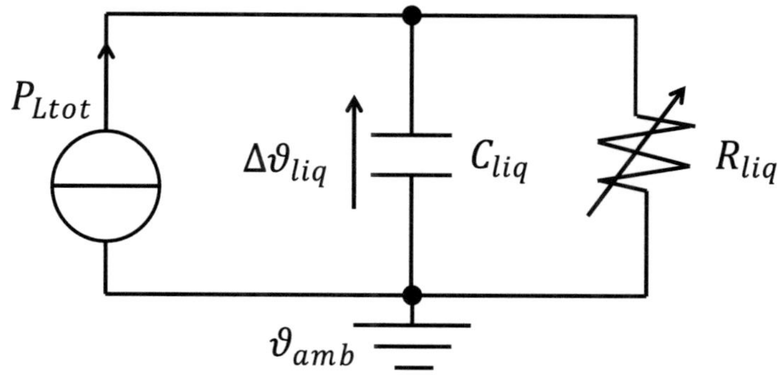

Figure 20. Radiator dynamic thermal model.

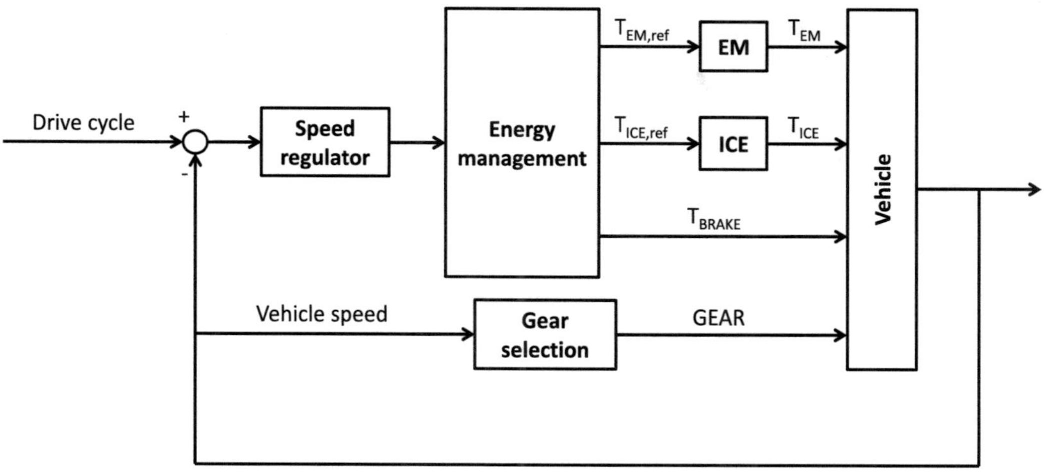

Figure 21. Driver and energy management control block scheme.

At last for the ICE gearbox a simple algorithm to set the correct ratio has to be implemented. The algorithm increase the gear if the ICE speed Ω_{ICE} exceed a certain threshold and decrease the gear if the speed Ω_{ICE} is below a different threshold. It is important to introduce an hysteresis zone on the speed Ω_{ICE} in order to avoid continuous gear shift.

11. Examples

In the current section some results compared with experimental data will be presented.

11.1. B segment car

In this section a B segment car will be considered; this car, originally propelled only with an ICE, has been transformed in a PHEV capable to run as a FEV up to 70 km/h and to cover a driving range of about $40km$. The main data of the vehicle are reported in Table 1.

Vehicle data			
Vehicle		**Internal Combustion Engine**	
Vehicle mass	$1100kg$	Fuel	Gasoline
Gearbox ratios	3.90 2.15 1.48 1.12 0.92	Max Torque	$102Nm$
Final ratio	4.071	Max Power	$50kW$
Wheel radius	$0.27m$	Total Displacement	$1200cc$

Table 1. Vehicle data.

11.1.1. Electrical power train simulation

First of all the validation of the vehicle behavior when run as a FEV will be presented. For this purpose it has been requested to the model to follow the same drive cycle executed using prototypal vehicle during experimental tests; this drive cycle is reproduced in Figure 22.

Electrical traction system data					
Battery		**Inverter**		**Motor**	
Element type	Li-Ion	V_{DC}	$80 - 400V$	Type	Induction
Number of elements	60	Typology	FOC	Peak Power	$30kW$
Rated Capacity	$50Ah$	Rated Current	$234A$	Rated Speed	$2950rpm$
Rated Voltage	$222V$	Max Current	$352A$	Rated Voltage	$105V$
Min. Voltage	$252V$	Aux Supply	$12V_{DC}$	Rated Current	$70A$
Max. Voltage	$192V$	Cooling	Water	No Load Curr.	$33.6A$
Total Energy	$11,1kWh$			Pole number	4
Max. Power	$30kW$			Cooling	Water

Table 2. Electrical traction system data.

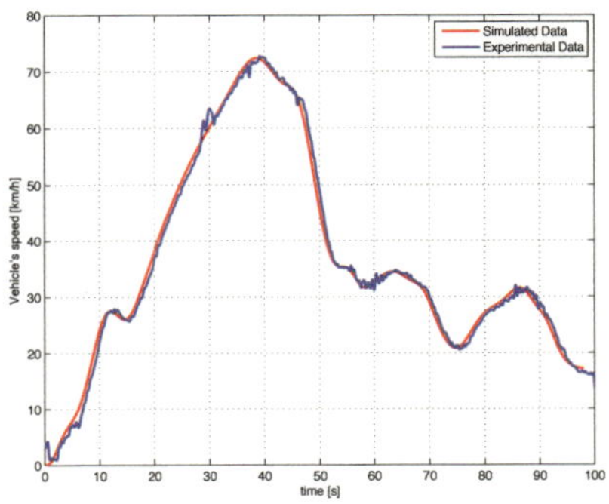

Figure 22. Electrical drive cylce.

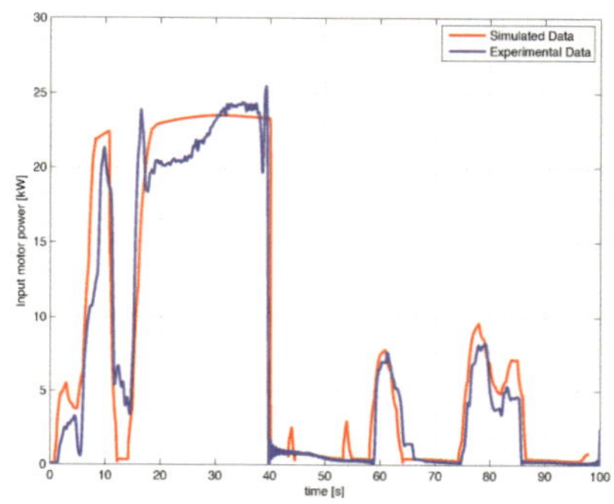

Figure 23. Electric motor power.

Using the cycle represented in Figure 22 it is possible to validate the battery model in terms of total voltage v_{batt} and in terms of current i_{batt}. The comparison between the model simulation results and the experimental data is shown in Figures 26 and 27. In the over mentioned figures it is also reported the energy consumption E evaluated through the acquired data and through the output of the vehicle's model. The comparison shows a good correspondence between the simulation and experimental data; as consequence the kilometric energy consumption is also well estimated by the model. Furthermore it is possible to validate the electrical motor model by numerical-experimental comparison performed considering the output power, as reported in Figure 23, the phase current and line to line voltage, as reported respectively in Figures 24 and 25.

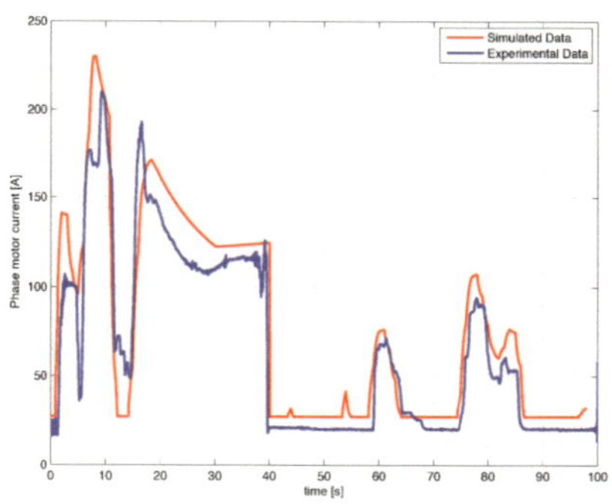

Figure 24. Motor phase current.

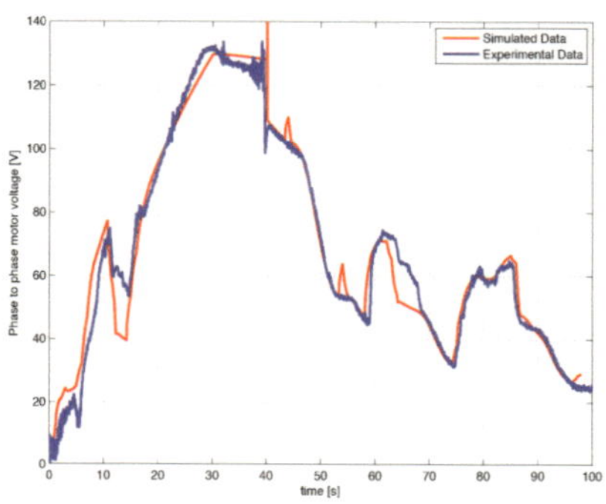

Figure 25. Motor phase to phase voltage.

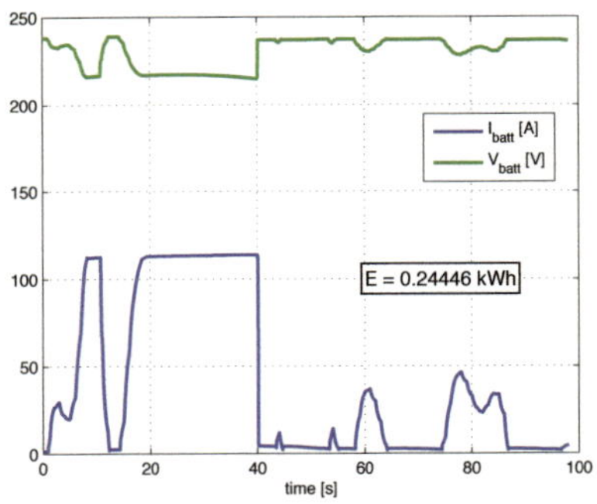

Figure 26. Simulated battery data.

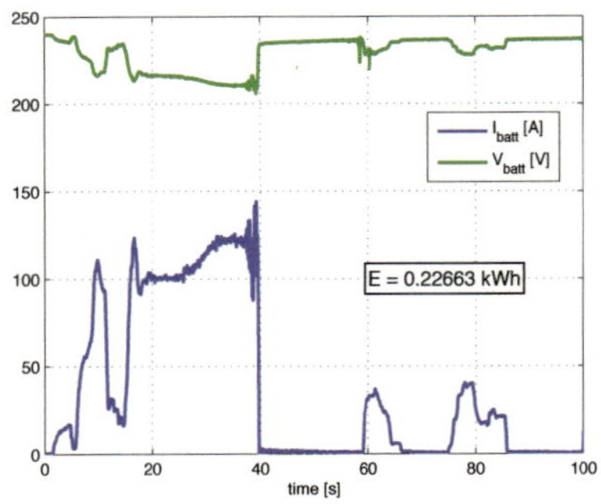

Figure 27. Real battery data.

11.1.2. Hybrid power train simulation

At last it has been implemented a Start&Stop strategy on the prototypal vehicle. This very simple strategy ask to the electrical drive traction system to propel the vehicle up to a speed threshold set to 32 km/h; above this speed threshold the vehicle is propelled by the ICE motor.

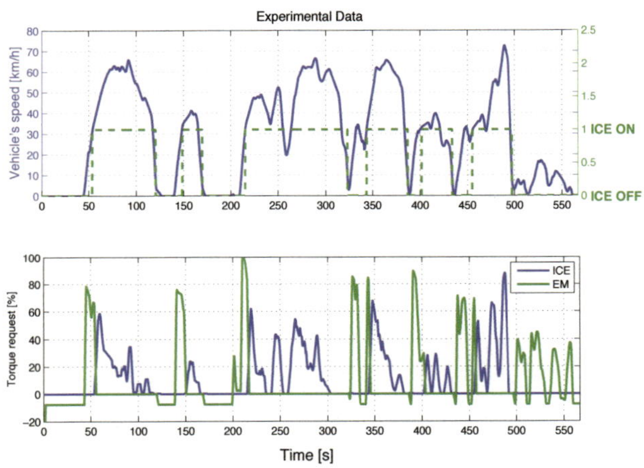

Figure 28. Drive cycle with superimposed the ICE status

In the upper part of Figure 28 it is shown the drive cycle used to validate the model in the Start&Stop mode and in the lower part it is shown the torque request repartition between the electrical motor and the ICE motor.

Finally in Figures 29 and 30 it is reported the comparison of experimental data and simulation results obtained using the drive cycle and the strategy reported in Figures 28.

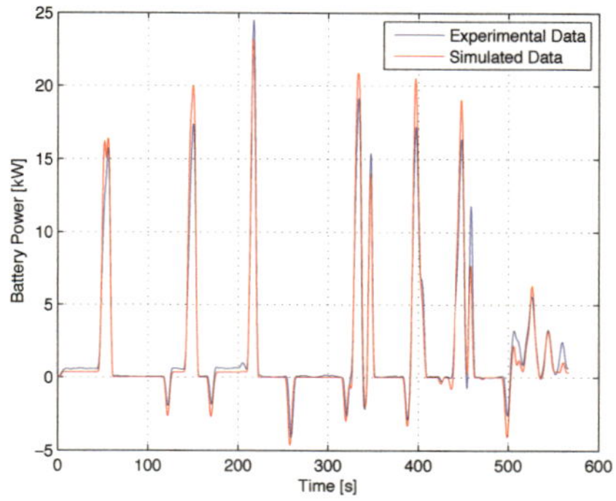

Figure 29. Battery power.

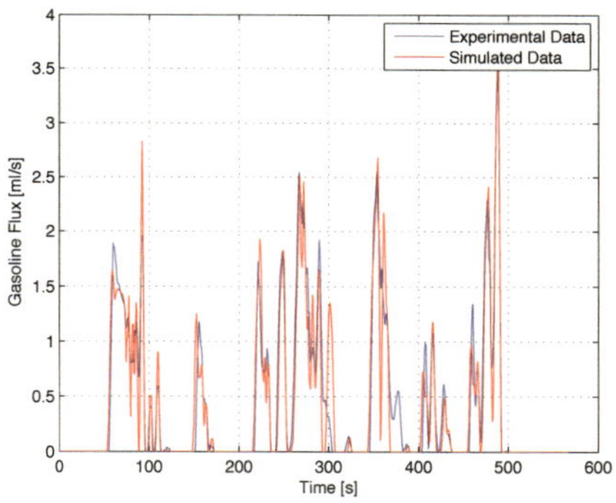

Figure 30. ICE gasoline flux.

11.2. Commercial vehicle

In this section a full electric commercial van will be considered. Its main characteristics are reported in Table 3.

As done for the previously described PHEV it has been requested to the simulation model to cover the same driving cycle executed by the prototypal vehicle (Figure 31).

Finally in Figures 32 and 33 are reported some comparison between simulated data and experimental ones; in particular Figure 32 refers to the EM torque and Figure 33 refers to the total battery current i_{batt}.

Full Electric truck data			
Vehicle		**Battery**	
Vehicle mass	$2500kg$	Element Type	Li-Ion
Final ratios	3.75	Number of elements	68
Wheel radius	0.325	Rated Capacity	$90Ah$
Max weight	$3500kg$	Rated Voltage	217
Inverter		**Electrical Motor**	
V_{DC}	$80 - 400V$	Type	Induction
Typology	FOC	Peak Power	$60kW$
Rated Current	$240A_{RMS}$	Rated Speed	$2400rpm$
Max Current	$350A_{RMS}$	Rated Voltage	$115V$
Aux Supply	$12V_{DC}$	Rated Current	$200A$
Cooling	Water	No Load Curr.	$95A$
		Pole number	4

Table 3. Electrical traction system data.

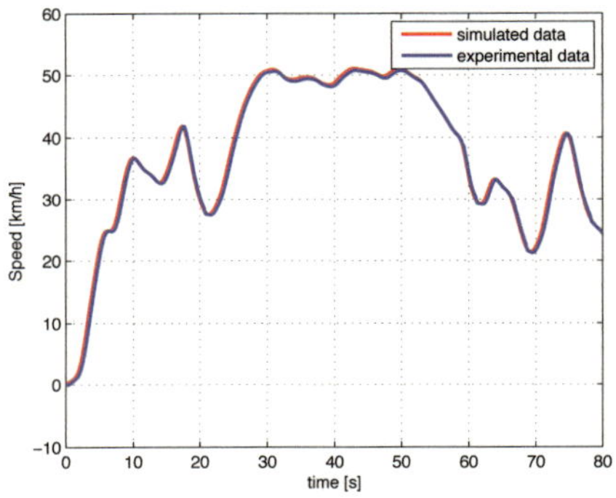

Figure 31. Drive cycle for the full electric commercial vehicle.

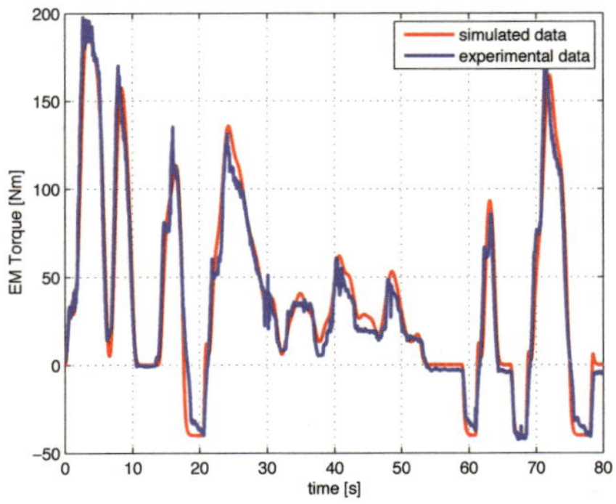

Figure 32. EM torque.

http://dx.doi.org/10.5772/53570

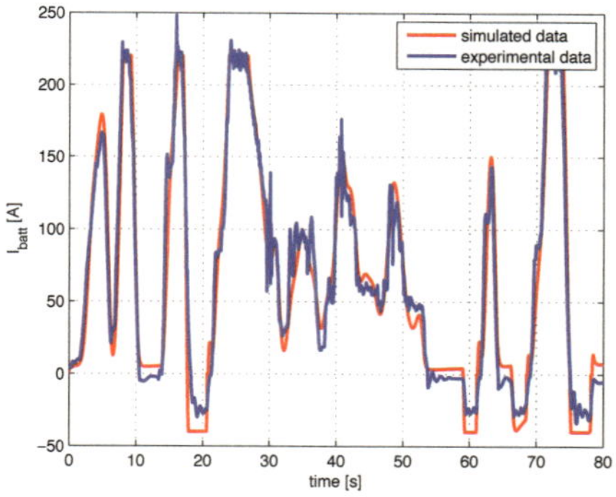

Figure 33. Total battery current i_{batt}.

12. List of Acronyms

HEV Hybrid Electrical Vehicle

RMS Root Mean Square

PHEV Plug-in Hybrid Electrical Vehicle

FEV Full Electrical Vehicle

SOC State Of Charge

IGBT Insulated Gate Bipolar Transistor

EM Electrical Motor

ICE Internal Combustion Engine

CAN Controller Area Network

ECM Equivalent Circuit Model

ODE Ordinary Differential Equation

FOC Field Oriented Control

AC Alternating Current

PMSM Permanent Synchronous Magnets Motor

PWM Pulse Width Modulation

Acknowledgements

The authors thank Davide Annese and Alberto Bezzolato for their precious help.

Author details

Ferdinando Luigi Mapelli and Davide Tarsitano

Mechanical Department, Politecnico di Milano, Milan, Italy

References

[1] D. W. Gao, C. , Mi, and A. Emadi. Modeling and simulation of electric and hybrid vehicles. *Proceedings of the IEEE*, 95(4):729–745, 2007.

[2] F. Cheli, F.L. Mapelli, R. Manigrasso, and D. Tarsitano. Full energetic model of a plug-in hybrid electrical vehicle. In *SPEEDAM 2008 - International Symposium on Power Electronics, Electrical Drives, Automation and Motion*, pages 733–738, Ischia, 2008.

[3] F. L. Mapelli, D. Tarsitano, and M. Mauri. Plug-in hybrid electric vehicle: Modeling, prototype realization, and inverter losses reduction analysis. *IEEE Transactions on Industrial Electronics*, 57:598–607, 2010.

[4] F.L. Mapelli, D. Tarsitano, and A. Stefano. Plug-in hybrid electrical commercial vehicle: Modeling and prototype realization. In *2012 IEEE International Electric Vehicle Conference, IEVC 2012*, Greenville, SC, 2012.

[5] T. Huria, M. Ceraolo, J. Gazzarri, and R. Jackey. High fidelity electrical model with thermal dependence for characterization and simulation of high power lithium battery cells. In *2012 IEEE International Electric Vehicle Conference, IEVC 2012*, Greenville, SC, 2012.

[6] A. Fratta and F. Scapino. Modeling inverter losses for circuit simulation. In *Conference of 2004 IEEE 35th Annual Power Electronics Specialists Conference, PESC04;*, volume 6, pages 4479–4485, Aachen, 2004.

[7] R. Manigrasso and F.L. Mapelli. Design and modelling of asynchronous traction drives fed by limited power source. In *Conference of 2005 IEEE Vehicle Power and Propulsion Conference, VPPC*, volume 2005, pages 522–529, Chicago, IL, 2005.

[8] http://www.infineon.com/cms/en/product/index.html

[9] http://www.infineon.com/cms/en/product/index.html

[10] R. Manigrasso and F.L. Mapelli. Design and modelling of asynchronous traction drives fed by limited power source. In *IEEE Vehicle Power and Propulsion Conference, VPPC*, volume 2005, pages 522–529, Chicago, IL, 2005.

[11] P. Vas. *Electrical machines and drives: a space-vector theory approach*. Clarendon Press, 1992.

[12] P. Vas. *Vector Control of AC Machines*. Clarendon Press, 1990.

[13] M. Mauri, F.L. Mapelli, and D. Tarsitano. A reduced losses field oriented control for plug-in hybrid electrical vehicle. In *19th International Conference on Electrical Machines, ICEM 2010*, Rome, 2010.

[14] G. Rizzoni, L. Guzzella, and B.M. Baumann. Unified modeling of hybrid electric vehicle drivetrains. *IEEE/ASME Transactions on Mechatronics*, 4(3):246–257, 1999.

[15] M. M. Rathore and R. Kapuno. *Engineering Heat Transfer*. Jones & Bartlett Publishers, 2010.

Chapter 8

Multiple Energy Sources Hybridization: The Future of Electric Vehicles?

Paulo G. Pereirinha and João P. Trovão

Additional information is available at the end of the chapter

http://dx.doi.org/10.5772/53359

1. Introduction

Energy availability and cost is at the heart of today's political and scientific agenda involving many economic, ecologic and geopolitical aspects. For instance, the European Council has established the objectives of reducing greenhouse gas emissions by 20%, of increasing the share of renewable energy to 20% and of improving energy efficiency by 20% by 2020 [1]. On March 2011, the European Commission adopted its new White Paper on Transport policy with a roadmap of 40 initiatives for the next decade to reduce Europe's dependence on imported oil and decrease the carbon emissions in transport by 60% by 2050, and in December 2011, has communicated the Energy Roadmap 2050 to pave the way to those objectives.

According to the New Policies Scenario, the central scenario of World Energy Outlook 2011 which supposes if recent government policies on energy and climate change are implemented in a cautious manner, the International Energy Agency, IEA, forecasts that the world primary demand for energy will increase by one-third between 2010 and 2035 [2]. The world Total Primary Energy Supply, TPES (with nearly 87 % coming from fossil fuels in 2009 [3]), has to fulfill this Demand. It should be noted that in 2009, around 31 % of the TPES was spent on energy transformation, leaving only about 69 % of TPES for Consumption.

The Total Final Consumption of energy, TFC, in the modern world is also mainly in the form of fossil fuels and according to the IEA, oil will remain the single largest fuel in the fuel shares of total final consumption (43.4 % in 2005, 41.3 % in 2009) with transport and power generation sectors absorbing a growing part of global energy. Indeed, the transports sector alone was responsible for 60.3% of the World Oil Consumption in 2005, 60.5 % in 2006, 61.4% in 2008, and 61.74% in 2009, against 45.4% in 1973 [3] [4]. This increasingly fuel consumption and the existent or latent conflicts mainly in the Middle East lead to oil shortage fear and price rise [5, 6], confirmed by the July 2008 crude oil peak prices, around US$150. Besides the price problem, there are also very important energy dependence and security concerns as the crude oil come mainly from Middle East unstable countries [3]. Indeed, even if the past 2008 economic crisis with the resulting consumption decline, led to a decrease in crude prices, the development and mobility levels sought and somehow already felt by many developing countries, namely the Big Emerging Market (BEM) economies, in particular China and India, will put a long term increasing pressure on the oil consumption, availability and prices [2-4]. We might very likely be already entering "the last trillion barrels of oil", as Non-OPEC oil production might have already peaked, and OPEC production could follow around 2020 [6].

Another issue is that the mass utilization of Internal Combustion Engine (ICE) vehicles in the transportation sector also increases pollution emissions, especially Greenhouse Gas emissions, which must be prevented for the sustainability of the planet and for life quality. The emissions of ICE vehicles are also one of the major sources of urban pollution, especially in medium-size and large cities. The high incidence of respiratory problems, allergies, asthmas, and some cancers is an increasing problem leading to public health concerns, as air pollution contributes definitively to mortality and morbidity. A study conducted in Austria, France, and Switzerland estimated the impact of outdoor (total) and traffic-related air pollution on public health [7]. It concluded that air pollution caused 6% of total mortality or more than 40000 attributable cases per year. About half of all mortality caused by air pollution was attributed to motorized traffic, accounting also for: more than 25 000 new cases of chronic bronchitis (adults); more than 290 000 episodes of bronchitis (children); more than 0.5 million asthma attacks; and more than 16 million person-days of restricted activities. Living in a polluted environment will undoubtedly lead to a lifetime decrease [8]. However, even living in a usually non polluted environment, a pollution peak can cause an "unexpected" increase in deaths and illnesses, like the ones in Europe during summer 2003.

2. Perspectives for sustainable transportation solutions

Due to the prior mentioned issues, there is now a general public awareness for the need for more economic, ecological and efficient transportation, namely electric vehicles (EVs) and hybrid electric vehicles (HEVs). Indeed, electric traction is the key to advanced and sustainable transports as the electric motor (EM) is much more efficient (typically with 70-90 % efficiency) than the ICE (10-30 %). This allows a much smaller *in vehicle* (or *Tank-To-Wheel*) energy consumption in vehicles driven by EMs comparatively to ICE vehicles (Fig 1, [9]), even with those complying with Euro 5 requirements, and makes the HEVs more energy efficient and cleaner (Fig. 2) than the ICE vehicles using the same engine technology. The Diesel HEVs promise to be a very effective option.

Here it should be pointed out that even though biomass fuels have much smaller Greenhouse Gas (GHG) emissions than fossil fuels (Fig. 2) its Source-to-Service fuel consumption is very high (Fig. 1). This should preclude the farmed biomass large scale utilization, contrarily to the news and hopes that have come to public mainly in 2006 and 2007, but had already lead to serious food price problems in 2008 due to the food-for-fuel dilemma: to use land to produce biomass for fuel instead of for food.

http://dx.doi.org/10.5772/53359

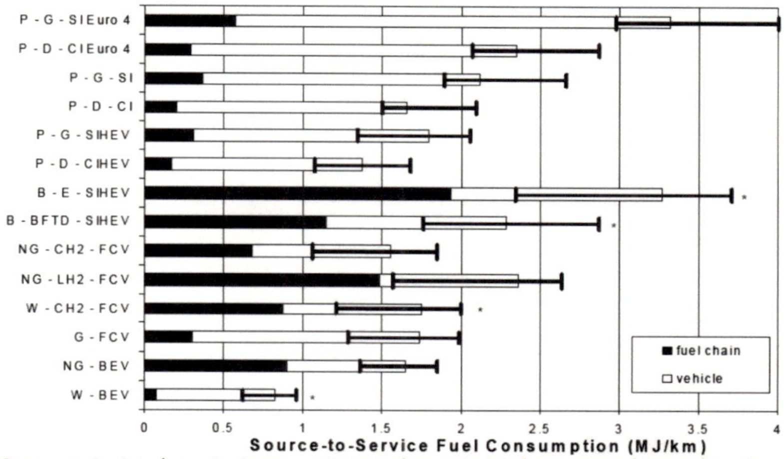

- ***Resources:** P = Petroleum; B = Biomass; NG=Natural Gas; W = Wind; GT = Natural Gas Turbine; * = renewable*
- ***Energy Carriers:** G = Gasoline; D = Diesel; E = Ethanol; BFTD = Bio Fischer-Tropsch Diesel; CH2 = Compressed Hydrogen; LH2 = Liquid Hydrogen; E = Electricity*
- ***Powertrains:** SI = Spark Ignition; CI = Compression Ignition; SIHEV = Spark Ignition Hybrid Electric; CIHEV; Compression Ignition Hybrid Electric; FCV = Fuel Cell; BEV = Battery Electric*

Figure 1. Vehicle and Source-to-Service fuel consumption (based on higher heating values of all chemical energy carriers). [9]

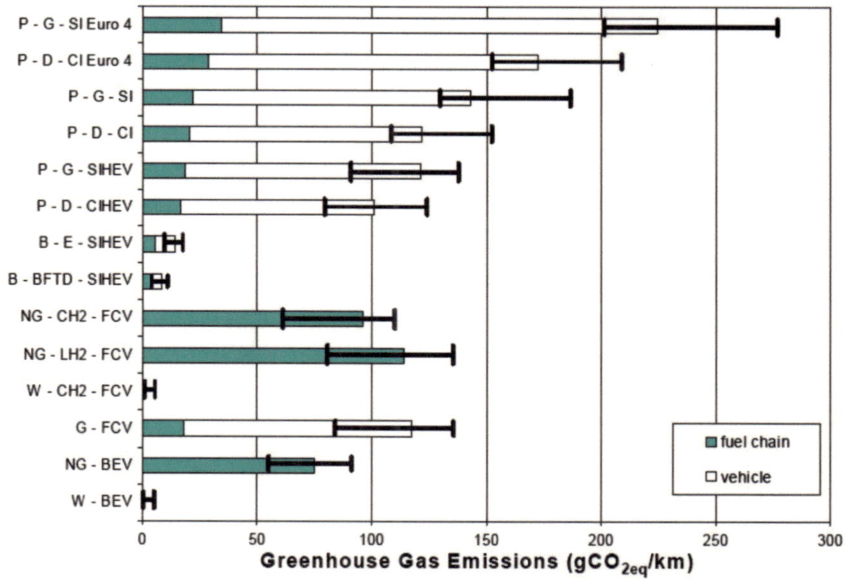

Resources: P = Petroleum; B = Biomass; NG=Natural Gas; W = Wind; GT = Natural Gas Turbine; * = renewable
Energy Carriers: G = Gasoline; D = Diesel; E = Ethanol; BD = Bio Fischer-Tropsch Diesel; CH2 = Compressed Hydrogen; LH2 = Liquid Hydrogen; E = Electricity
Powertrains: SI = Spark Ignition; CI = Compression Ignition; SIHEV = Spark Ignition Hybrid Electric ; CIHEV; Compression Ignition Hybrid Electric; FCV = Fuel Cell; BEV = Battery Electric

Figure 2. Greenhouse Gas Emission for various fuels and powertrain technologies [9].

The problem here is that the efficiency of photosynthesis is only about 0.40% [10], which is extremely low. Taking into account the other losses in the fuel chain, the net energy available for electricity production or transportation utilization is even much smaller. For electricity it would be much better to use photovoltaic panels (PV): considering a 80% land use, a 12% efficiency in the DC generation from PV array, and 85% for the DC/AC conversion and transmission, the overall efficiency for photovoltaics is 8.16%, much higher than the overall efficiency of 0.137% obtained using bio-methane or the 0.074% from bio-hydrogen [10]. Concerning the energy harvest for transportation, the distance that can be driven with the annual energy extracted from one hectare of land is nearly 21500 km for Biodiesel, 22500 km for Bioethanol, and 3250000 km with Electricity from PV (151 and 144 times more, respectively). So, the problem is not the substitution of gasoline by biofuels but the replacement of inefficient ICE by efficient electric motors. [10]

Furthermore, while being basically non pollutant during its lifetime (considering proper battery recycling) and highly efficient, EVs are silent and gentle to drive.

Particularly, battery electric vehicles (BEVs) present zero emission of pollutants locally, which is very important for urban driving. If batteries are recharged using electricity from some renewable sources, namely wind (Fig. 1 and Fig. 2) or PV, then the differences, relative to ICE cars, increase and all the potential of the BEVs is shown. For these reasons, it is the authors' belief that the future of Sustainable Mobility passes surely by the BEVs supplied from wind, hydro and PV, or other high efficient and clean renewable energy sources.

It should also be noted that as nowadays only a small percentage of the world electricity is produced from oil (5.8% in 2006, decreasing to 5.1% in 2009 [3]), there is another advantage in shifting the share of transports primary energy from oil to electricity, especially if the electric energy storage devices are charged during the night, using the energy surplus usually available in the grid and that can be increased by fostering the public lightning efficiency.

By all that has been presented, at the moment, the future perspectives for Sustainable Transportation Solutions seem to be [10]-[12]:

- Efficient electric or hybrid-electric cars for commuting and local transport ("Wind-to-Wheel" efficiency up to 70%);
- Avoidance of hydrogen for ICE and fuel cell vehicles ("Wind-to-Wheel" efficiency of 20% to 25%);
- Distant land, air and ocean transport with oil or biofuels.

A transition step while looking forward to the ideal solution of Zero Emissions Vehicles, are the Low Emissions Vehicles, as the HEV, specially the Plug-In Hybrid Electric Vehicles, PHEV. Several projects and models of EV, HEV and PHEV, including buses, vans and cars, have been developed in the last few years, resulting in cleaner, more economic and less noisy vehicles, some of them already available commercially. [13][14]

However, some incentives are still needed to allow electric vehicle (EV) technologies to develop and become more competitive.

3. Energy storage in electric vehicles

3.1. The energy storage issue in electric vehicles

To allow EVs to become the effective sustainable transportation solution, a great effort has to be done in R&D to overcome the major technical issue in EVs: the energy storage.

Typically, EVs store energy in batteries (usually Lead-Acid, NiMH and, more recently, Li-Ion) that are bulky, heavy and expensive. The specific energy of gasoline is about 12500 Wh/kg (of which only 2000-3000 can be considered useful energy, due to the very low efficiency of ICE) against typically 40-50 Wh/kg in good lead-acid batteries or 70 Wh/kg in NiMH, which gives an idea of the volume and weight necessary to store the energy needed to do the same work. Li-ion batteries have higher specific energy, around 150 Wh/kg but they are still expensive and some particular Li-ion technologies have safety issues that have to be carefully addressed. Due to these problems, with current battery technologies it is very difficult to make a general purpose EV that effectively competes with ICE cars. For massive deployment of EV, its driving range problem must be solved. [15]-[17]

3.2. Main available energy sources

At present and in the foreseeable future, the viable EVs energy sources are batteries, fuel cells, SuperCapacitors (SCs) and ultrahigh-speed flywheels.

Batteries are the most mature source for EV application. But they offer either high specific energy (HSE) or (relatively) high specific power (HSP). Fuel cells are comparatively less mature and expensive for EV application. They can offer exceptionally HSE, but with very low specific power. In spite of some quite expensive prototypes, such low specific power poses serious problems in their application to EVs that desire a high acceleration rate or high hill climbing capability. Also, they are incapable of accepting the high peaks of regenerative energy during EV braking or downhill driving. SCs have low specific energy for standalone application. However, they can offer exceptionally HSP (with low specific energy). Flywheels are still technologically immature for EV application. [18]-[20]

Some recent information on the energy sources can be found for example in [17], [21], [22].

3.3. Multiple energy sources hybridization

For the "full electric" EV the solutions pass by significant progresses in battery technology and by using different energy sources with optimized management of the energy flow as none of the available energy sources can easily fulfill alone all the demand of EVs to enable them to compete with gasoline powered vehicles. In essence, these energy sources have a common problem: they have either HSE or HSP, but not both. A HSE energy source is favorable for long driving range, whereas a HSP energy source is desirable for high acceleration rate and high hill climbing capability. The concept of using and coordinating multiple energy sources to power the EV is typically denominated *hybridization*. Hence, the specific advantages of the various EV energy sources can be fully utilized, leading to optimized energy economy while satisfying the expected driving range and maintaining other EV performances. [22]-[25]

The basic operation of the hybridized system is shown in Fig. 3. In operations that require high power, as is required during a hard acceleration or traveling up slopes, the two energy sources provide power to the powertrain system, as shown in Fig. 3 a).

Moreover, in operations that require less power, for example, during travel at constant speed (cruising), the source with characteristics of high specific energy provides power to the drive system while simultaneously recharges the second source that only has characteristics of high specific power, as shown in Fig. 3 b), to prepare it for new high power demand situations. In braking and deceleration mode, the regenerative energy will essentially be stored in the source with high specific power characteristics, particularly the peaks and only a small, limited to its maximum power value, is absorbed by the source with high specific energy (see Fig. 3 c)). Thus, to try feeding an EV with only one of these sources with the same responsiveness as the one described above, the volume, weight and cost of the unique source would be so large that the system would be incapable of operating properly.

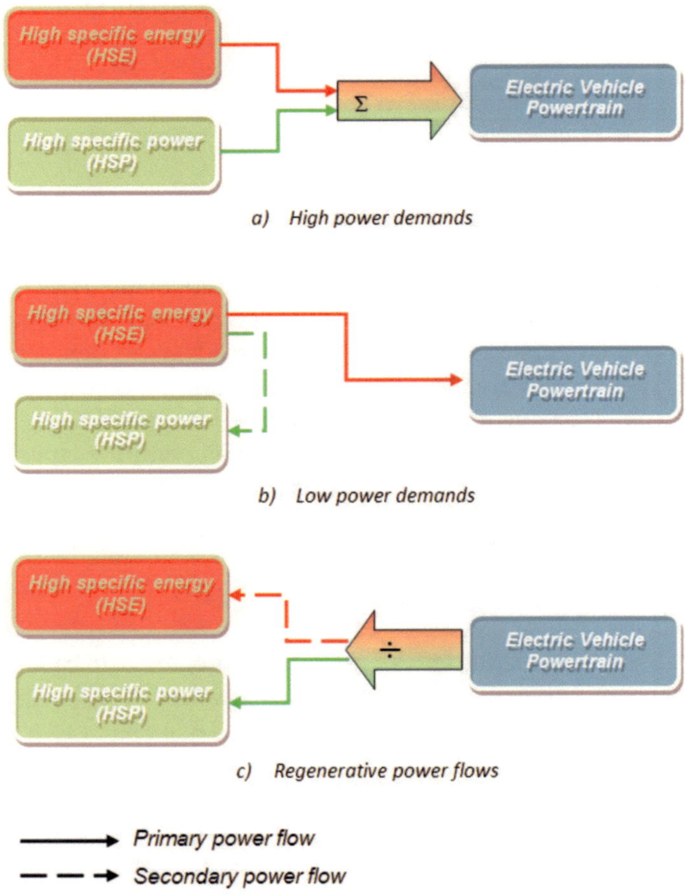

Figure 3. Concept of energy sources hybridization: a) shared power supply; b) power supply and recharging the high specific power source; c) regenerative energy shared storage.

Therefore, the hybridization concept presents a scale economy using complementary feeding systems fusing the sources' advantages and better responds to the drive requests. For this purpose, any work related to the hybridization concept of EVs should start with an optimized sizing of the on-board vehicle energy sources, meeting the minimum characteristic requirements aimed for the EV.

As mentioned in Section 3.2, with the current state of technological development, the future of EVs seems to go through the hybridization of various energy sources. This strategy seeks to benefit from the best qualities of each available energy source and is especially useful in urban driving. In [23], a methodology to optimize the sizing of the energy sources for an electric vehicle prototype, using different driving cycles, maximum speed, a specified acceleration, energy regeneration and gradeability requests is presented. The possibility of using a backup system based on solar energy is also studied, which may be considered in the design or as an extra to cope with unforeseen routines and to minimize the recharge of energy sources.

3.4. Generated, stored, demanded and available energies

The total energy generated or stored ($W_{ge.st}$) and demanded (W_{dem}) over a time period can be written in terms of the generated solar, regenerative break and storage powers and the power demand as follows:

$$W_{ge.st} = \int_0^t \left(P_{PV} + P_{Bat} + P_{reg\_SC}\right) dt$$
$$W_{dem} = \int_0^t P_{dem} dt \tag{1}$$

where P_{Bat} is the power supplied from (or to) the batteries, P_{PV} is the power generated by a specified PV array, and P_{reg_SC} is the regenerative break power to be stored by the SCs. At any moment the available energy, W_{avail}, is given by

$$W_{avail} = W_{ge.st} - W_{dem} \tag{2}$$

The values of $W_{ge.st}$ and W_{dem} and W_{avail} should be updated for small time steps (for the EV VEIL [26, 27] case study presented later in Section 5, time steps of 1 s were considered). The W_{avail} evolution can be plotted and used to analyze the storage capacity and the EV autonomy for a specific drive journey, as will be shown in Section 5.3.3.

4. EV dynamical model

For the EV model, the mechanical parts, including body and transmission units, and the dynamic and aerodynamic vehicle characteristics have to be considered.

Considering a vehicle of mass, m, (see Fig. 4), the opposing forces to the vehicle motion are: the rolling resistance force (F_{rr}) due to friction of the vehicle tires on the road; the aerodynamic drag force (F_{ad}) caused by the friction of the body moving through the air; and the climbing force (F_{hc}) that depends on the road slope. The F_{ad} force is directly derived from aerodynamic theory ignoring the lateral forces.

The total tractive effort is equal to F_R and is the sum of the resistive forces, as in (3):

$$F_R = F_{rr} + F_{ad} + F_{hc} \tag{3}$$

The F_{rr} force is the sum of the rolling resistance force of each wheel, depending on the coefficient of rolling resistance (μ_{rr}) and of the vehicle mass, as presented in (4). The typical values for μ_{rr} may vary between 0.015, for conventional tires, and 0.005 for tires developed specially for EV [28].

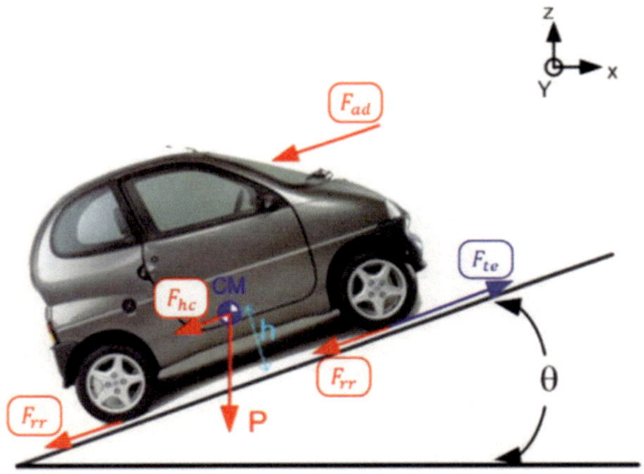

Figure 4. Forces applied to the vehicle.

The aerodynamic drag force is given by the second term on the right side of equation (4), where the symbol ρ represents the air density, C_D the drag coefficient, A_F the frontal projection area and V_V the vehicle speed relative to the wind [28]. It must also be noted that air density is variable, as a function of the atmospheric pressure, temperature and hygrometric conditions, and that the aerodynamic drag is proportional to the square of vehicle velocity. Thus the power applied to the motor, necessary to overcome F_R, increases with the cube of the speed.

The weight component of the vehicle relative to the rolling plane angle, expressed in the last term of (4), corresponds to a force that opposes the motion when climbing, and is a function of the climbing angle θ and the vehicle mass m.

$$F_R = \mu_{rr} mg + \frac{1}{2}\rho C_D A_F V_V^2 + mg\sin\theta \tag{4}$$

The dynamic behavior of the electrical motor, in the motor referential, considering an ideal mechanical transmission, is given by

$$T_m - \frac{r}{i}\cdot F_R = J_T \cdot \frac{d\omega_m}{dt} \tag{5}$$

The load torque results from a set of vehicle motion resistant forces (F_R) in the motor referential, considering the wheel radius r, the transmission gearbox ratio, i, ω_m is the motor angular speed and T_m is the motor torque.

The total moment of inertia associated to the vehicle (J_T), in the motor referential, is given by (6), and is equal to the sum of the moments of inertia from electric motor (J_m), wheel (J_r) and the one associated with the vehicle that is a function of the road characteristics [28].

$$J_T = J_m + J_r + \frac{1}{2}m\left(\frac{r}{i}\right)^2 (1-\varepsilon) \tag{6}$$

The moment of inertia corresponding to the mass of the vehicle is the last term in (6), where ε represents the slipping of the wheels.

The mechanical power needed on the wheels (P_u) is then in the motor referential:

$$P_u = T_m \cdot \omega_m \tag{7}$$

The formulation presented in (1) to (7) can be and is usually used to study the vehicle power and energy need from a high-level energy management and sources comparison points of view [30]-[33].

However, to correctly size the energy sources, all the energy chain with the corresponding losses need to considered. That is, the efficiency of each one of the components has to be considered [34]. Therefore, the required electric power (P_e), considering the total efficiency η_{tot} of the powertrain (the total efficiency of all the components used between the energy sources and the wheels see Fig. 8) is given by:

$$P_e = \eta_{tot} \cdot P_u \tag{8}$$

5. Case study

5.1. Hybridization project for VEIL prototype

As previously described, the utilization of multiple energy sources is a well suited solution to overcome current EV barriers. To study the utilization of multiple energy sources in electric vehicles a small electric vehicle is used: at the Electrical Engineering Department of the Engineering Institute of Coimbra (DEE-ISEC), the authors' team started the on-going VEIL project to convert a small Ligier 162 GL, initially with an internal-combustion engine (ICE), into an electric vehicle (Fig. 5) [26] [27].

Figure 5. VEIL during road tests at ISEC campus.

For the VEIL project prototype, the hybridization of three energy sources was considered to be viable: a HSE storage system – Batteries –, a HSP system – SCs – and photovoltaic panels, PV. Fig. 6 shows this hybridization configuration.

Considering the available space, 1300x1100 mm on the rooftop, and 550x1100 mm over the hood, it is possible to implant 5 selected PVs (cf. Table 2), 4 on the rooftop and 1 on the hood.

http://dx.doi.org/10.5772/53359

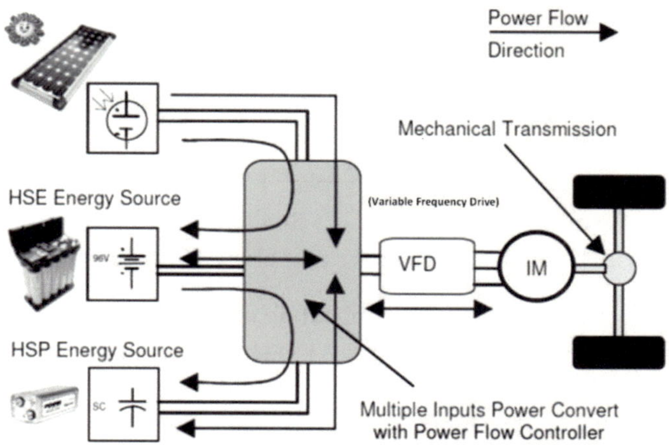

Figure 6. EV project power scheme.

5.2. Test cycles, scenarios and sources

In this work, three different scenarios have been used to study the utilization of different combinations of energy sources, corresponding to typical possible utilizations of a small electric vehicle and in particular of the VEIL [35]. In the three scenarios three different time periods were considered: a first displacement in the morning, starting at 7:30 and taking 1.5 h (to get to work, for instance), a second period where the car is parked outdoor and lasting 8 h, and a third period equal to the first one, corresponding to the return back home, from 17:00 to 18:30.

The first scenario, Scenario 1, corresponds to a typical routine for mobility in big European cities with low average speed and very frequent stops and goes. To simulate this behavior the ECE 15 cycle presented in Fig. 7 a) was used. The travel in the morning consists of a sequence of 27 ECE 15 cycles, corresponding to nearly 27.35 km. The same distance has to be travelled in the evening to make the way back. Scenario 1 consists then of 27 ECE 15 cycles during 1.5 h, followed by a period of 8 h parked outdoor, and then again 27 ECE cycles. The total journey distance is 54.7 km (2 times 27.35 km).

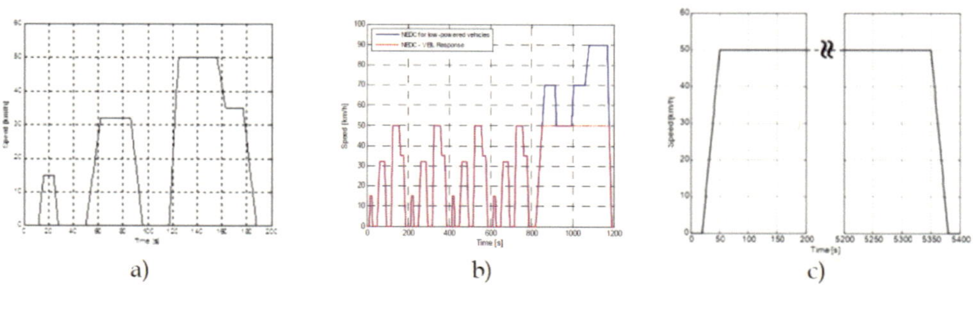

Figure 7. Driving cycle speed versus time: a) ECE 15 urban; b) NEDC and VEIL speed and c) constant 50 km/h

The second scenario, Scenario 2, corresponds to a mix use of urban and sub-urban/extra urban driving. It uses the New European Driving Cycle (NEDC), represented in Fig. 7 b), which consists of the combination of 4 ECE 15 cycles, repeated without interruption, followed by one EUDC cycle, which is limited to 90 km/h for low-powered vehicles. For the VEIL, which by law is limited to 45 km/h (free driving license car), the maximum vehicle speed was considered as 50km/h on the flat road. To fulfill the 1.5 h travel in the morning and in the evening, Scenario 2 uses 4.5 NEDCs totalizing 40.24 km followed by 8 h parked and again 4.5 NEDCs. The total journey distance is about 80.48 km.

Scenario 3 corresponds to an extra urban utilization at the VEIL full speed that for this kind of vehicle is limited to 45 km/h. Nevertheless, the study was done considering the slightly higher speed of 50 km/h, as in Fig. 7 c). In this case, 1.5 h is able to cover a 73.59 km distance between home and work in the morning and the same distance in the evening, to return back home (almost 149.18 km, in total).

For the considered test cycles, the request initial-acceleration performances are defined as accelerating the EV from standstill to 15km/h in 4s, to 32km/h in 22s and 50km/h in 8s, for the ECE 15 cycle, and from standstill to 50km/h in 14s, for NEDC cycle. The most demanding situation considered is then the 3rd period of acceleration in the ECE-15 cycle, where it is necessary to reach 50 km/h in 8s.

For the sources, several different types of batteries were considered with the main characteristics shown in Table 1, and SCs and PV panels with the characteristics shown in Table 2. The presented prices in Tables 1 and 2 are only indicative, as they correspond to the market prices obtained for the project quantities. They might decrease significantly for big quantities.

Battery	Manufacturer	NVxBS	Capacity[Ah]	Total energy[kWh]	Mass [kg]	Vol. [dm³]	Approx. Cost [€]
Pbacid	-	12 x 8 = 96V	27.5	3.95	159.2	69.5	1200
Li-ion	SAFT	10.8x9=97.2V	80	7.8	72.0	50.9	(>>18000)
NiMH-VH module[1]	SAFT	12 x 8 = 96 V	2 x 13.5(@ 2C)	2.9	48.0	27.2	2200
NiMH-VH module[2,3]	SAFT	12 x 8 = 96 V	4 x 13.5(@ 2C)	5.8	96	54.4	4400
Li-ion[3]	Thunder Sky	3.2 x 30= 96 V	60 (@ 0.3C)	5.76	75.0	45.25	3671
Li-ion[3]	Thunder Sky	3.2 x 30= 96 V	90 (@ 0.3C)	8.64	96.0	65	4235

NV: nominal voltage; BS: number of batteries in series. For Li-ion, the value ">>18000", was a 2006 quotation for a specific battery model; the other values are for new products in the market, end of 2008 prices.

[1] Values presented for two battery banks in parallel; [2]Values presented for four battery banks in parallel; [3]New options considered for the VEIL project.

Table 1. Previously [30] and new considered batteries to obtain 96 V

Source	Manufacturer	Capacitance Voltage	Series	Parallel	Capacity	Operation Voltage [V]	Total Mass [kg]	Approx. Cost [€]
SCs	Maxwell **BMOD0500–E16**	500 F / 16 V	5	2	200.0 F	40-65	50	6150
Source	**Manufacturer**	**Panel Dimensions L x W [mm]**	**Series**	**Parallel**	**Warranted Power [W]**	**Operation Voltage [V]**	**Total Mass [kg]**	**Approx. Cost [€]**
PV Array	BP Solar **BP MSX 30**	616 x 495	5	1	135	84-105	15	1200

Table 2. Characteristics of SCs and PV panels

One particular case corresponds to each type of batteries. Each one of these cases, Case A to Case F, considers different possible combinations for Batteries (Bat), SCs and PVs, taking into account the different weights of the sources used, as shown in Table 3.

Case	Bat. Type	Only Bat.	Bat.+SC [1]	Bat.+PV	Bat.+PV+SC [1]
A	NiMH VH module (2 banks)	432.0 kg	485.0 kg	447.0 kg	500.0 kg
B	Pbacid	543.2 kg	596.2 kg	558.2 kg	611.2 kg
C	Li-ion	456.0 kg	509.0 kg	471.0 kg	524.0 kg
D	NiMH VH module (4 banks)	480.0 kg	533.0 kg	495.0 kg	548.0 kg
E	Li-ion	459.0 kg	512.0 kg	474.0 kg	527.0 kg
F	Li-ion	480.0 kg	533.0 kg	495.0 kg	548.0 kg

[1]A 3 kg weight increase was considered for the SCs DC/DC converter and other associated equipment. Cases D-F represent new energy source considered.

Table 3. Vehicle mass with different sources

5.3. Calculation and results

Using the previously presented formulation, several relevant quantities can be calculated. To implement the model, Matlab/Simulink® was used [36] and the characteristics of the electrical drive, the transmission ratio of the gearbox (i = 10), the wheel radius (r = 26 cm), the load vehicle mass (m = 500 kg), the air density (ρ=1.204 kg/m^3 @ 20ºC), the drag coefficient (C_D=0.51), the frontal projection area (A_F=2.4 m^2), the coefficient of rolling resistance (μ_{rr} =0.015), and the total moment of inertia associated to the vehicle (J_T=0.53 kg/m^2) were taken into account. The slipping of the wheels, for the control purposes, is not considered.

5.3.1. Power on the wheels and power in the sources

To properly size the power source to supply the VEIL prototype, the global power chain should be considered, as shown in Fig. 8, i.e., including the corresponding losses, considering the η_{tot} of the powertrain (all components' efficiency, between the power supply and the wheels), given by (8).

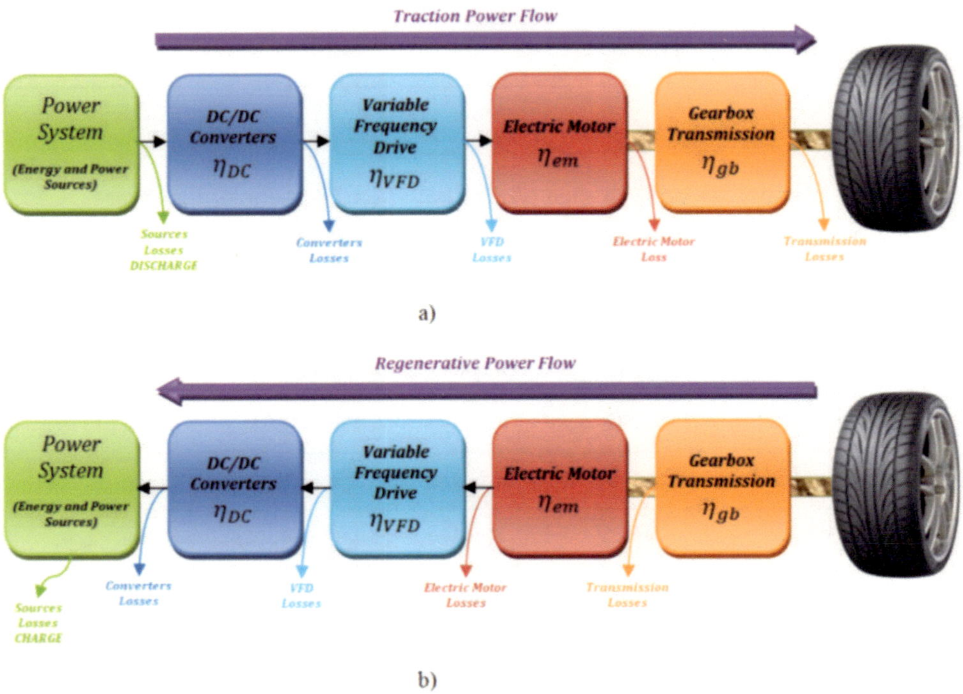

Figure 8. Power flow diagrams for global system EV: a) Traction Mode; b) Regenerative Mode.

To calculate the power needed from the electric sources, *Pe*, given by (9), for the VEIL case, the efficiency of the gearbox (η_{gb}), the efficiency of the electric motor (η_{em}), the efficiency of the Variable Frequency Drive (VFD) (η_{VFD}) and the efficiency of the DC/DC converter (η_{DC}) had to be considered. In traction mode, η_{tot} in (8) becomes η_{totT} given by (9)

$$\eta_{totT} = \eta_{gb} \times \eta_{em} \times \eta_{VFD} \times \eta_{DC} \tag{9}$$

And when the EV is in the breaking mode, to calculate the energy that can be stored in and recovered from the SCs, the powertrain power (or energy) recover efficiency η_{totR} has to take into account also the efficiency of the SCs (η_{SC}), both for charge and discharge

$$\eta_{totR} = \eta_{gb} \times \eta_{em} \times \eta_{VFD} \times \eta_{DC} \times \eta_{SC}^{2} \tag{10}$$

For the VEIL components, the average efficiencies in (9) and (10) are respectively η_{totT} = 90%*85%*96%*95% ≈ 70% and η_{totR} = 90%*85%*96%*95%*(96%)^2 ≈ *65%*. These values are not very different from those expected of a near future typical EV.

This clearly shows the need to carefully choose the components, as all the energy chain efficiency is strongly influenced by the less efficient component. For example, in Fig. 9 for the cycle ECE-15, the mechanical power needed on the wheels ("Power Demand", P_u), given by (7), the electric power to be supplied by the electric sources ("Electric Power Demand", P_e), given by (8) and also the "Total Regenerative Power" available on the wheels and the power that for the present case study can be recovered from the SCs, the "Effective Regenerative Power" (about 65% of the Total Regenerative Power available on the wheels), is presented.

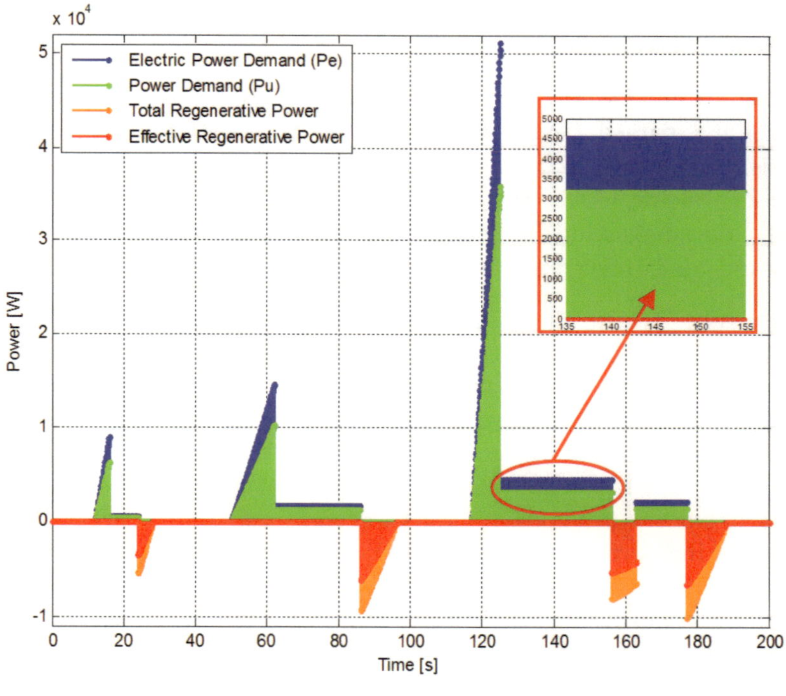

Figure 9. ECE 15 power demand and available regenerative power on the wheels and on the power sources.

It is also important to notice that, even though there are not experimental results for all the cases considered, the simulated results for the electric power demand at 50 km/h constant speed (zoom on Fig. 9), 4.5 kW, are in very good accordance with the measured ones, 4.3 kW [27], which validates the model used, for a high level daily energy study. Nevertheless, it should also be pointed out that to study the system response, a much more detailed study has to be performed with more accurate component models and smaller time step scales [38], and to manage the energy sources, a real time multiple energy source monitoring system has to be used [39].

5.3.2. Study of the efficiency powertrain influence

Using the presented formulation with the different electrical energy source combinations of PV array, SC and Batteries, and the different scenarios for typical drive journeys, as explained in Section 6, the available energy, W_{avail}, was calculated with (2) for 1 s steps for a typical 24 h period. As initial conditions, the batteries were considered fully charged and the SCs completely discharged.

For the solar energy, the average hourly statistics for direct normal solar radiation [Wh/m^2] for the last 30 years at the project location, Coimbra, was used. The PV array efficiency model was used to compute the global generated energy by the panels on a typical day of two different months, November and August, with the minimum and maximum solar radiation, respectively. To account for the near horizontal position of the PV panels, as well as for some undesirable aspects like the different solar panels orientation (giving origin to non-uniform irradiation), and the effects of the buildings and trees shadows, a depreciation of 25% on the normal irradiation was considered. When the car is moving, the PV energy can be directly used by the powertrain, decreasing the amount of energy supplied by the batteries. When the car is parked, the PV energy is stored in the batteries. As the charge current, around 1 A, is much smaller than any of the considered batteries typical charge currents, the batteries' losses were neglected. The expected accumulated energy varies between 900 Wh and 1350 Wh a day, depending on the considered month, and supposing that the driver can find a sunny and good oriented parking place.

In Fig. 10 the results for Case A and Case C are compared for scenario 1 (2x27xECE 15 cycles) and with or without considering the efficiency of the components on the power/energy chain. From the energy management and sources comparison points of view sometimes it is only considered the energy at the wheels [30]-[33]. However, as can be seen from these two graphs, for the sources or autonomy sizing it is fundamental to consider the energy efficiency of all the energy chain. For example, from Fig. 10a) it could be said that using only the NiMH batteries and SCs the travel of work-return home could be accomplished (curve 1, with a slightly positive value at the end of the journey) but considering the η_{tot} it can be seen that it is not possible in any case, not even with the help of the PV panels (curve 2). Comparing curves 1 and 3, it can also be seen that the influence of the regenerative breaking energy is much smaller (only nearly 65% – η_{totR} – reach and can be extracted from the SCs, and only 70% – η_{totT} – of this energy return back to the wheels, which gives an overall recovery of 45.5%, from wheel-to-wheel for the present EV components).

5.3.3. Study of possible suitable solutions for the autonomy objectives: 4 NiMH banks (Case D) vs new Li-ion batteries (Cases E&F)

To study possible suitable solutions for the autonomy objectives, three new solutions were considered: Case D, where the duplication of the present two NiMH battery banks to four banks was considered, and Case E and F, using Li-ion batteries that more recently appeared in the market. The relevant quantities were calculated for all the six cases in Table 3 for each one of the three considered displacement scenarios, and with and without considering the components efficiency.

http://dx.doi.org/10.5772/53359

The W_{avail} evolution considering η_{tot} (η_{totT} and, when applicable, η_{totR}), is presented in Fig. 11 and 12, for Case D and Case F, respectively. These figures contain a lot of information from where some important conclusions could be extracted. Some of them will now be presented.

The Scenario 1 is the one corresponding to the most likely utilization of a small urban electric vehicle. From the W_{avail} evolution in Fig. 11 a), it can be concluded that even using the four NiMH small battery packs, it is only possible to drive the vehicle back home for Scenario 1, using the combination Bat+SCs+PVs. The other two scenarios (Scenarios 2 with 80.5 km, and 3 with nearly 150 km) are not possible to carry out with these batteries quantity.

For Case F, using a different combination of Li-ion batteries, PV panels and SCs, it can be concluded from Fig. 12 a) that for Scenario 1, the batteries alone are sufficient to drive the 55 km planned for the journey. However, the batteries would be almost depleted (SOC below 15%), which is dangerous in terms of autonomy reserve or in case of unbalanced batteries, besides being severe for the batteries life time.

Furthermore, for Case F the values presented for the Li-ion batteries' capacity are for 0.3C that implies a 27 A discharge, which is a quite low value; for bigger discharge rates the capacity will certainly be significantly lower. It is then concluded that the utilization of PVs or SCs could overcome these issues. The SCs also will increase the EV dynamic performances; the batteries' efficiency and life time are also improved. For the NEDCs in Fig. 12 b), it can be seen that the chosen batteries alone do not have enough energy and so it is clear that the best option with that batteries is to use Bat+PV+SCs (even for November, the SOC at the end of the trip would be around 17 %). However not even with Bat+PV+SCs, is it possible to drive Scenario 3 with this battery pack.

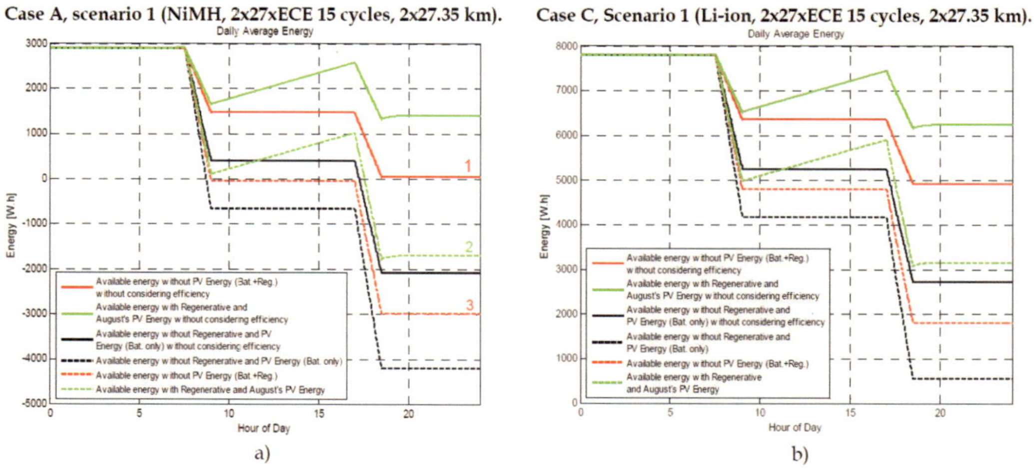

Figure 10. a) Available energy for Case A (NiMH batteries) and different efficiency consideration; b) Available energy for Case C (Li-ion batteries) and different efficiency consideration.

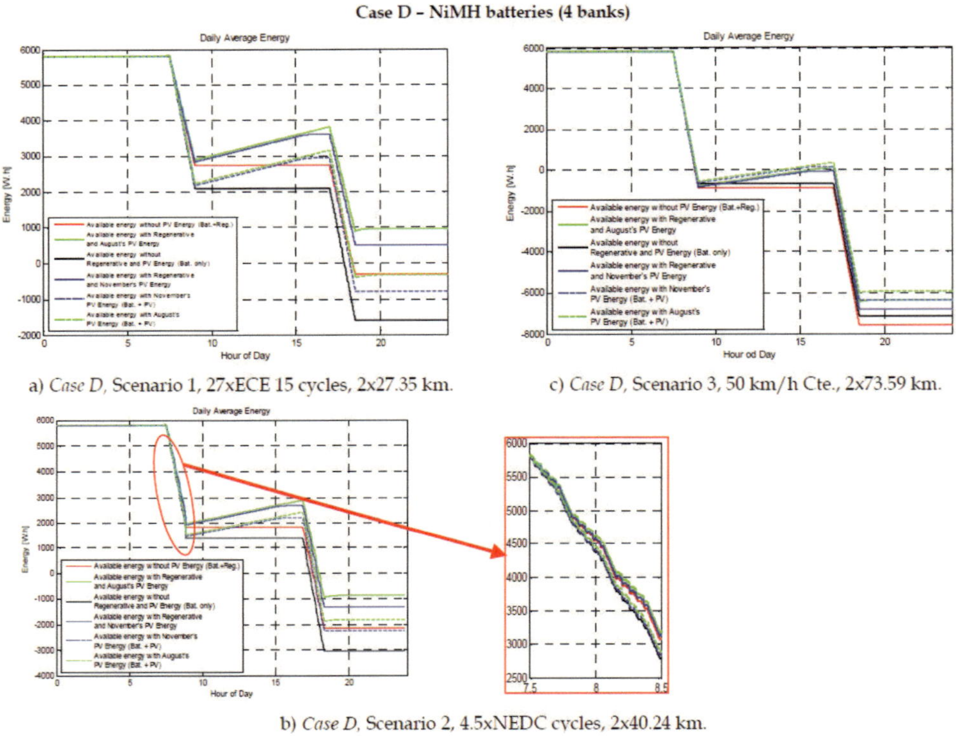

a) *Case D*, Scenario 1, 27xECE 15 cycles, 2x27.35 km.

c) *Case D*, Scenario 3, 50 km/h Cte., 2x73.59 km.

b) *Case D*, Scenario 2, 4.5xNEDC cycles, 2x40.24 km.

Figure 11. Available energy for Case D (NiMH batteries – Four banks) and 3 different mobility scenarios.

Comparing the graphics a), b) and c) in Fig. 11 and 12, it can also be seen that the relative importance of the regeneration, i.e. of the SCs, decreases. Indeed, as for the 50 km/h cte (graphics c), the regenerative braking energy is negligible, and it is also clear that the SCs do not bring any advantage; in reality, it is the contrary: the weight increase due to the SCs, associated electronics and support structures, increases the energy consumption, decreasing the W_{avail} relative to the batteries-only solution. It can also be concluded that for extra urban utilization the correct choice is to add more battery packs. This conclusion is regardless the batteries or SCs prices.

http://dx.doi.org/10.5772/53359

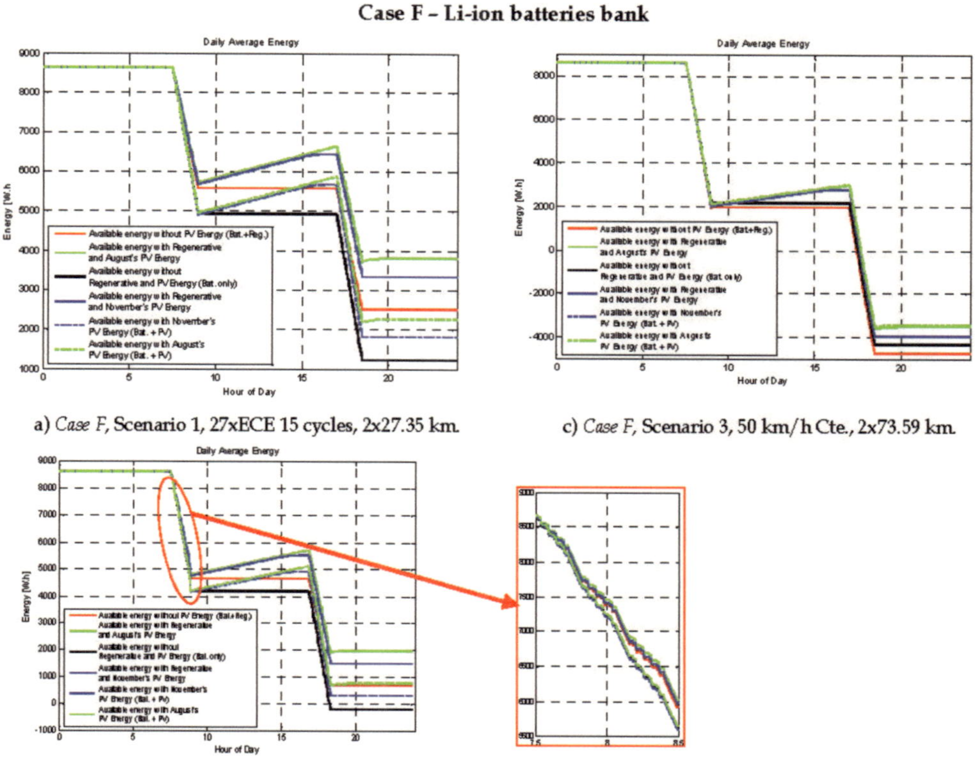

a) *Case F*, Scenario 1, 27xECE 15 cycles, 2x27.35 km.

c) *Case F*, Scenario 3, 50 km/h Cte., 2x73.59 km.

b) *Case F*, Scenario 2, 4.5xNEDC cycles, 2x40.24 km.

Figure 12. Available energy for Case F (Li-ion 90 Ah@ 0.3C) and 3 different mobility scenarios.

Finally, a very important part of the regenerative breaking energy can be recovered using SCs. In spite of their present high cost, which is expected to decrease in a near future, a general conclusion is that the regenerative braking recovering is particularly important for urban traffic. This fundamental aspect implies multiple energy sources hybridization and a global energy management system.

6. Hierarchical management concept suitable for multiple sources EVs

Recently, some authors [40]-[45] have studied the energy storage management in EVs focused on online control and optimization. In [40], the authors use the Energetic Macroscopic Representation to define and implement different strategies for hybrid energy storage systems for EVs. In [42] and [44], stochastic dynamic programming is used to determine an energy management strategy for online control of the power flows during operation considering the stochastic influences of traffic and driver behavior. In [45], an optimal online power management strategy is developed using machine learning and fuzzy logic in order to minimize energy sources' power losses.

Thereby analyzing the results presented in Section 5.3, it is clear that hybridization of a multiple energy sources for electric vehicles presents a set of requirements for a global energy management system resulting essentially in an energy and power management problem, with several time scales to define implementable solutions for sharing energy and power between the selected sources with different power and energy characteristics.

The management concept is based on the use of all available resources to obtain the desired result efficiently, which implies an effective use and especially the coordination of all available resources to achieve the objectives with maximum efficiency. Thus, the management concept focuses on the organization of all processes, from the point of view of long term-action, always considering the short-term one. Hence, a global process management system it is necessary that involves different levels to form a strategic vision, defining objectives and identifying a strategy, which will then be implemented and imposed. From this definition, there is the applicability of this concept to the energy and power management systems using multiple energy sources.

6.1. Classic hierarchical management structure

Typically, an overall management structure consists of several levels or layers with a hierarchically well-defined chain, as shown in Fig. 13, or even a hierarchical command chain. To achieve a common goal, several agents or decision processes, at each level of this hierarchy, are to receive and perform very different tasks, especially with different time, but always with a collaborative point of view. It is recognized that the highest level of this hierarchy is mainly responsible for the high level guidelines that influence long-term objectives of the process. The directives listed in high level management are therefore referred to the decision process at intermediate level. However, the long-term guidelines, achieved through the implementation of these directives, do not need to be well defined or known in the subsequent levels. But it is essential that the intermediate level receive enough information from the high-level process to make a decision that meets its objectives while respecting the guidelines of the particular level of superior guidance. Likewise, the higher hierarchical level management does not require detailed information on the particular objectives of the intermediate level, and how to execute their long-term orientations.

According to the guidelines and restrictions imposed by management level, the medium level takes decisions almost continuously affecting the system operations based on pre-established policies management. The content of the tactical management level decisions are more interventionist, which leads to a shorter periodicity decision as compared to the high level management. Therefore, at a periodic frame, the upper level may rethink its current strategy and its long-term goals, and as a result, amend its guidelines, which are communicated to the medium level so that its decisions take place. The update rate of these decisions is greater than the change rate of the high level guidelines.

The tasks that carry out the implementation of very specific guidelines for a global system are defined as low level management in classic management hierarchy. The low level management takes quick local decisions which directly influence the process using as boundaries the decisions handed down by the medium level management. The action frequency at the operational level, the lowest management level, is much higher than the frequency of renewal decisions of the higher levels.

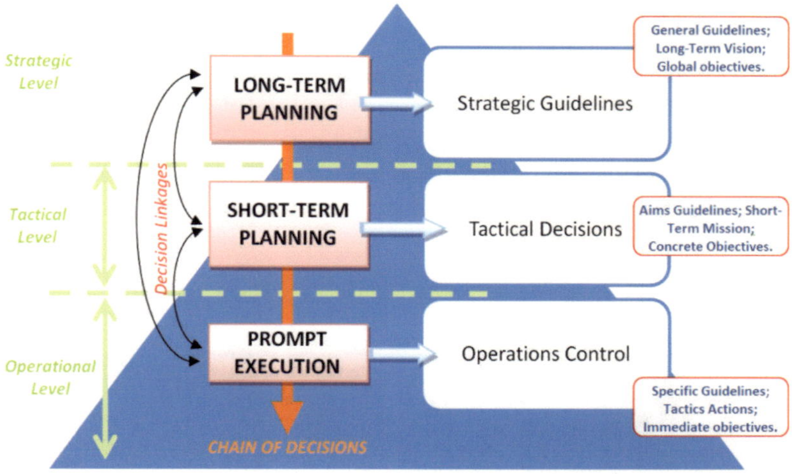

Figure 13. Classic hierarchical management.

The different decision levels in the hierarchical management structure, shown in Fig. 13, for a given system or process, enables to clearly define the overall objectives, the frequency of decisions (long, medium, and short term) and forms of interaction between the various levels. The modular organization of this strategy approach allows an easy and rapid modification in each module independently without the process management restructuration. Each management module is responsible for a specific purpose defined and is strictly responsible for its decisions and sending decisions/orientations to the down module of the decision chain.

This hierarchical management structure concept provides a systematic dissemination and simultaneously evaluates the process to manage. Given their different nature, the guidelines, decisions and implementations will inevitably have different cadence decision, in which each level shall take its decisions with different cadences, and must be synchronized in time so that no mismatch can occur between the various modules. Thus, the high level responsible for dictating the long-term strategy will have a refresh rate of its objectives slower than the decision level intermediate tactical maker that will decide several times during one cycle of the upper module. The lower level has a responsibility to produce reactions in a very short time, and scarcely have an instant response to any change in the behavior of the process within the guidelines of decision's modules located hierarchically above. Therefore, the overall process of management is divided into three decision modules with different responsibilities and different times for the revaluation of its decisions.

The differentiated step concept of the decision based on each decision module is illustrated in Fig. 14. As shown in this Figure, multiple implementations of the low level module occur before the intermediate level module makes a further decision, and consequently, several decisions of the same level occur before a new guideline consideration is provided by the high level module.

6.2. Architecture and hierarchy of the energy management system for hybridized EV

It should be noticed that when applying the hierarchical management to organizations or companies, a strict schedule time to make decision is not fundamental. However, for the applicability of this concept to the particular problem of multiple energy sources management, the concept of the classic hierarchical management methodology for a modular management, where decisions are made in a discrete and deterministic way, is mandatory. This is presented in Fig. 14.

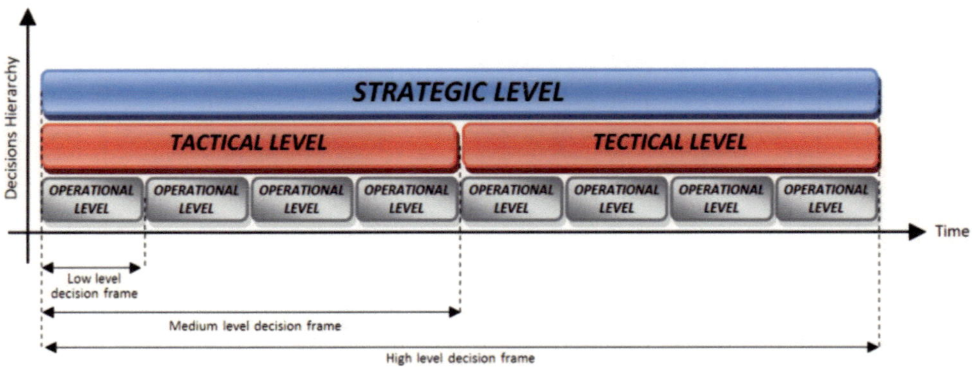

Figure 14. Temporal organization of the decisions in a hierarchical management system.

As evidenced, the modular hierarchical management methodology has various concepts that can be modeled and adapted to project energy management systems in general and particularly to the management of multiple sources. This approach clearly demonstrates that a global management process with simultaneous objectives (long, medium and short term) can be divided into several smaller processes, where each process has one or more well definable tasks. The fact to distinguish perfectly the natural interconnections between the several management modules with different time scales should also be stressed. This structure has particular interest to the energy management problem of the EV multiple sources, and this question cannot be dissociated from a correct power sharing of the embedded sources or energy storage. The closed relationship between the power (*P*) and the energy (*W*) due to those parameters are simply related to each other by a single parameter, the time (*t*). The relationship between two quantities is characterized by (12).

$$P=\frac{dW}{dt} \Leftrightarrow W=\int P\,dt \tag{11}$$

From (12), the energy is simply the cumulative use of power over a period of time. Thus, in a direct way and searching for a solution for the energy and power management, the management of energy can be associated to the hierarchically higher one than the power one.

http://dx.doi.org/10.5772/53359

Based on the presented hierarchical management concept and its adaptability to the multiple energy sources EV management, a complete on-line energy management system architecture for dual-source EV is presented in Fig. 15, with the introduction of different management levels.

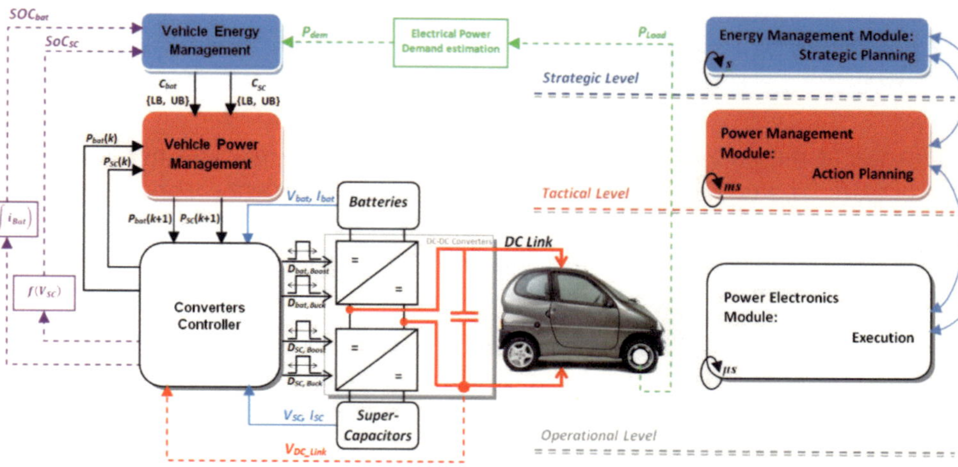

Figure 15. Architecture and hierarchy of the energy management system for a dual-source EV system.

The formulation of the EV energy management problem with multiple sources, with particular emphasis on urban circuits, is primarily based on three fundamental objectives for the correct EV operations. The global results of the management have to maximize the use of the source that best suits the powertrain power demand answering the driver and route requirements. The listed objectives for this problem are: Long-Term Planning (energy management), responsible for the definition of an overall management strategy to produce a set of guidelines to consider in the decisions of lower management levels; Short-Term Planning (power management), whose main function is the definition of actions that will lead the lower level to produce the reference signals to control and perform the wanted operations, and finally, the Prompt Execution (operational control), responsible for the control signals generation in order to implement the guidelines and directives of the two higher levels and command the power electronic converters. Thus, using a top-down approach, the first objective defines a global strategy and therefore defines the guidelines and restrictions that restrict the decision space of the second management level, which together dictate rules to produce control signals that will control the DC/DC converters [46] [47], as presented in the blocs diagram of Fig. 15.

Only an approach based on energy and power management through a hierarchical structure, using various management modules with different responsibilities, may lead us to obtain the good energy efficiency results presented in Section 5.3. These results were achieved considering an optimized management system for on-line energy and power management.

Although hybridization of multiple sources and energy management topics are still open to further study, in the present work we attempted to suggest some wide challenges and describe new research opportunities in order to obtain an effective energy management system for multiple energy sources electric vehicle.

7. Conclusions

The emphasis of the presented work is on the multiple energy sources hybridization for EVs. A comparative study on the impact of the utilization of different energy sources, namely different types of batteries, SCs and PV panels, for different common scenarios of daily use was done. The importance of considering the efficiency of all the energy chain in the EV was also clearly shown.

Simulation results for the VEIL Project powered by a mix of energy sources were presented and analyzed. At project start, small NiMH modules seemed to be a good option for a typical urban utilization. However, with the Li-ion price decrease, some apparently more interesting solutions appeared in the market. It was shown that the regenerative braking energy can be quite important in urban driving, together with the PV utilization when a typical home-work-home journey is forecast, with long outdoor parked periods. This leads to the need of sources hybridization for urban utilization. Besides the range extension, the PV utilization can also supplement the long-term batteries self-discharge, and in some cases avoid the need of a charge during the day, which can be particularly relevant in terms of energy cost for the EV owner and for grid energy management, especially by decreasing the need for fast charges. The presented methodology, which is quite simple to apply and extend to any EV, can be followed for the correct sizing and choice of the energy sources to use, depending on the assumed utilization, and also to estimate what will be the EV autonomy or its ability to perform different utilizations. This can be used to customize the EV energy sources for the client needs and desires.

Finally, to allow effective multiple energy sources hybridization, the architecture of a three level hierarchic energy management system for a dual-source EV was proposed. This system, with its simulation and hardware implementation has been under development by the authors showing promising results.

Acknowledgements

This work was supported in part by the Science and Technology Foundation under Grant SFRH/BD/36094/2007 and project Grant PTDC/EEA-EEL/121284/2010 and FCOMP-01-0124-FEDER-020391.

Author details

Paulo G. Pereirinha[1,2,3] and João P. Trovão[1,2]

1 Department of Electrical Engineering, Polytechnic Institute of Coimbra, IPC-ISEC, Rua Pedro Nunes, Coimbra, Portugal

2 Institute for Systems and Computers Engineering at Coimbra - R&D Unit INESC Coimbra, Rua Antero de Quental 199, Coimbra, Portugal

3 Portuguese Electric Vehicle Association, Lisbon, Portugal

References

[1] Presidency Conclusions, European Council of 8-9 March 2007 (Available in July 2012 at http://ec.europa.eu/archives/european-council/index_en.htm).

[2] IEA, "World Energy Outlook 2011 Factsheet", OCDE, 2011.

[3] IEA, "Key World Energy Statistics", 2011 edition.

[4] IEA "World Energy Outlook 2010", OCDE, 2010.

[5] James L. Williams, "Oil Price History and Analysis". (Available in July 2012 at www.wtrg.com/prices.htm).

[6] Peter R. A. Wells, "The Peak in World Oil Supply", September 2008. (Available at "The Last Trillion Barrels", EV World, Open Access Article Originally Published: September 27, 2008 www.evworld.com/article.cfm?storyid=1535).

[7] N. Künzli, R. Kaiser, S. Medina, et al., "Public-health impact of outdoor and traffic-related air pollution: a European assessment", The Lancet, vol. 356, Issue 9232, pp. 795-801, 2 September 2000.

[8] L. Int Panis, R. Torfs, "Health effects of traffic related air pollution", Proc. European Ele-Drive Transportation Conference 2007, EET-2007, 30th May - 1st June 2007, Brussels, Belgium, in CD-ROM.

[9] Jan H. J. Thijssen, "Viable and Sustainable Energy Strategies Grounded on Source-to-Service Analyses: A Perspective of the Role of Fuel Cells in Transportation", Presented at the Lucerne Fuel Cell Forum 2004.

[10] U. Bossel, "Phenomena, Facts and Physics of a Sustainable Energy Future", Presentation at the European Sustainable Energy Forum, 3 July 2007, Lucerne / Switzerland.

[11] U. Bossel, "Does a Hydrogen Economy Make Sense?", Proc. of the IEEE, October 2006, pp. 1826-1836.

[12] J. Van Mierlo, G. Maggetto, Ph. Lataire, "Which energy source for road transport in the future? A comparison of battery, hybrid and fuel cell vehicles", Energy Conversion and Management, 47, pp. 2748–2760, October 2006.

[13] P.G. Pereirinha, J.C. Quadrado, J. Esteves, "Sustainable Mobility: Part II – Some Possible Solutions Using Electric and Hybrid Vehicles", CEE´05 – Inter. Conf. on Electrical Engineering, in CD-ROM, October 2005, Coimbra, Portugal.

[14] Buying guide, Available in July 2012 at www.thechargingpoint.com/buying-guide.html

[15] Iqbal Husain, "Electric and Hybrid Vehicles. Design Fundamentals", 2nd Edition, CRC Press, 2010.

[16] Paulo G. Pereirinha, João P. Trovão, Alekssander Santiago; "Set Up and Test of a LiFePO4 Battery Bank for Electric Vehicle," Electrical Review - Przeglad Elektrotechniczny, Warsowa, Polland, ISSN PL 0033-2097, R. 88 NR 1a/2012, pp. 193-197.

[17] Hugo Neves de Melo, João P. Trovão, Paulo G. Pereirinha, "Study of Lithium-Ion Batteries Usability for Electric Vehicle Powertrain", Proceedings of the 2011 3rd International Youth Conference on Energetics (IYCE), pp.1-7, 7-9 July 2011.

[18] C. C. Chan, Y.S. Wong, A. Bouscayrol and K. Chen, "Powering Sustainable Mobility: Roadmaps of Electric, Hybrid and Fuel Cell Vehicles", Proceedings of the IEEE, April 2009.

[19] A. Burke, "Batteries and Ultracapacitores for Electric, Hybrid, and Fuel Cell Vehicle, Proceedings of the IEEE, Vol. 95, No. 4, April 2007.

[20] Lukic, S.M., Jian Cao, Bansal, R.C., Rodriguez, F.; Emadi, A., "Energy Storage Systems for Automotive Applications", IEEE Transactions on Industrial Electronics, vol. 55, pp. 2258, Jun 2008.

[21] Mehrdad Ehsani, Ali Emadi, Yimin Gao, Modern Electric, Hybrid Electric, and Fuel Cell Vehicles Fundamentals, Theory, and Design (2nd Edition), CRC Press, ISBN 978142005398-2, 2009.

[22] A. Khaligh and Z. Li, "Battery, ultracapacitor, fuel-cell, and hybrid energy storage systems for electric, hybrid electric, fuel cell, and plug-in hybrid electric vehicles: Stat-of-art," IEEE Transactions on Vehicular Technology, vol. 59, no. 6, pp. 2806-2814,July 2010.

[23] Trovão, J. P., P. G. Pereirinha, H. M. Jorge. "Design Methodology of Energy Storage Systems for a Small Electric Vehicle", World Electric Vehicle Journal Vol. 3 - ISSN 2032-6653, 2009.

[24] R.M. Schupbach, J.C. Balda, M. Zolot; B. Kramer, "Design methodology of a combined battery-ultracapacitor energy storage unit for vehicle power management", Power Electronics Specialist Conf., PESC ´03, IEEE 34th Annual; vol.1, June 2003.

[25] M.J. West, C.M. Bingham, N. Schofield, "Predictive control for energy management in all/more electric vehicles with multiple energy storage units", Electric Machines and Drives Conf., 2003. IEMDC'03. IEEE International; vol. 1, June 2003.

[26] P. G. Pereirinha; J. Trovão; A. Marques; A. Campos; F. Santos; J. Silvestre; M. Silva; P. Tavares, "The Electric Vehicle VEIL Project: A Modular Platform for Research and Education", Proc. of the European Ele-Drive Transportation Conference 2007, EET-2007, 30th May - 1st June 2007, Brussels, Belgium, in CD-ROM.

[27] Paulo G. Pereirinha, João P. Trovão, L. Marques, M. Silva, J. Silvestre, F. Santos: "Advances in the Electric Vehicle Project-VEIL Used as a Modular Platform for Research and Education", EVS24 International Battery, Hybrid and Fuel Cell Electric Vehicle Symposium, Stavanger, Norway, 13-16 May 2009.

[28] Hodkinson, R., Fenton, J.: "Lightweight Electric/Hybrid Vehicle Design", Society of Automotive Engineers, 2001.

[29] Hori, Y, Toyoda, Y., Tsuruoka, Y: "Traction Control of Electric Vehicle: Basic Experimental Results Using the Test EV – UOT Electric March", IEEE Transactions on Industry Applications, Vol. 34, n. 5, September/October 1998.

[30] Paulo G. Pereirinha, João P. Trovão, "Comparative study of multiple energy sources utilization in a small electric vehicle", 3rd European Ele-Drive Transportation Conference EET-2008 - Geneva, March 11-13, 2008.

[31] Wu, Y.; Gao, H., "Optimization of Fuel Cell and Supercapacitor for Fuel-Cell Electric Vehicles", IEEE Trans. on Veh. Technol., Vol. 55, No. 6, Nov. 2006, pp. 1748-1755.

[32] R. Schupbach and J. Balda, "The role of ultracapacitors in an energy storage unit for vehicle power management," Proc. IEEE Veh. Technol. Conf., Orlando, FL, 2003, p. 3236.

[33] J. Bauman, M. Kazerani, "A Comparative Study of Fuel-Cell–Battery, Fuel-Cell–Ultracapacitor, and Fuel-Cell–Battery–Ultracapacitor Vehicles" IEEE Trans. Veh. Technol., vol. 57, pp. 760, March 2007.

[34] W. Gao, "Performance comparison of a fuel cell-battery hybrid powertrain and a fuel cell–ultracapacitor hybrid powertrain", IEEE Trans. Veh. Technol., vol. 54, pp. 846, May 2005.

[35] João P. Trovão; Paulo G. Pereirinha; Humberto M. Jorge; "Analysis of operation modes for a neighborhood electric vehicle with power sources hybridization," 2010 IEEE Vehicle Power and Propulsion Conference (VPPC), pp.1-6, 1-3 Sept. 2010.

[36] João P. Trovão; Paulo G. Pereirinha; Humberto M. Jorge; "Simulation model and road tests comparative results of a small urban electric vehicle," 35th Annual Conference of IEEE Industrial Electronics, 2009. IECON '09, pp.836-841, 3-5 Nov. 2009.

[37] Paulo G. Pereirinha, João P. Trovão, L. Marques, M. Silva, J. Silvestre, F. Santos, "Advances in the Electric Vehicle Project-VEIL Used as a Modular Platform for Research

and Education", EVS24 International Battery, Hybrid and Fuel Cell Electric Vehicle Symposium, Stavanger, Norway, 13-16 May 2009.

[38] João P. Trovão, Paulo G. Pereirinha, Fernando J. T. E. Ferreira, "Comparative Study of Different Electric Machines in the Powertrain of a Small Electric Vehicle", 18th International Conference on Electrical Machines, ICEM'08, Vilamoura, Portugal, 6-9 September 2008.

[39] Marco Silva, João P. Trovão, Paulo Pereirinha, Luís Marques, "Multiple energy sources monitoring system for electric vehicle", 19th International Symposium on Power Electronics, Electrical Drives, Automation and Motion, SPEEDAM 2008, Ischia, Italy, 11-13 June 2008.

[40] A. L. Allègre, R. Trigui, A. Bouscayrol, Different energy management strategies of Hybrid Energy Storage System (HESS) using batteries and supercapacitors for vehicular applications, 6th IEEE Vehicle Power and Propulsion Conference, VPPC 2010, September 1-3, 2010, Lille, France.

[41] S. Caux, D. Wanderley-Honda, D. Hissel, M. Fadel; On-line energy management for HEV based on particle swarm optimization, 6th IEEE Vehicle Power and Propulsion Conference, VPPC 2010, September 1-3, 2010, Lille, France.

[42] C. Bordons, M. A. Ridao, A. Pérez, A. Arce, D. Marcos; Model predictive control for power management in hybrid fuel cell vehicles, 6th IEEE Vehicle Power and Propulsion Conference, VPPC 2010, September 1-3, 2010, Lille, France.

[43] C. Romaus, K. Gathmann, J. Böcker; Optimal energy management for a hybrid energy storage system for EVs based on stochastic dynamic programming, 6th IEEE Vehicle Power and Propulsion Conference, VPPC 2010, September 1-3, 2010, Lille, France.

[44] Scott J. Moura, Duncan S. Callaway, Hosam K. Fathy, Jeffrey L. Stein, Tradeoffs between battery energy capacity and stochastic optimal power management in plug-in hybrid electric vehicles, Journal of Power Sources, Volume 195, Issue 9, 1 May 2010, Pages 2979-2988.

[45] Yi L. Murphey, ZhiHang Chen, Leonidas Kiliaris, M. Abul Masrur, Intelligent power management in a vehicular system with multiple power sources, Journal of Power Sources, Volume 196, Issue 2, 15 January 2011, Pages 835-846.

[46] R. de Castro, R.; João P. Trovão; P. Pacheco; P. Melo; Paulo G. Pereirinha; R. E. Araujo; "DC link control for multiple energy sources in electric vehicles," 2011 IEEE Vehicle Power and Propulsion Conference (VPPC), pp.1-6, 6-9 Sept. 2011.

[47] Mário A. Silva, João P. Trovão, Paulo G. Pereirinha; "Implementation of a multiple input DC-DC converter for Electric Vehicle power system," Proceedings of the 2011 3rd International Youth Conference on Energetics (IYCE), pp.1-8, 7-9 July 2011.

Chapter 9

Investigation and Analysis of the Mechanical Behaviors of the Electric Vehicles

Liang Zheng

Additional information is available at the end of the chapter

http://dx.doi.org/10.5772/53275

1. Introduction

An electric vehicle (EV), also referred to as an electric drive vehicle, uses one or more electric motors or traction motors for propulsion. Electric vehicles are being widely developed due to their better performance than the traditional fuel vehicles in terms of environmental pollution and energy consumption [1-3]. However, comparing with the traditional petroleum vehicles, the relatively short driving distance of the electric vehicle is the main hindrance to prevail. To improve the performance of the electric vehicle and excel the internal combustion engine vehicles, research and study need to be done. In this chapter, analyses and applications are developed to investigate the mechanical behaviors of the electric vehicles such that the performance of the electric vehicles can be considerably improved.

2. Investigation of the regenerative braking force in electric vehicles

In order to overcome the weakness of the short driving distance, the energy efficiency of the electric vehicle has to be substantially improved [4]. The regenerative braking technology is able to convert the excessive kinetic energy of the vehicle into another form, which can be either used immediately or stored until needed [5]. In this section, the regenerative braking force is analyzed and an optimized control strategy for the regenerative braking process was proposed based on the results of the analysis. With the implementation of this optimized control strategy, the energy efficiency of the electric vehicle can be appreciably improved.

2.1. Analysis of the regenerative braking force

The amount of recovered energy during the braking process depends on the magnitude of the regenerative braking force. In order to achieve the maximum amount of recovered energy, the forces in the front and rear tires of the vehicle during the braking process should be analyzed and distributed at the optimal level. Figure 1 represents a schematic of the vehicle during the braking process in which the deceleration of the vehicle is denoted as j, the gravitational acceleration is denoted by g and the braking forces in the front and the rear wheels are denoted as F_{xb1} and F_{xb2}, respectively.

Force Equilibrium yields:

$$F_{z1} = mg(\frac{b}{L} + \frac{j}{g}\frac{h_g}{L}) \text{ and } F_{z2} = mg(\frac{a}{L} - \frac{j}{g}\frac{h_g}{L}) \tag{1}$$

The braking rate z and the adhesion coefficient Φ_i are defined as:

$$z = \frac{j}{g} \text{ and } \Phi_i = \frac{F_{xbi}}{F_{zi}} \tag{2}$$

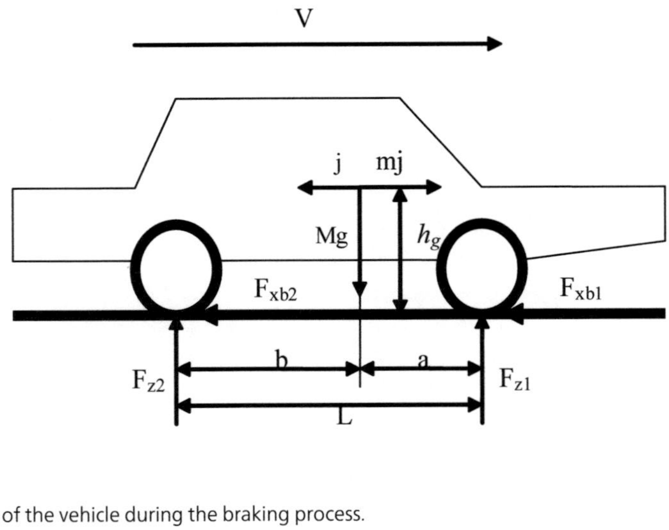

Figure 1. Schematic of the vehicle during the braking process.

In general, there are three different scenarios due to the various distribution of the braking forces in the front and rear wheels: i) both the front and rear wheels are locked at the same time; ii) the front wheels are locked whereas the rear wheels are unlocked; and iii) the rear wheels are locked whereas the front wheels are unlocked.

http://dx.doi.org/10.5772/53275

The first scenario is the ideal case that has the best braking effect. The relationship between the forces in the front and rear wheels can be derived and expressed as:

$$F_{xb2} = \frac{1}{2}\left[\frac{mg}{h_g}\sqrt{b^2 + \frac{4h_g L}{mg}F_{xb1}} - (\frac{mgb}{h_g} + 2F_{xb1})\right] \tag{3}$$

The second scenario is a dangerous case in which the vehicle loses the steering capability but it won't be rolled over. The relationship between the forces in the front and rear wheels in this scenario can be shown as:

$$F_{xb2} = F_{xb1}(\frac{L - \Phi h_g}{\Phi h_g}) - \frac{mgb}{h_g} \tag{4}$$

The third scenario is an extremely dangerous case in which a rollover may be occurred. The relationship between the forces in the front and rear wheels in this case can be derived and written as:

$$F_{xb2} = \frac{\Phi}{L + \Phi h_g}(mga - h_g F_{xb1}) \tag{5}$$

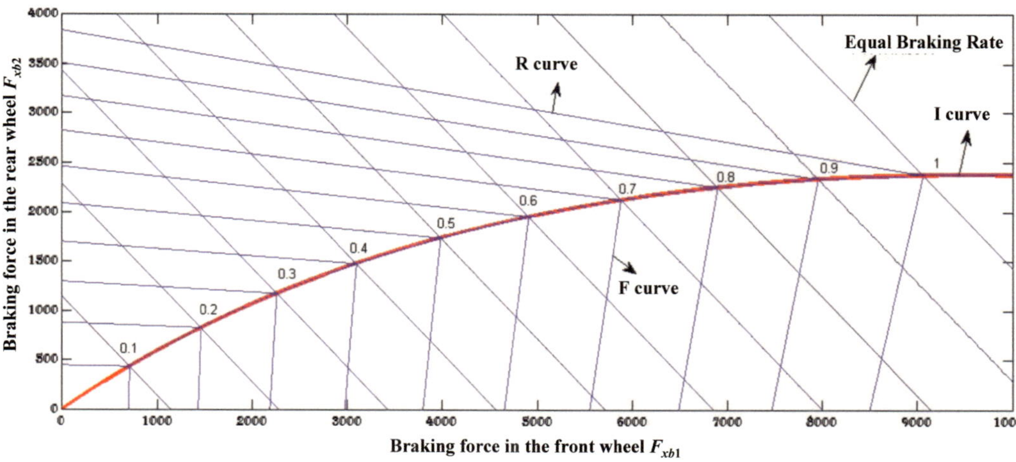

Figure 2. Plots represent the relationships between the braking forces in the front and rear wheels.

Based on Eqs. (3)~(5), the relationships between forces in the front and rear wheels of the electric vehicle in these three different scenarios can be plotted and shown in Fig.2. I curve, as illustrated with red dark color, represents the first scenario (i.e., the ideal case); the second scenario is represented by a series of F curves; and R curves represent the third scenario which is the extremely dangerous case.

The above analysis reveals that the distribution of the braking forces between the front and rear wheels plays the crucial role on the braking performance of the vehicle. In order to ensure the braking safety of the vehicle, United Nations Economic Commission for Europe (UNECE) established a regulation (ECE R13) which strictly regulates the distribution range of the braking forces between the front and rear wheels, as shown in Fig. 3 [6]. The horizontal axis *z* represents the braking ratio, and the vertical axis *k* is the adhesion coefficient between the tire and the road in the figure.

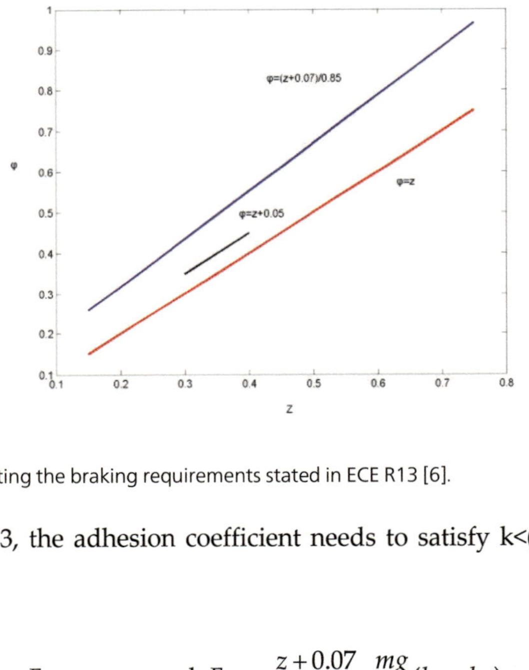

Figure 3. Diagram representing the braking requirements stated in ECE R13 [6].

According to ECE R13, the adhesion coefficient needs to satisfy k<(z+0.07)/0.85, which results in:

$$F_{xb1} + F_{xb2} = mgz \text{ and } F_{xb1} = \frac{z+0.07}{0.85} \cdot \frac{mg}{L}(b + zh_g) \tag{6}$$

The braking ratio *z* can be vanished and it yields:

$$\frac{(F_{xb1} + F_{xb2})^2}{mgL} + \frac{F_{xb1} + F_{xb2}}{L}(b + 0.07h_g) + \frac{0.07mgb}{L} - 0.085F_{xb1} = 0 \tag{7}$$

In addition, the adhesion coefficient of the rear wheel has to be less than that of the front wheel to avoid the occurrence of the extremely dangerous scenario. This means:

$$k_2 = \frac{F_{xb2}}{F_{z2}} \leq k_1 = \frac{F_{xb1}}{F_{z1}} \quad \Rightarrow \quad \frac{F_{xb1}}{F_{xb2}} \geq \frac{b + h_g z}{a - h_g z} \text{ and } F_{xb1} + F_{xb2} = mgz \tag{8}$$

The braking ratio z can be cancelled and it becomes:

$$F_{xb2} = \frac{1}{2}\left[\frac{mg}{h_g}\sqrt{b^2 + \frac{4h_g L}{mg}F_{xb1}} - (\frac{mgb}{h_g} + 2F_{xb1})\right] \tag{9}$$

which evidently matches with the ideal distribution I curve.

Using Eqs. (7) and (9), the distribution range of the braking forces in front and rear wheels can be plotted and shown in Fig. 4. It can be seen from Fig. 4 that I curve represents Eq. (9) and M curve represents Eq. (9). The acceptable braking forces in the front and rear wheels should be distributed between the range enclosed by I curve and M curve, according to the technical requirements stated in ECE R13.

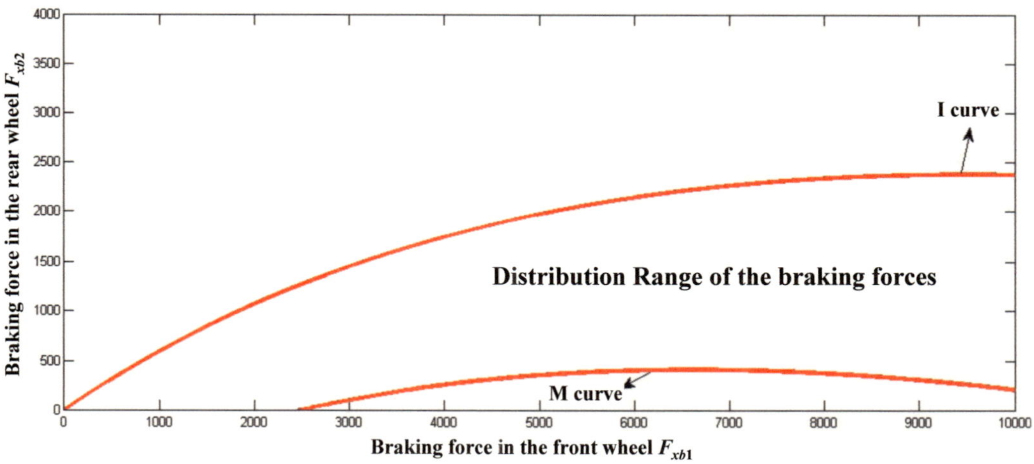

Figure 4. The distribution range of the braking forces in front and rear wheels, satisfying ECE R13.

2.2. An optimized control strategy of the regenerative braking

Based on the analysis developed in the previous section, an optimized control strategy of the regenerative braking process is proposed in this section. This proposed control strategy aims at the following three goals: 1) safety of the vehicle braking; 2) maximization of the recovered energy; 3) a simple control system with a low cost of manufacturing. A front-wheel drive pure electric vehicle is utilized in this research, with the parameters listed in Table 1.

Pure Electric Vehicle	Mass	1144 kg
	Drag coefficient	0.335
	Height of center of mass	0.5m
	Frontal area	2.0 m^2
	Wheelbase	2.6 m
	Coefficient of rolling resistance	0.009
	Distance between center of mass and front axle	1.04 m
	Radius of the wheel	0.282 m
AC Induction Motor	Nominal power	75 KW
	Nominal rotating speed	2640 r/min
	Maximum rotating speed	10000 r/min
	Maximum torque	271 N·M
	Overload factor	1.8
Gear	Gear reduction ratio	6.6732
	Main gear reduction ratio	1
Power	Lead acid battery pack	

Table 1. Parameters of the pure electric vehicle utilized in this research.

The energy due to the regenerative braking can be recovered when the braking rate is between 0 and 0.6. No energy will be recovered when the braking rate is higher than 0.6. The optimal distribution of the braking forces between the front and rear wheels is presented as the OGFC-I curve shown in Fig. 5.

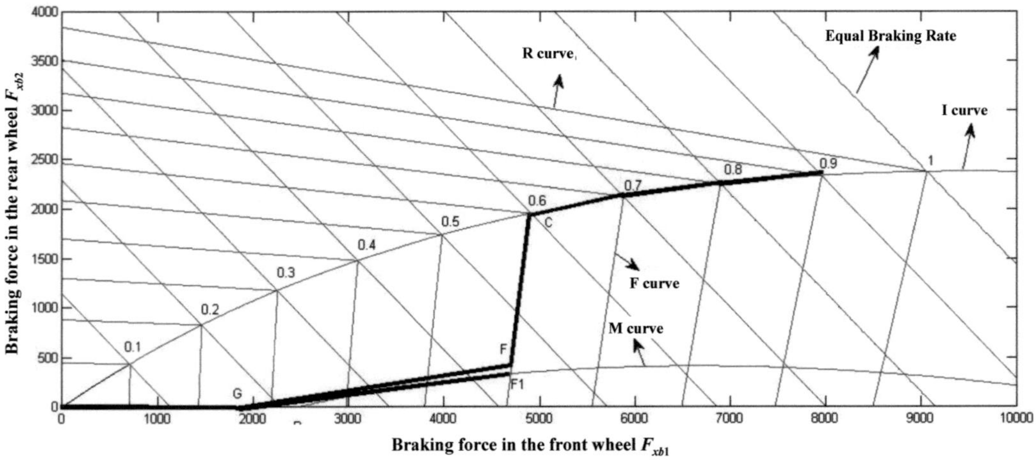

Figure 5. Optimized distribution curve of braking forces in front and rear wheels.

1. $0 < z \leq 0.16$

The braking forces in the front and rear wheels are distributed along the line segment OG.

2. $0.16 < z \leq 0.455$

The braking forces in the front and rear wheels are distributed along the line segment GF. The line GF_1 is tangent to M curve and the equation for GF_1 can be expressed as:

$$F_{xb2} = 0.12F_{xb1} - 219.6 \tag{10}$$

In order to keep distant from the M curve that the rear wheel may be locked before the front wheel which can result in a rollover, one can chose the line segment GF instead of GF_1. The line GF can be presented by:

$$F_{xb2} = 0.15F_{xb1} - 219.6 \tag{11}$$

3. $0.455 < Z \leq 0.6$

The braking forces in the front and rear wheels are distributed along the line segment FC. With the increase of the braking rate, it quickly approaches to I curve. The line segment FC can be described as:

$$F_{xb2} = \frac{23}{3}F_{xb1} - 35692.8 \tag{12}$$

4. $Z > 0.6$

It is the case of emergency braking, the braking forces in the front and rear wheels are distributed along the ideal case (as described in I curve).

2.3. Modeling and simulations

Implementing the optimized control strategy in Section III, the controlling module of the braking force was established and plugged into the commercial software ADVISOR. The distribution modules of the optimized braking forces in the front and rear wheels are shown in Fig. 6.

In order to validate this optimized controlling strategy, three drive cycles were used, i.e., CYC_ARTERIAL, CYC_LA92, and CYC_NYCC. The parameters of these three drive cycles are listed in Table 2.

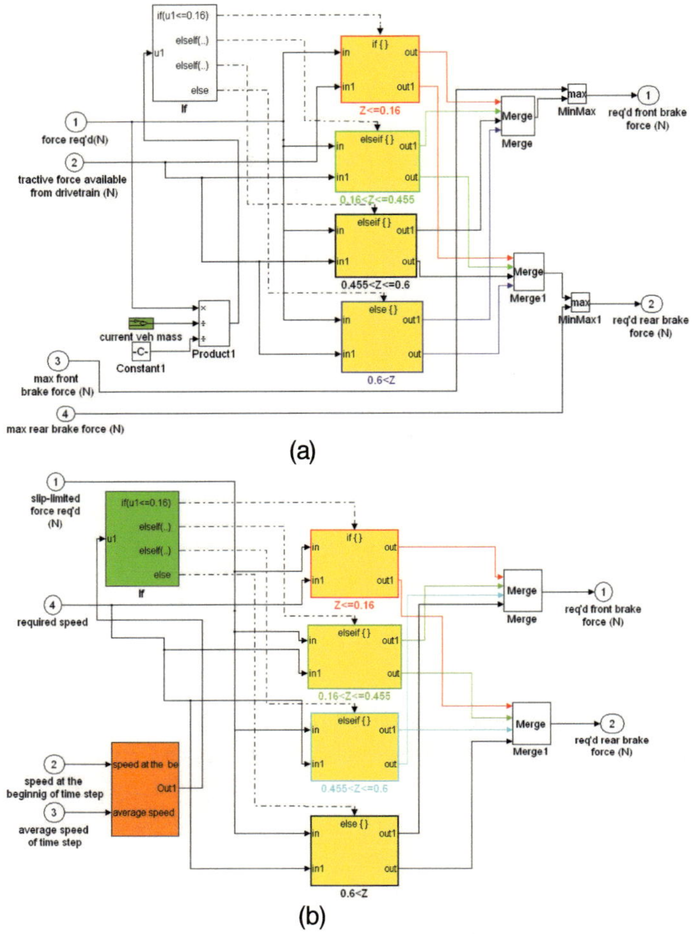

Figure 6. Distribution Modules of the optimized braking forces in the (a) front and (b) rear wheels.

The pure electric vehicle was simulated in these three drive cycles in ADVISOR, using the optimized control strategy. Then the obtained results were compared with the actual (original) results. Figure 7 represents the comparisons of the two sets of results in CYC_ARTERIAL and CYC_NYCC drive cycles. It can be seen from Fig. 7 that these two sets of results match very well, meaning that the optimized control strategy did not affect the safety of the vehicle.

The ADVISOR software has a default control strategy which is easy to operate but can not recycle much energy due to the regenerative braking. More importantly, the default control strategy is unable to guarantee a quick and safe stop for the high-speed running vehicle due to the limited amount of regenerative braking force generated during the braking process. Here, the results using the optimized control strategy are compared to the results using the default control strategy provided by ADVISOR.

Parameters	CYC_ARTERIAL	CYC_LA92	CYC_NYCC
Time (s)	291	1435	598
Max. speed (km/h)	64.3	108.1	44.58
Avg. speed (km/h)	39.7	39.6	21.4
Max. acceleration (m/s^2)	1.07	3.08	2.68
Max. deceleration (m/s^2)	-2.01	-3.93	-2.64
Avg. acceleration (m/s^2)	0.6	0.67	0.62
Avg. deceleration (m/s^2)	-1.8	-0.75	-0.61
Stops	4	16	18
Driving distance (km)	3.2	15.8	1.9

Table 2. Parameters of three drive cycles

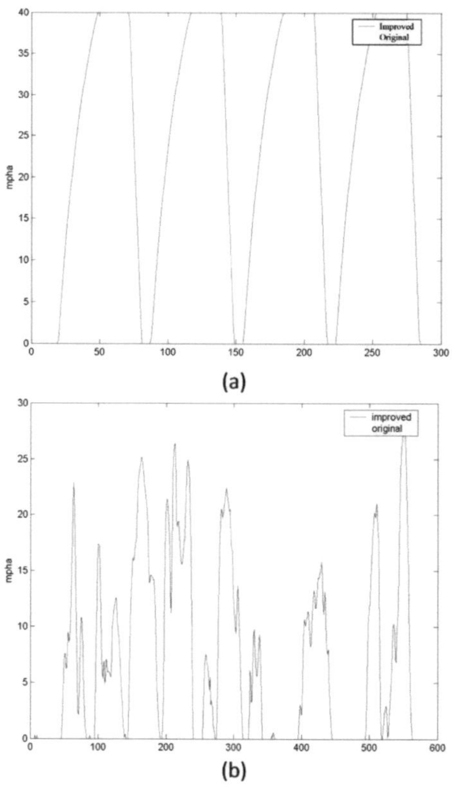

Figure 7. Comparisons of the vehicle speed in (a) CYC_ARTERIAL and (b) CYC_NYCC cycles.

Firstly, the SOC of the battery is selected to be compared by using these two different control strategies. Figure 8 illustrates the comparisons of SOC in CYC_LA92 and CYC_NYCC drive cycles, using the default and optimized control strategies. It can be seen that the SOC curve using the optimized control strategy has a flatter slope than the SOC curve using the default one. It means more electric energy is reserved in the battery and consequently more kinetic energy can be recovered during the regenerative braking process.

Secondly, the recovered energy during the braking process will be compared using these two different control strategies. As listed in Tables 3 to 5, the parameters for comparison include the consumed energy, the braking energy, the total recovered energy, the energy recovery efficiency, the effective energy recovery efficiency, and the energy efficiency of the entire vehicle. The optimized control strategy and the default control strategy in ADVISOR are employed in the three drive cycles.

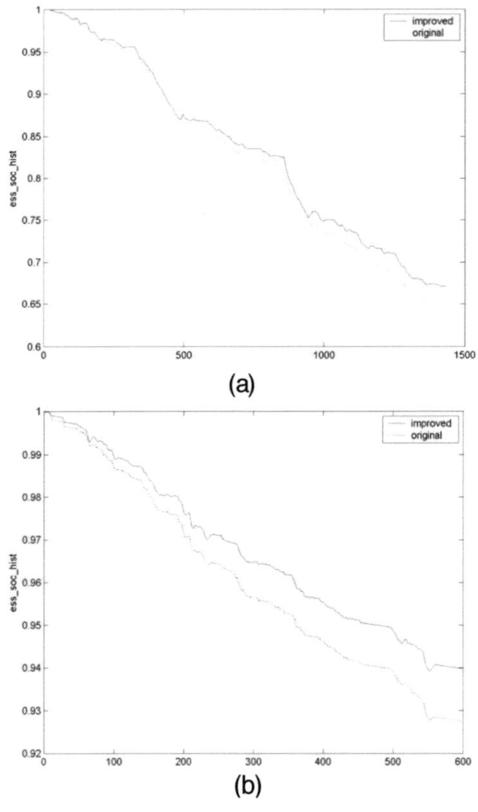

Figure 8. Comparisons of SOC in (a) CYC_LA92 and (b) CYC_NYCC cycles, using two different control strategies.

Parameters	Optimized Control Strategy	Default Control Strategy
Total Energy Consumption (KJ)	2198	2172
Braking Energy (KJ)	671	671
Recovered Energy (KJ)	409	299
Energy Recovery Efficiency	60.95%	44.56%
Effective Energy Recovery Efficiency	18.6%	13.77%
Energy Efficiency of the entire vehicle	36.5%	34.8%

Table 3. Comparison of the recovered energy in the CYC_ARTERIAL cycle

Parameters	Optimized Control Strategy	Default Control Strategy
Total Energy Consumption (KJ)	11168	11209
Braking Energy (KJ)	2891	2891
Recovered Energy (KJ)	1740	1066
Energy Recovery Efficiency	60.2%	36.87%
Effective Energy Recovery Efficiency	15.6%	9.5%
Energy Efficiency of the entire vehicle	42.6%	39.6%

Table 4. Comparison of the recovered energy in the CYC_LA92 drive cycle

Parameters	Optimized Control Strategy	Default Control Strategy
Total Energy Consumption (KJ)	2040	2212
Braking Energy (KJ)	565	565
Recovered Energy (KJ)	277	116
Energy Recovery Efficiency	49.02%	20.53%
Effective Energy Recovery Efficiency	13.6%	5.2%
Energy Efficiency of the entire vehicle	13.5%	11.3%

Table 5. Comparison of the recovered energy in the CYC_NYCC drive cycle

It can be seen from Tables 3 to 5 that with the use of the optimized control strategy, the parameters of all categories have been improved considerably, comparing with the acquired results using the default control strategy. With the same amount of braking energy, the energy recovered from the regenerative braking process has been increased dramatically via the use of the optimized control strategy, for all three drive cycles. For instance, in the CYC_NYCC drive cycle, the amount of recovered energy using the optimized control strategy (i.e., 277 KJ) is more than doubled, comparing to the number (116 KJ) utilizing the default control strategy provided by ADVISOR. In addition, the energy efficiency of the entire vehicle has also been improved for all three drive cycles, via the use of the optimized control strategy. The largest improvement occurs in the CYC_NYCC drive cycle with the amount of almost 20% (13.5% vs. 11.3%), with less total energy consumption (2040 KJ vs. 2212 KJ). The optimized control strategy also prevails over the default control strategy in other categories, as shown in Tables 3 to 5.

2.4. Summary on the analysis of the regenerative braking Force

In this section, the regenerative braking force was analyzed and the distribution range of the braking forces in the front and rear wheels was determined to meet the requirements in UN-ECE R13. An optimized control strategy of the regenerative braking process was proposed based on the results of the analysis. The regenerative braking process with the optimized control strategy is simulated in the commercial software ADVISOR. Three different drive cycles with real data and the default control strategy in ADVISOR were employed to conduct the comparison. Results of the comparison reveal that the optimized control strategy prevails over the default control strategy in every category of energy recovery. Comparing to the default control strategy, this optimized control strategy can improve the energy recovery efficiency by more than 100% for one drive cycle. This optimized control strategy can also be used for the hybrid vehicles.

3. Simulation and analysis of Series Hybrid Electric Vehicle (SHEV)

Series-hybrid vehicles are driven by the electric motor with no mechanical connection to the engine. Unlike piston internal combustion engines, electric motors are efficient with exceptionally high power-to-weight ratios providing adequate torque over a wide speed range. In a series-hybrid system, the combustion engine drives an electric generator instead of directly driving the wheels. The generator provides power for the driving electric motors [7]. However, the efficiency of the engine and the working condition of the battery have to be considerably improved to make the series hybrid electric vehicles more competitive in the industry. In order to improve the efficiency of the engine and make the battery work in the better condition, analyses and simulations need to be developed and an optimized control strategy has to be obtained and utilized.

3.1. Control strategy for SHEVs

3.1.1. Operation mode for SHEVs

A typical power control flow for SHEVs can be described by the following four kinds of operation modes, as shown in Fig. 9 [8].

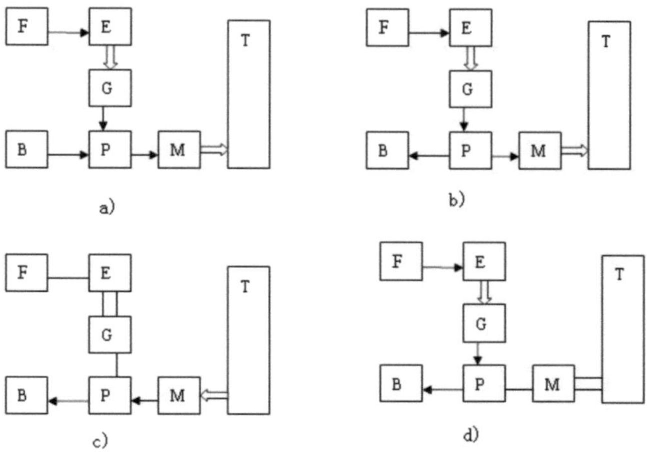

Figure 9. A typical power control flow for SHEVs.

When a SHEV starts to run, the engine outputs the electrical energy via the generator and the battery, transmits the energy to the converter, which drives the motors, ultimately drive wheels through the mechanical transmission devices; When the vehicle is underloaded, the power outputted by the engine is more than the vehicle needs, then the excessive energy is utilized to recharge the battery via the generator, until the SOC reaches the predetermined limit; When the vehicle is braking or decelerating, the motor converts the kinetic energy into the electrical energy, recharging the battery through the power converter; When the vehicle stops, the engine can recharge the battery through generators and power converters.

3.1.2. Control strategy for SHEVs

During the SHEV automobile working process, it often happens that energies from two or more sources superimpose. The main purpose of the correct matching of SHEV powertrain is to rationally determine the operating characteristics of the engine and its power distribution and energy balance with the energy storage device. By controlling the working status of various energy sources, the energy conversion efficiency can be improved, the loss due to energy transmission can be reduced, and ultimately the energy can be utilized in a maximum way for SHEVs.

The general control strategy for SHEVs is developed normally based on the parameters such as battery SOC, the driver's accelerator pedal position, the wheel speed and the average power of the driving wheel; then the engine and electric motor generate the corresponding torque to meet the requirements of the driving torque for the driving wheel. Commonly used control strategies include "thermostat" and "power follower" [9], where the power follower strategy are described as follows:

1. When the battery SOC is greater than the upper limit of SOC_{max}, the engine stops working, but when the power demand of vehicle is too large, the engine needs to be restarted;

2. When the battery SOC is less than the lower limit SOC_{min}, the engine needs to get to work;

3. When the engine works, its power output should not only follow the changes of the power demand of the vehicle, but keep the battery SOC around the middle value of its working range;

4. While he engine is working, its output power should not be too small or too large in order to ensure high efficiency.

In this research, parameters are controlled to determine the energy flow of SHEVs, so that the control strategy of "power follower + thermostat" is designed. The general logic control flow is shown in Fig. 10.

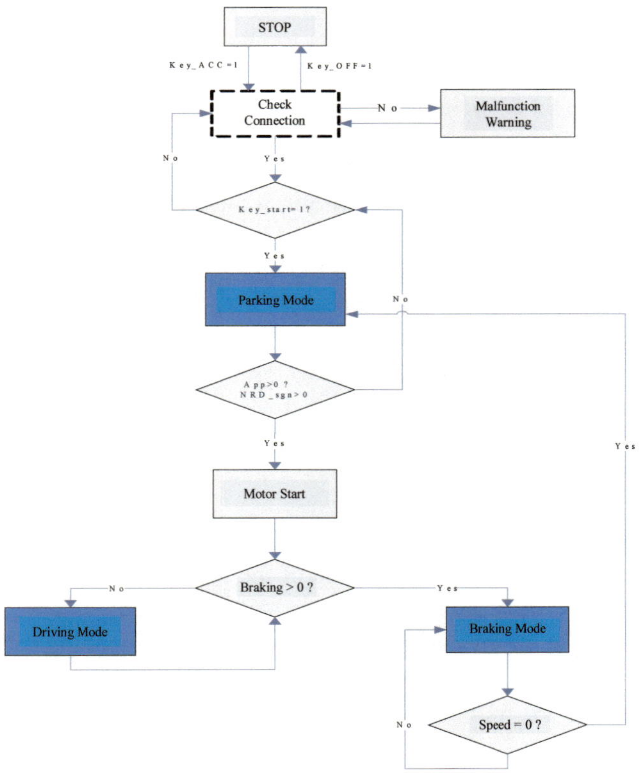

Figure 10. General logic control.

http://dx.doi.org/10.5772/53275

3.1.3. SIMULINK/stateflow control process

Based on the above analysis, one can establish the vehicle control systems in the SIMULINK/Stateflow based on the energy flow of vehicles. Modeling in the SIMULINK/Stateflow, one can not only avoid the complex programming in MATLAB, but omit the direct construction of control frame in the SIMULINK. Furthermore, the status flow in Stateflow can explicitly represent the control process with figures, so that the control can be precisely achieved. Based on the above analysis and the energy flow analysis and control chart, the control flow designed in the SIMULINK/Stateflow is shown in Fig. 11.

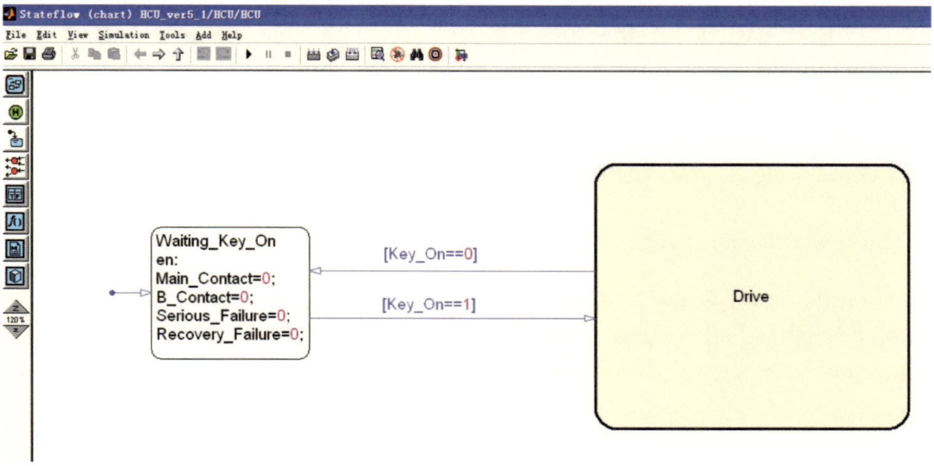

Figure 11. Stateflow control process.

The kernel part of this design is the driving component, which is shown Fig. 12.

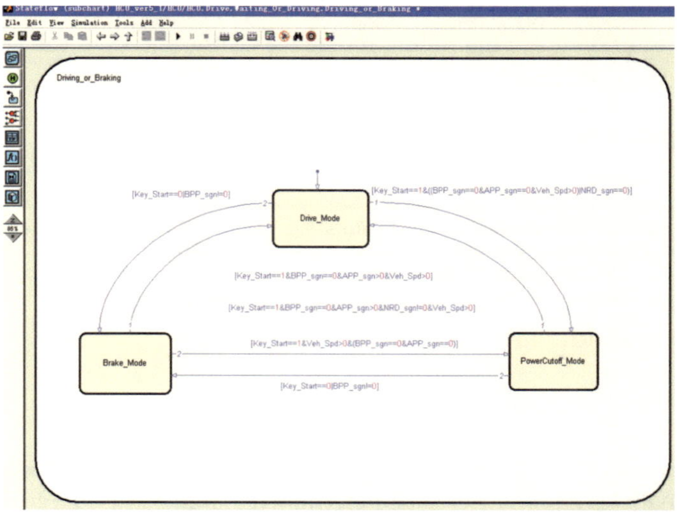

Figure 12. Status Flow of the Driving Component.

3.2. Simulations of SHEVs

ADVISOR employs a unique hybrid simulation method, which mixes the backward simulation and the forward simulation, with the backward simulation as the primary one and forward simulation as the auxiliary one [10]. It first carries on the backward simulation along with the opposite direction to the actual power flow, making the request of the needed speed and torque to the vehicle model according to requirement of the path circulation. Then the vehicle model delivers the request to the wheel and the axle module, to the main gearbox module, to the transmission gearbox module and so on, until the request is delivered to the power supply module, then the power provided by the power supply can be calculated. Next, it carries on the forward simulation; a stepwise computation will be developed along the actual direction of the power flow, until the automobile's real speed is determined.

3.2.1. Power matching of various parts of vehicles

3.2.1.1. Power selection for engines

SHEVs have the similar requirements to ordinary vehicles on dynamic performance, and continuous driving distance has to be independent of battery capacity, thus the engine's maximum output power needs to satisfy [11]:

$$P_{e\max} = \frac{1}{3600\eta_{gc}\eta_{mc}}(mgf + \frac{C_D A v_{a\max}^2}{21.55})v_{a\max} \tag{13}$$

where P_{emax} is the maximum output power of the engine; η_{gc} is the power conversion efficiency; η_{mc} is motor efficiency; v_{amax} is average driving speed; m is the vehicle mass; g is the acceleration due to gravity; f is the rolling resistance coefficient; C_D is the air resistance coefficient; A is the windward area.

Table 6 lists the parameters for the SHEV engine.

Maximum Rotating Speed	Normal Rated Power / Rotating Speed	Maximum Torque / Rotating Speed
4200 rpm	96 kw / 4200 rpm	300 N-m / 3000 rpm

Table 6. SHEV Engine Parameters

3.2.1.2. Selection of parameters for motors

Parameters of the motor should be chosen to meet the requirements of a wide range of vehicle speed and load variation. The to-be-determined parameters mainly include the low-speed constant power P_{mr}, rated speed of motor n_{mr}, max. speed of motor $v_{m\max}$ and rated voltage of motor V_{mr}.

Rated speed and maximum speed of motor: the maximum speed has an impact on the size of the transmission system and rated torque of motor, where the rated speed of the motor can be expressed as:

$$n_{mr} = n_{m\max} / \beta \tag{14}$$

where β is the ratio between the maximum speed and rated speed of the motor, also known as the coefficient of expanded constant power zone of the motor. Usually this coefficient is chosen from 4 ~ 6.

Rated power, maximum power of Motor: for SHEVs, the selection of power parameters for the motor must meet the performance requirements. In a road with good condition, the vehicle's power balance equation can be written as follows [12]:

$$T = \frac{1}{3.6}\int_0^v \frac{dt}{dv}dv = \frac{1}{3.6}\int_0^v \frac{\delta m}{F_t - F_w - F_f}dv \tag{15}$$

where T is the time for the speed accelerated from 0 to v; δ is the spinning mass conversion factor; F is the driving force; F_w is the air resistance; and F_f is rolling resistance.

The climbing capacity of a vehicle can be represented by the angle of the slope that the vehicle can climb under a certain speed, which is described as [13]:

$$\alpha = \arcsin\left(\frac{F_t - F_w}{mg\sqrt{1+f^2}}\right) - \arctan(f) \tag{16}$$

where, α is the slope angle; mg is weight of the vehicle, and f is rolling resistance coefficient.

According to the situation with acceleration and the situation with the highest speed, the needed power can be calculated for the running vehicle. Next, according to the maximum calculated value, considering the transmission efficiency, the power for the electric motor can be preliminarily determined. Then this value can be substituted back to the above two equations for verification. Once the maximum power of the motor is determined, the rated power of the electric motor can be calculated according to the equation below:

$$P_{on} = \frac{P_{m\max}}{\lambda} \tag{17}$$

in which, P_{on} is the rated power of the electric motor; $P_{m\max}$ is the maximum power of the electric motor; λ is overload factor of the electric motor.

Table 7 lists the parameters of the motor for SHEVs.

Peak Torque	Peak Power	Rated Power	Rated Torque	Base Rotating Speed
618 N·m	170 KW	60 KW	218N·m	2000 rpm

Table 7. Parameters of the motor for SHEVs

3.2.1.3. Selection of parameters for batteries

For the hybrid vehicles, the larger of battery capacity, the longer the vehicle can drive under the pure electric condition. However, larger capacity means larger weight and volume. Therefore, we should choose the battery based on several key factors, including the technical specifications of the vehicle, the specific energy and specific power of the battery. The selection of the number of batteries should be based on the needs of electric power and the mileage under pure electric condition.

First, from the aspect of the power need, the battery capacity must meet the maximum requirements of the output power of the motor. The number of batteries needed can be determined by the following formula [14]:

$$n = \frac{P_{b\max}}{(2E_b^2 / 9R_b)\eta_{mc}} \tag{18}$$

where n is the number of batteries, $P_{b\max}$ is the power needed from the battery under the worst condition (with the lowest SOC value), E_b is the electromotive force of the battery; R_b is the equivalent resistance of the battery.

As far as the mileage under pure electric condition is concerned, the number of batteries should meet the following criterion as [15]:

$$n' = \frac{P_{bc}(s / V_{bc})}{U_{\text{model}}C} \times 1000 \tag{19}$$

where, n' is the number of batteries; P_{bc} is the power under pure electric condition; s is the driving distance under pure electric condition; V_{bc} is driving speed under pure electric condition; U_{model} is the output voltage of a single battery; C is the capacity of the battery.

Finally, one can compare the values of n' and n, and the larger number should be chosen as the number of batteries.

Table 8 lists the selection results for the battery.

Capacity	Number	Rated Total Energy
93 Ah	40	44.64 kwh

Table 8. Parameters of hybrid batteries.

3.2.2. Simulation of the performance of the entire vehicle

Figure 13 illustrates the simulation of a SHEV system built in ADVISOR. A city bus (route R36) of Nuremberg, Germany is utilized here as the example.

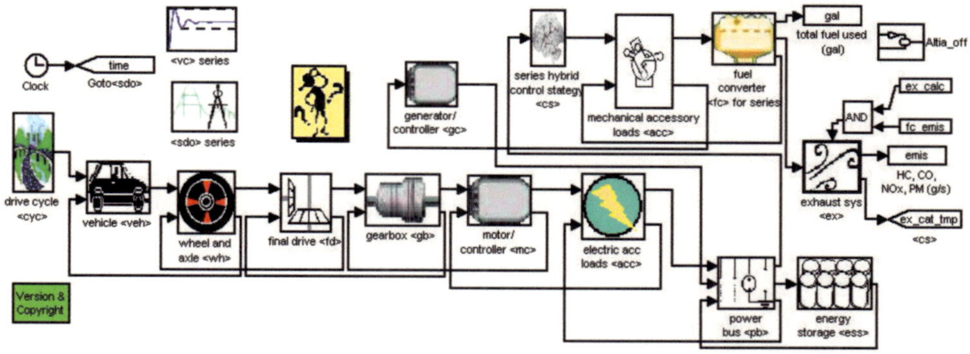

Figure 13. Simulation of a SHEV system in ADVISOR.

The cycle status of this SHEV is shown in Fig. 14. Embedding the control frame illustrated in Fig. 11 into the system shown in Fig. 13, and setting the values for the parameters of various parts that were calculated in previous sections, one can obtain the simulation results as shown in Figures 15~18, with Fig. 15 representing the SOC value of the battery, Fig. 16 illustrating the output speed of the engine, Fig. 17 presenting the speed of the entire vehicle, and Fig. 18 showing the output torque of the engine, respectively

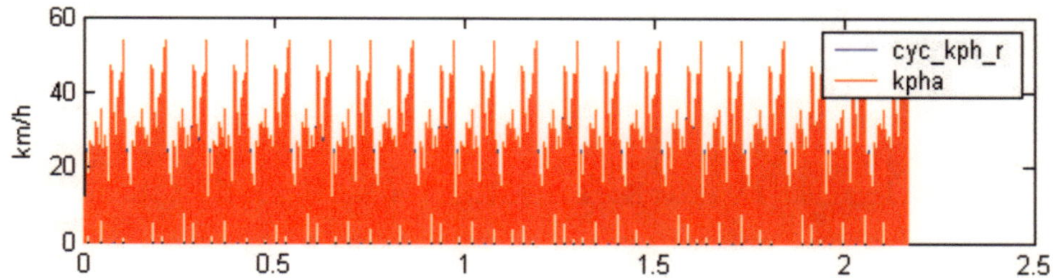

Figure 14. Cycle Status of a R36 bus in Nuremberg.

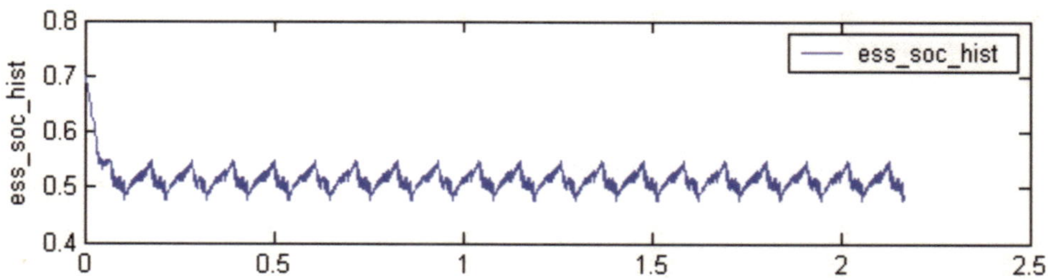

Figure 15. SOC value of the Battery.

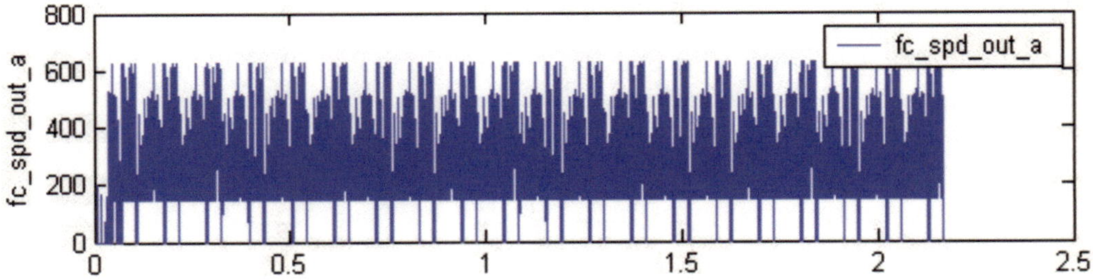

Figure 16. Output Speed of the Engine.

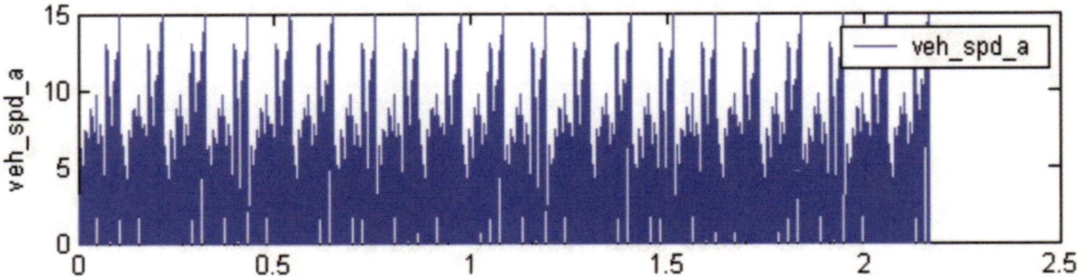

Figure 17. Vehicle Speed.

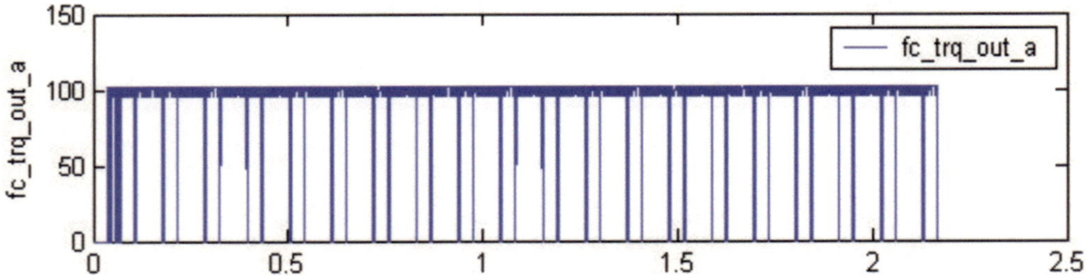

Figure 18. Output Torque of the Engine.

The simulation results indicate that through the control strategy designed in this research, the battery SOC value can be maintained at about 0.5, which completely meet the safety requirement of the battery; the average speed of the vehicle remained at the value around 40km / h, which conforms to the road and traffic conditions in most cities; from the output torque of the engine in Fig. 18 one can see that, under the control strategy designed in this research, the engine works in the best economic zone.

3.3. Summary on analysis of series hybrid electric vehicle

In this section, the simulations and analysis of the series hybrid electric vehicles (SHEVs). According to the technical specifications of SHEVs, a control strategy was designed utilizing the power follower and thermostat, while parameters were selected for various part of the SHEV. Simulation results revealed that this control strategy can not only meet the performance requirements of the vehicle, but make the engine and batteries work in a reasonable range. The study and optimization of the control strategies are the hotspots in the field of hybrid power system. The optimization of control strategies is not only to achieve the best fuel economy, but to adjust to various operating conditions. In particular, since all the countries are implementing the environmental protection to have a greener earth, the minimum vehicle emission becomes a crucial measurement for the superiority of a control strategy for SHEVs.

4. Investigation of the temperature distribution in the Li-ion battery for pure electric vehicles

The lithium-ion batteries have much more excellent performance than the other types of batteries so that they are widely used in the electric vehicles (EVs) [16-18]. However, the Li-ion battery is very sensitive to the temperature variation. One needs to consider not only the temperature range of the batteries, but the temperature difference of each battery in the pack. Low temperature can result in insufficient electrochemical reaction inside the battery, whereas high temperature will affect the working condition so that it shortens the life of service of the battery. Furthermore, the overshoot or overdischarge may be happened in some batteries if the temperature distribution among the battery pack is nonuniform. In order to minimize the heat generated in the battery pack, the temperature distributions in the battery and the battery pack utilized in EVs are investigated and parameters are optimized to lower the temperature and non-uniformity of the temperature distribution.

4.1. FE formulations of the heat conduction equation of the li-ion battery

The heat conduction equation in the Cartesian coordinates is given as [19]:

$$\rho C\frac{dT}{d\tau}=\frac{\partial}{\partial x}(k_x\frac{\partial T}{\partial x})+\frac{\partial}{\partial y}(k_y\frac{\partial T}{\partial y})+\frac{\partial}{\partial z}(k_z\frac{\partial T}{\partial z})+\phi_v \quad (20)$$

where ρ is the density and C is the heat capacity.

The Galerkin method is employed to obtain the finite element forms of the heat conduction equation. For the steady-state heat conduction equation, the finite element form is expressed as [20]:

$$K\phi=P \quad (21)$$

and the finite element form for the transient heat conduction equation is written as [20]:

$$K\phi + C\dot{\phi} = P \tag{22}$$

where:

$$K_{ij} = \sum_{e} \int_{\Omega} \left(k_x \frac{\partial N_i}{\partial x}\frac{\partial N_j}{\partial x} + k_y \frac{\partial N_i}{\partial y}\frac{\partial N_j}{\partial y} + k_z \frac{\partial N_i}{\partial z}\frac{\partial N_j}{\partial z}\right) d\Omega + \sum_{e} \int_{\tau_3} h N_i N_j d\tau \tag{23}$$

$$P_i = \sum_{e} \int_{\tau_2} N_i q d\tau + \sum_{e} \int_{\tau_3} h N_i \varphi_a d\tau + \sum_{e} \int_{\Omega} N_i \rho Q d\Omega \tag{24}$$

In Eqs. (21)-(24), K is the heat conduction matrix, ϕ is the nodal temperature matrix, P is the temperature load matrix, C is the heat capacity matrix, and $\dot{\phi}$ represents the derivative of ϕ with respect to time. In addition, N_i denotes the shape function, and k_x, k_y, k_z represents the thermal conductivity along three axes.

4.2. Finite element simulations of the li-ion battery

4.2.1. Equivalent material properties of the battery

Figure 19 illustrates the rectangular Li-ion battery employed in this research. It has the height of 280 mm, the depth of 71 mm, and the width of 182 mm. The internal structure of the Li-ion battery mainly includes three parts: the shell, the contact layer and the central area. The central area is the heating zone of the battery since most reactions are occurred in the central area.

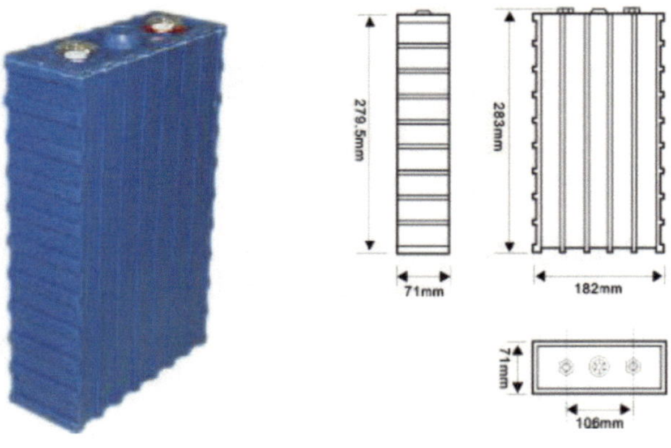

Figure 19. Internal Structure of the rectangular Li-ion battery.

The central area is the most complex area of the battery since it is stacked with hundreds of layers made of different materials. Therefore it is essentially infeasible to generate and implement finite element models of the entire battery incorporating the actual structures. Consequently, equivalent modeling techniques must be developed to investigate the temperature distribution of the battery while circumventing the time and intense computational requirements. In other words, equivalent material properties need to be calculated and utilized in the finite element models. Eqs. (25)-(28) represent the calculations of the equivalent material properties of the battery:

$$\text{Density:}\quad \rho = \frac{m}{V} = \sum_{i=1}^{n} \frac{m_1 + m_2 + m_3 + \cdots + m_n}{V_1 + V_2 + V_3 + \cdots + V_n} \tag{25}$$

$$\text{Specific heat:}\quad C = \sum_{i=1}^{n} \frac{C_1 V_1 + C_2 V_2 + C_3 V_3 + \ldots + C_n V_n}{V_1 + V_2 + V_3 + \ldots + V_n} \tag{26}$$

Thermal Conductivity:

$$\text{Serial Connection:}\quad K = \frac{\sum_{i=1}^{n} l_i}{\sum_{i=1}^{n} \frac{l_i}{k_i}} \tag{27}$$

$$\text{Parallel Connection:}\quad K = \frac{\sum_{i=1}^{n} A_i k_i}{\sum_{i=1}^{n} A_i} \tag{28}$$

where l_i denotes the length of each battery along the heat flux, whereas A_i denotes the cross-sectional area of each battery along the heat flux. Both serial and parallel connection cases have been considered and the thermal conductivity of both cases were calculated (as shown in Eq. (27)).

The material properties of the lithium-ion battery used in this research are listed in Table 9, including thickness, thermal conductivity, specific heat, density, etc.

Based on the values of the parameters provided in Table 9, the equivalent material properties of the central area in the battery can be determined:

Equivalent Density:ρ=2123.83 kg/m^3

Equivalent specific heat:C= 1165.91 J/(kg×K)

Thermal conductivity:k_x = 1.022 W/(m×K) k_y= 25.746 W/(m×K) k_z= 26.364 W/(m×K)

	Material	Thickness	Thermal conductivity	Specific heat	Density
Anode	$LiCoO_2$	0.014	1.58	1269.21	2328.5
Anode current collector	AL	0.002	238	903	2702
Negative electrode	Graphite	0.0116	1.04	1437.4	1347.33
Negative collector	Cu	0.0014	398	385	8933
Diaphragm	Polyethylene	0.0035	0.3344	1978.16	1008.98
Contact layer	Electrolyte	0.05	0.6	2055.1	1129.95
Shell	AL	0.07	238	903	2702

Table 9. Material properties of the Lithium-ion battery.

4.2.2. Discussion of boundary conditions

The Li-ion battery used in this research has a rectangular shape, which means it is under the natural convection if there is no dissipation source. Under the natural convection, the heat transfer is mainly generated by the buoyancy of the air, so different boundary surfaces may have different convection coefficients. The convection coefficient for each boundary surface needs to be calculated separately. In this study,the constant temperature of the wall is set to be 30 ºC, the thermal conductivity is set as k = 2.63×10^{-2} W/(m k), the kinematic viscosity is ν = 15.53×10^{-6} m^2/s, and the Prandtl number is Pr = 0.702.

For the top surface of the battery, the heat can be easily dissipated so that the heat convection is strong. The calculations of the convection coefficient can be shown as follows:

Grashof number: $Gr=\frac{g\alpha\Delta t l^3}{v^2}=2.26\times10^4$

⇒ Raleigh number: $Ra=Gr\times\Pr=1.5865\times10^4$

⇒ Nusselt number: $Nu=0.54Ra^{\frac{1}{4}}=6.06$

⇒ Convection coefficient: $h=Nu\frac{k}{l}=6.25W\Big/(m^2\cdot K)$

For the bottom surface of the battery, the heat convection is weak.

$$Nu=0.27Ra^{\frac{1}{4}}=3.03\Rightarrow h=Nu\frac{k}{l}=3.125W\Big/(m^2\cdot K)$$

For the vertical surfaces, they should have the same convection coefficients:

$$Gr=2.9772296\times10^7 \quad Nu=39.89 \quad h=Nu\frac{k}{l}=3.75W\Big/(m^2\cdot K)$$

http://dx.doi.org/10.5772/53275

4.2.3. Finite element simulations

There are two types of most commonly used Li-ion batteries, one with a cylindrical shape and the other with a rectangular shape. Figure 20 illustrates the temperature distributions in these two different types of Li-ion batteries.

It can be seen that under the same condition (1C electric charge and at the temperature of 293K), the maximum temperature in the cylindrical battery is 326.12K and the largest temperature difference in the battery is 8.57K; while the maximum temperature in the rectangular battery is 320.58K and the largest temperature difference in the battery is 6.33K. It indicates that the rectangular battery is better than the cylindrical one in terms of heat dissipation.

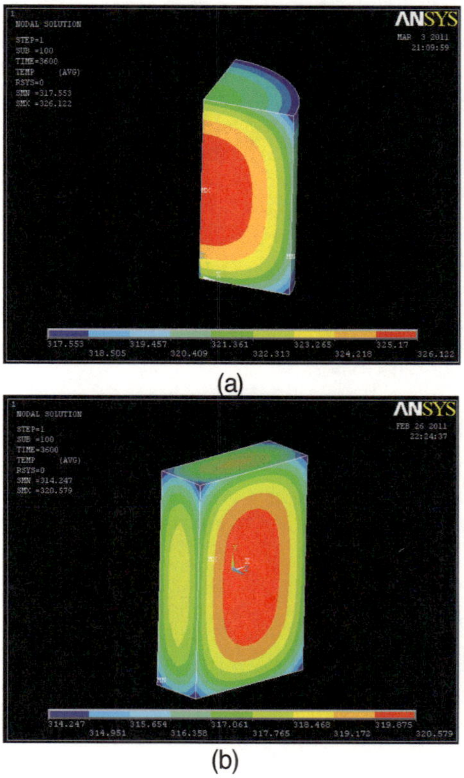

Figure 20. Temperature Distribution of (a) cylindrical battery and (b) rectangular battery at 1C293K.

The effect of the discharge rate of the battery on the temperature distribution of the battery pack was also investigated. Figure 21 illustrates the temperature distribution in the battery with different discharge rates. The comparison between Fig. 21(a) and Fig. 21(b) reveals that the temperature of the battery with a faster discharge rate is higher than the temperature of the battery with a slower discharge rate. A slower discharge rate should be chosen to lower the temperature and decrease the temperature difference in the battery. Due to the demand of high power, EVs usually employ battery packs in which each pack consists of tens of batteries. Figure 22 shows a typical Li-ion battery pack used in the EVs [21].

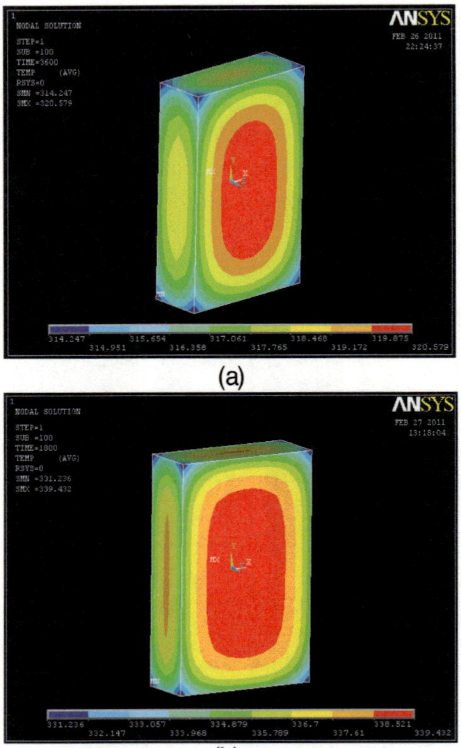

(b)

Figure 21. Temperature distribution in the battery at (a) 1C discharge and (b) 2C discharge.

Figure 22. A typical Li-ion battery pack utilized in the EVs [21].

The cooling strategy of the battery pack is investigated by considering two different strategies. In the first strategy, the cooling gas (air) is blowing along the longitudinal direction of the battery pack, it is called the serial cooling; in the second strategy, the cooling direction is perpendicular to the longitudinal direction of the pack, it is called the parallel cooling.

Figure 23 illustrates the temperature distributions in the battery pack using the serial cooling strategy with two wind speeds (4 m/s and 10 m/s). The choice of these two values is based on the critical wind speed (4.2 m/s), calculated based on the critical Reynolds number of 500000 to differentiate the laminar flow and the turbulent flow.

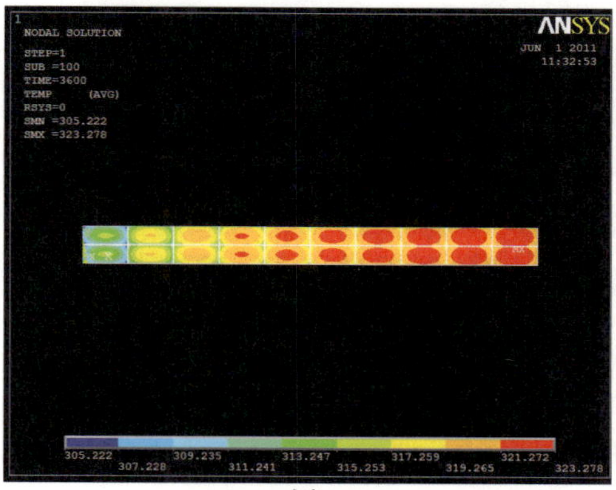

(a)

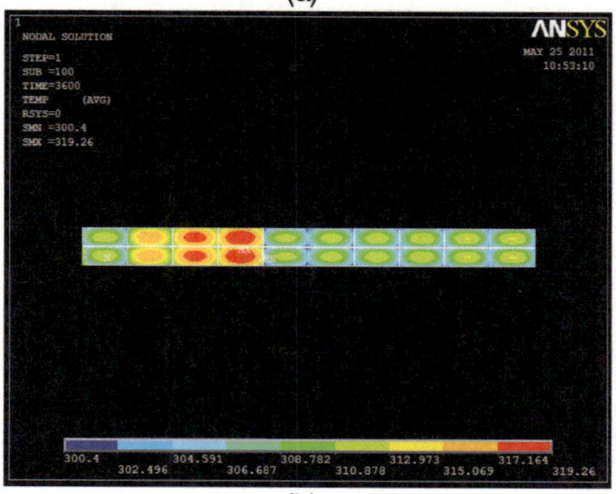

(b)

Figure 23. Temp. Distributions in the battery pack with the wind speed of (a) 4 m/s and (b) 10 m/s.

It can be seen in Fig. 23(a) that the air flow will turn to be the turbulent flow at the speed of 10 m/s which is much larger than the critical value of the wind speed (4.2 m/s). Furthermore, it can be seen that there are quite large temperature differences (about 19 degrees) in the battery pack for both wind speeds, which will deteriorate the batteries in the long term.

The temperature distributions in the battery pack using the parallel cooling strategy is presented in Fig. 24. The battery pack is under the condition of 1C electric charge and 293K temperature. The convection coefficient is 50 $W/m^2 K$. It can be seen that the maximum temperature difference in the battery pack is less than 7 degrees. The temperature distribution is the same for each battery. Comparing Figs. 23 and 24, one can easily see that the parallel cooling strategy is much better than the serial cooling strategy in this category.

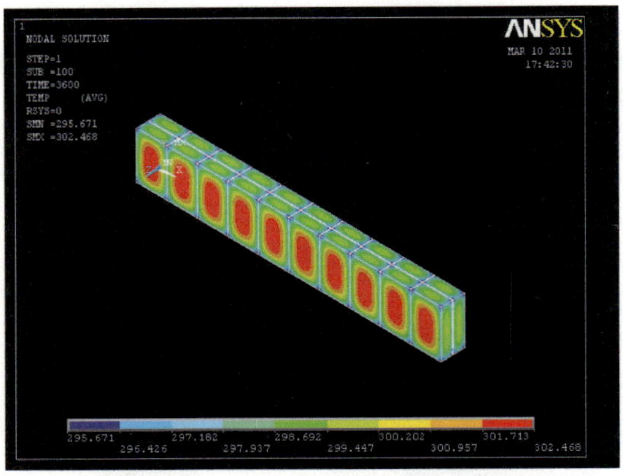

Figure 24. Temperature distribution in the battery pack utilizing the parallel cooling.

4.3. Summary on investigation of the temperature distribution in the Li-ion battery

In this section, the finite element formulations of the heat conduction equation of the Lithium-ion battery are established for both steady state and transient sate cases. Then the boundary conditions of the battery were discussed and the heat convection coefficients were determined for each surface of the battery. A parametric study was developed to investigate and optimize the temperature distribution in the battery. Finite element results indicate that a rectangular shape battery should be chosen to minimize the temperature difference in the battery. Results also revealed that a slower discharge rate needs to be selected to not only lower the temperature inside the battery, but make the temperature distribution in the battery as uniform as possible. Finally, the Li-ion battery pack was investigated and different cooling strategies were studied. Investigation indicates that the parallel cooling should be chosen to accelerate the heat dissipation and minimize the temperature difference in the battery pack. To acquire more accurate results on the temperature distribution in each battery and the entire battery pack, much more sophisticated finite element models need to be established to simulate the real structure inside the battery.

5. Summary and conclusions

In this chapter, the mechanical behaviors of the electric vehicles were investigated to improve the performance of the electric vehicles such that they can prevail over the conventional petroleum vehicles in the near future. Three different aspects of the electric vehicles were analyzed and substantially improved via the use of various methods.

Firstly, the regenerative braking force in the pure electric vehicle was analyzed and the distribution range of the braking forces in the front and rear wheels was determined to meet the requirements in UNECE R13. An optimized control strategy of the regenerative braking process was proposed based on the results of the analysis. Simulation results indicate that this optimized control strategy can considerably improve the energy efficiency of the regenerative braking process, as well as that of the entire vehicle. This optimized control strategy can also be used for hybrid vehicles.

Secondly, series-hybrid electric vehicles (SHEVs) were investigated to improve the efficiency of the engine and the working condition of the battery. According to the technical specifications of SHEVs, a control strategy was designed utilizing the power follower and thermostat, while parameters were selected for various part of the SHEV. Simulation results revealed that this control strategy can not only meet the performance requirements of the vehicle, but make the engine and batteries work in a reasonable range. The study and optimization of the control strategies are the hotspots in the field of the hybrid power system.

Finally, the temperature distribution in the lithium-ion battery for pure electric vehicles are studied and analyzed. Finite element formulations of the heat conduction equation of the Lithium-ion battery; boundary conditions of the battery were discussed and the heat convection coefficients were determined for each surface of the battery; then a parametric study was developed to investigate and optimize the temperature distribution in the battery. Results indicate that a rectangular shape battery should be chosen to minimize the temperature difference in the battery and a slower discharge rate is needed to make the temperature distribution in the battery as uniform as possible.

The research and study developed in this chapter have not only substantial theoretical values but also significant practical values. The results obtained in this chapter provide clear guidelines for the electric vehicle manufacturers to improve the quality and competitiveness of the electric vehicles such that they will be able to prevail the automobile industry in the near future.

Author details

Liang Zheng

Address all correspondence to: liangzheng@hitsz.edu.cn

Harbin Institute of Technology Shenzhen Graduate School, University Town of Shenzhen, Shenzhen, China

References

[1] Kawashima K., Uchida T., and Hori Y. Rolling stability control of in-wheel electric vehicle based on two-degree-of-freedom control. International Workshop on Advanced Motion Control, AMC, 2008 (1): 751-756.

[2] Weissinger C., Buecherl D., and Herzog H.-G. Conceptual design of a pure electric vehicle. IEEE Vehicle Power and Propulsion Conference, 2010.

[3] Wang R.-R., Chen Y., Feng D.-W., Huang X.-Y., and Wang J.-M. Development and performance characterization of an electric ground vehicle with independently actuated in-wheel motors, Journal of Power Sources, 2011, 196(8): 3962-3971.

[4] Younis A., Zhou L., and Dong Z.-M. Application of the new SEUMRE global optimization tool in high efficiency EV/PHEV/EREV electric mode operations, International Journal of Electric and Hybrid Vehicles, 2011, 3(2): 176-190.

[5] http://en.wikipedia.org/wiki/Regenerative_brake

[6] Official website of United Nations Economic Commission for Europevhttp://www.unece.org/trans/main/wp29/wp29regs1-20.html

[7] http://en.wikipedia.org/wiki/Hybrid_vehicle#Series_hybrid

[8] Xiong J., Wang G. Hybrid electric bus development and industrialization. Bus Technology and Research (in Chinese), 2002, 3: 4-6.

[9] Li X.-Y., Yu X.-M., and, Wu Z.-X. Series Hybrid Electric Vehicle Control Strategy. Journal of Jilin University: Engineering Science, 2004, 35: 250-254.

[10] Wipke K. B., Cuddy M. R. and Burch S. B. ADVISOR 2.1: A User Friendly Advanced Powertrain Simulation Using a Combined Backward / Forward Approach. NREL/JA-540-26839, 1999.

[11] Wang P. and Sun J. Design and Simulation of Series Hybrid Transit Bus". Journal of Technology and Research (in Chinese), 2007(11): 42-43.

[12] Yu Z.-S., editor, Automotive theory. Mechanical Industry Press, 2003.

[13] Zhu S.-S., Zhu W.-D. Study on the matching of generator and Motor in Series Hybrid Electric Bus. Journal of R & D of Automobiles (in Chinese), 2001(6): 46-48.

[14] Yang X., Qian L.-J. Series Hybrid Electric Bus parameter selection and simulation. Bus Technology (in Chinese), 2006(1): 25-27.

[15] Khanipour A., Ebrahimi K., and Seale W. J. Conventional Design and Simulation of an Urban Hybrid Bus, Engineering and Technology. 2007(28): 26-32.

[16] Einhorn M., Roessler W.; and Fleig J. Improved performance of serially connected Li-ion batteries with active cell balancing in electric vehicles. IEEE Transactions on Vehicular Technology, 2011, 60(6): 2448-2457.

[17] Bandhauer T. M., Garimella S., and Fuller T. F. A critical review of thermal issues in lithium-ion batteries. Journal of the Electrochemical Society, 2011, 158(3), R1-R25.

[18] Guerrero C. P., Li J.-S., Biller S., and Xiao G.-X.: Hybrid/electric vehicle battery manufacturing: The state-of-the-art, 2010 IEEE International Conference on Automation Science and Engineering, 2010, 281-286.

http://dx.doi.org/10.5772/53275

[19] Holman J. P. Heat Transfer, ninth edition, p. 4, McGraw-Hill (2002).

[20] Cook R. D., Malkus D. S., Plesha M. E. and Witt R. J. Concepts and Applications of Finite Element Analysis, fourth edition, p. 464-466, John Wiley (2001).

[21] http://en.wikipedia.org/wiki/File:Lithium-Ionen-Accumulator.jpg.

Chapter 10

Z-Source Inverter for Automotive Applications

Omar Ellabban and Joeri Van Mierlo

Additional information is available at the end of the chapter

http://dx.doi.org/10.5772/52724

1. Introduction

In a context of volatile fuel prices and rising concerns in terms of energy security of supply and climate change issues, one of the discussed technological alternatives for the transportation sector are electric based vehicles. Advanced technology vehicles such as hybrid electric vehicles (HEVs), plug-in hybrid electric vehicles (PHEVs), fuel cell hybrid electric vehicles (FCHEVs), and electric vehicles (EVs) require power electronics and electrical machines to function. These devices allow the vehicle to use energy from the battery to assist in the propulsion of the vehicle, either on their own or in combination with an engine [1]. Therefore, many research efforts have been focused on developing new converters and inverters suitable for electric vehicles applications. One of the most promising topologies is the Z-source inverter (ZSI). The ZSI, as shown in Figure 1, is an emerging topology for power electronics DC/AC converters. It can utilize the shoot-through (ST) state to boost the input voltage, which improves the inverter reliability and enlarges its application field [2]. In comparison with other power electronics converters, it provides an attractive single stage DC/AC conversion with buck-boost capability with reduced cost, reduced volume and higher efficiency due to a lower component number. Therefore, the ZSI is a very promising and competitive topology for vehicular applications [3].

As a research hotspot in power electronics converters, the ZSI topology has been greatly explored from various aspects, such as: ST control methods [2], [4]-[8], designing of the Z-network elements [9], modeling of the ZSI [10], [11], feedback control strategies [12]-[26], motor control algorithms [27]-[31] and automotive applications [32]-[37].

This chapter starts by presenting a summary for ZSI operation modes and modeling. Then, a review and a comparison between four ST boost control methods, which are: simple boost control (SBC), maximum boost control (MBC), maximum constant boost control (MCBC), and modified space vector modulation (MSVM) boost control, is presented based on simulation and experimental results. Control strategies of the ZSI are important issue and several feedback control strategies have been investigated in recent publications. There are four methods for controlling the dc-link voltage of the ZSI, which are: capacitor voltage control, indirect dc-link voltage control, direct dc-link control and unified control. A review of the above mentioned control methods with their drawbacks will be presented. Two new proposed control methods, which are dual-loop capacitor voltage control and dual-loop peak dc-link voltage control, will be presented and demonstrated by simulation and experimental results. Then, this chapter presents a comparative study of the most significant control methods, which are: scalar control (V/F), indirect field oriented control (IFOC) and direct torque control (DTC), for an induction motor fed by a ZSI for automotive applications. These control techniques are implemented using PWM voltage modulation. Finally, this chapter proposes three applications of the ZSI for automotive applications.

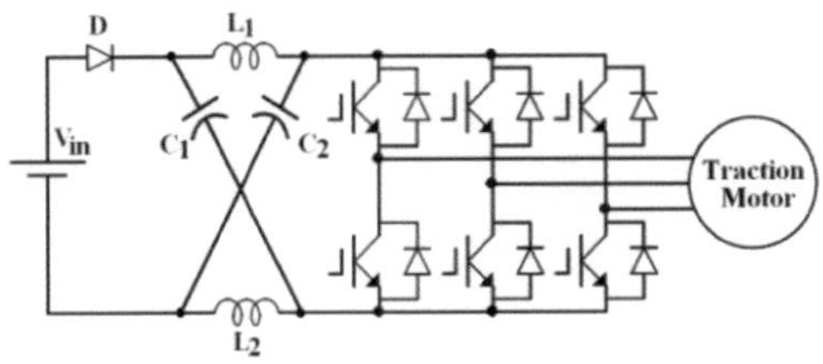

Figure 1. ZSI basic structure

2. ZSI Operation Modes and Modeling

To design a controller for the ZSI, a proper dynamic model for its switching operation is needed. An accurate small signal model of the ZSI gives not only a global also a detailed view of the system dynamics and provides guidelines to system controllers design since the transfer functions could be derived accordingly. Figure 1 shows the basic ZSI topology, which consists of two inductors (L1 and L2) and two capacitors (C1 and C2) connected in X shape to couple the inverter to the dc voltage source. The ZSI can produce any desired ac output voltage regardless of the dc input voltage. Because of this special structure, the ZSI has an additional switching state, when the load terminals are shorted through both the upper and lower switching devices of any phase leg, which called the shoot-through (ST) state besides the eight traditional non-shoot through (NST) states. General operation of a ZSI can be illustrated by simplifying the ac side circuit by an equivalent RL load in parallel with a switch S_2 and the input diode D is represented by a switch S_1, as shown in Figure 2. Where, $R_l$is given by $R_l = 8 \mid Z_{ac} \mid / 3\cos\phi$ and$L_l$is determined so that the time constant of the dc load is the same as the ac load [16]. Two operation modes involving two different circuit topologies can be identified in the ZSI operation as shown in Figure 3. In Mode 1, Figure 3-a, the energy transferred from source to load is zero because the load side and source side are decoupled by the ST state and the open status of S1. In Mode 2, Figure 3-b, real energy transfer between source and load occurs.

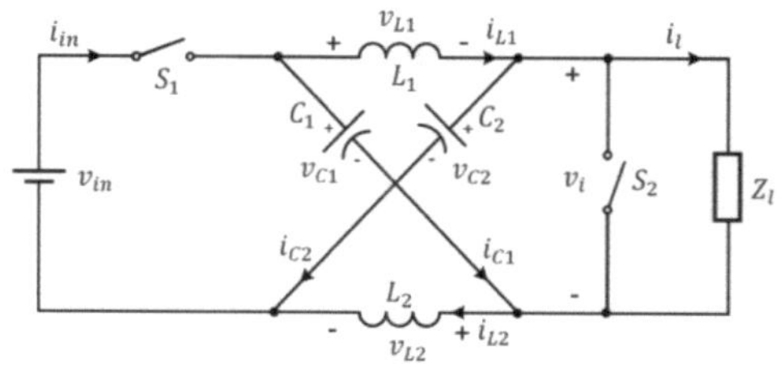

Figure 2. A simplified equivalent circuit for the ZSI

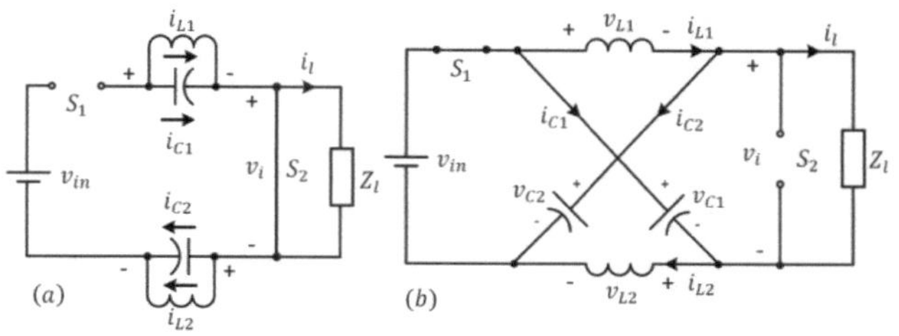

Figure 3. The basic two equivalent operation modes: (a) ST state, (b) NST state

Equations (1-4) represent: the third order small signal model, the steady state values of the state variables, the control to capacitor voltage $G_{vd}(s)$and control to inductor current $G_{id}(s)$ small signal transfer functions of the ZSI, where V_{in}, R_l, L_l, I_L, V_C, I_l, $D_0$are the input voltage, the equivalent dc load resistance, the equivalent dc load inductance, and the steady state values of inductor current, capacitor voltage, load current and ST duty ratio at certain operating point, respectively, and L , $C$are the Z-network inductor and capacitor, respectively.

$$\frac{d}{dt}\begin{bmatrix}\tilde{i}_L(t)\\ \tilde{v}_c(t)\\ \tilde{i}_l(t)\end{bmatrix}=\begin{bmatrix}0 & \frac{2D_0-1}{L} & 0\\ \frac{1-2D_0}{C} & 0 & \frac{-(1-D_0)}{C}\\ 0 & \frac{2(1-D_0)}{L_l} & \frac{-R_l}{L_l}\end{bmatrix}\cdot\begin{bmatrix}\tilde{i}_L(t)\\ \tilde{v}_c(t)\\ \tilde{i}_l(t)\end{bmatrix}+\begin{bmatrix}\frac{1-D_0}{L}\\ 0\\ \frac{-(1-D_o)}{L_l}\end{bmatrix}\cdot\tilde{v}_{in}(t)+\begin{bmatrix}\frac{2V_C-V_{in}}{L}\\ \frac{-2I_L+I_l}{C}\\ \frac{-2V_C+V_{in}}{L_l}\end{bmatrix}\cdot\tilde{d}_0(t) \tag{1}$$

$$\begin{aligned}V_C&=\frac{1-D_o}{1-2D_0}V_{in}\\ I_L&=\frac{1-D_o}{1-2D_0}I_l\\ I_l&=\frac{V_C}{R_l}\end{aligned} \tag{2}$$

$$G_{vd}(s)=\frac{(-2I_L+I_l)L_lLs^2+[(-2I_L+I_l)R_lL+(1-D_0)(2V_C-V_{in})L+(1-2D_0)(2V_c-V_{in})L_l]s+(1-2D_0)(2V_c-V_{in})R_l}{L_lLCs^3+R_lLCs^2+[2L(1-D_0)^2+L_l(2D_0-1)^2]s+R_l(2D_0-1)^2} \quad (3)$$

$$G_{id}(s)=\frac{(2V_C-V_{in})L_lCs^2+[R_lC(2V_C-V_{in})+(1-2D_0)(-2I_L+I_l)L_l]s+(1-D_0)(2V_C-V_{in})+(1-2D_0)(-2I_L+I_l)R_l}{L_lLCs^3+R_lLCs^2+[2L(1-D_0)^2+L_l(2D_0-1)^2]s+R_l(2D_0-1)^2} \quad (4)$$

Predicting the right half plane (RHP) zeros of the related transfer functions is one of the major advantages of the small-signal modeling. By considering the control to capacitor voltage transfer function, given by Eq. 3 as an example, the numerator is a quadratic equation. As known, for a quadratic equation$ax^2+bx+c=0$, there will be two different poles α and β if the discriminant$b^2-4ac>0$. Regarding this case, it can be acquired that $a=(-2I_L+I_l)L_lL<0$ and$c=(1-2D_0)(2V_c-V_{in})R_l>0$, therefore, this transfer function has two zeros: one is negative while the other is a positive one, which is called RHP zero. This identifies a non-minimum phase characteristic in the capacitor voltage response that is known to potentially introduce stability issues in the closed loop controlled system. The design of a feedback controller with an adequate phase margin becomes more difficult when RHP zeros appear in the transfer function, since it tends to destabilize the wide bandwidth feedback loops, implying high gain instability and imposing control limitations.

3. Review of PWM control methods for ZSI

3.1. Simple ST Boost Control (SBC)

The SBC method [2], uses two straight lines equal to or greater than the peak value of the three phase references to control the ST duty ratio in a traditional sinusoidal PWM, as shown in Figure 4. When the triangular waveform is greater than the upper line, V_p, or lower than the bottom line, V_n, the circuit turns into ST state. Otherwise it operates just as traditional carrier based PWM. This method is very straightforward; however, the resulting voltage stress across the switches is relatively high because some traditional zero states are not utilized.

3.2. Maximum ST Boost Control (MBC)

Reducing the voltage stress under a desired voltage gain becomes more important to control the ZSI; this can be achieved by making the ST duty ratio as large as possible. The MBC control, [4], turns all the traditional zero states into ST state. As shown in Figure 5, the circuit is in ST state when the triangular carrier wave is either greater than the maximum curve of the references (V_a, V_b and V_c) or smaller than the minimum of the references. The ST duty ratio varies at six times of the output frequency. The ripples in the ST duty ratio will result in ripple in the inductor current and the capacitor voltage. This will cause a higher requirement of the passive components when the output frequency becomes very low. Therefore, the MBC method is suitable for applications that have a fixed or relatively high output frequency.

http://dx.doi.org/10.5772/52724

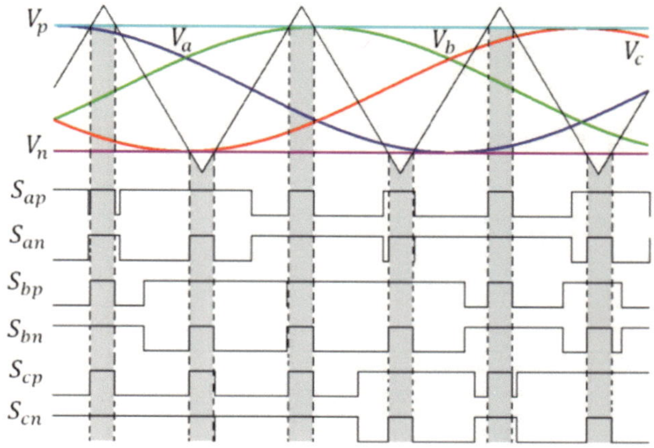

Figure 4. SBC method waveforms

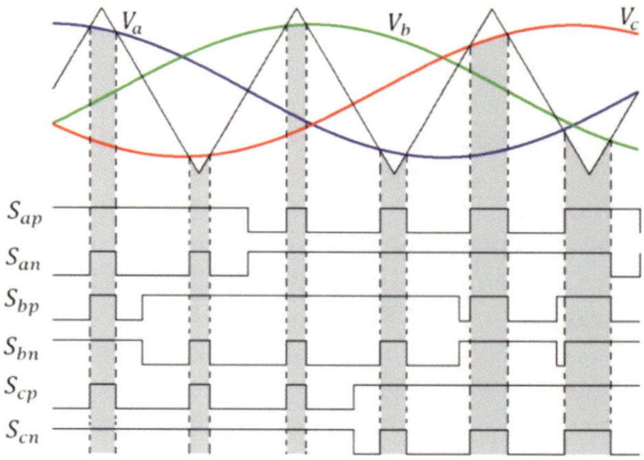

Figure 5. Waveforms of MBC method

3.3. Maximum Constant ST Boost Control (MCBC)

In order to reduce the volume and the cost, it is important always to keep the ST duty ratio constant. At the same time, a greater voltage boost for any given modulation index is desired to reduce the voltage stress across the switches. The MCBC method achieves the maximum voltage gain while always keeping the ST duty ratio constant [5]. Figure 6 shows the sketch map of the maximum constant ST boost control with third harmonic injection. Using the third harmonic injection, only two straight lines, V_p and V_n, are needed to control the ST time with 1/6 of the third harmonic injected.

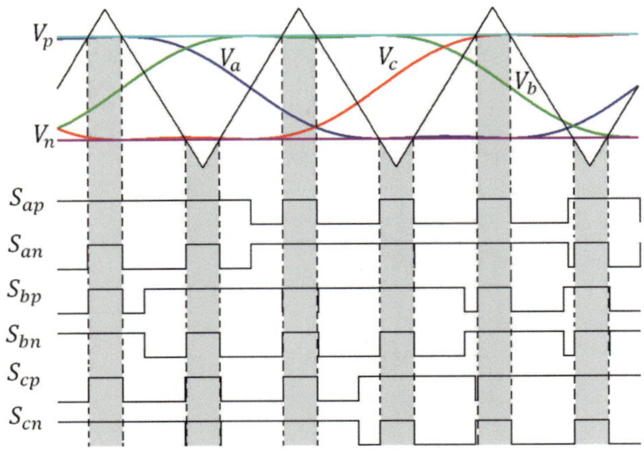

Figure 6. MCBC method with third harmonic injection

3.4. Modified Space Vector Modulation ST Control Method

The space vector PWM (SVPWM) techniques are widely used in industrial applications of the PWM inverter because of lower current harmonics and a higher modulation index. The SVPWM is suitable to control the ZSI. Unlike the traditional SVPWM, the modified space vector modulation (MSVM) has an additional ST time T_0 for boosting the dc-link voltage of the inverter beside the time intervals T_1, T_2 and T_z. The ST states are evenly assigned to each phase with $T_0/6$ within zero voltage period T_z. The zero voltage period should be diminished for generating a ST time, and the active states T_1 and T_2 are unchanged. So, the ST time does not affect the PWM control of the inverter, and it is limited to the zero state time T_z. The MSVM can be applied using two patterns. The MSVM1 as shown in Figure 7-a, at this switch pattern, the ST time T_0 is limited to $(3/4)T_z$, because the period $(T_z/4\text{-}2T_s)$ should be greater than zero. The MSVM2 as shown in Figure 7-b, where the distribution of zero state time is changed into $(T_z/6)$ and $(T_z/3)$. The maximum ST time is increased to the zero state time T_z [6].

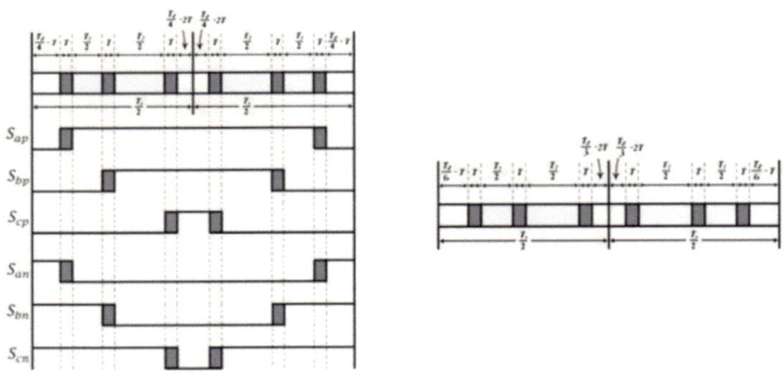

(a) (b)

Figure 7. Switching pattern for the MSVPWM: (a) MSVM1, (b) MSVM2

Table 1 shows a summary of all relations for the different ST boost control methods, where D_0 is the ST duty ratio, M is the modulation index, B is the boost factor, G is the voltage gain, and Vs is the voltage stress across the switch. Figure 8-a shows the voltage gain versus the modulation index and Figure 8-b shows the voltage stress versus the voltage gain for different ST boost control methods. At high voltage gain, the MSVPWM1 has the highest voltage stress.

ST method	SBC	MBC	MCBC	MSVM1	MSVM2
D_0	$1-M$	$\frac{2\pi-3\sqrt{3}M}{2\pi}$	$\frac{2-\sqrt{3}M}{2}$	$\frac{3}{4}\cdot\frac{2\pi-3\sqrt{3}M}{2\pi}$	$\frac{2\pi-3\sqrt{3}M}{2\pi}$
B	$\frac{1}{2M-1}$	$\frac{\pi}{3\sqrt{3}M-\pi}$	$\frac{1}{\sqrt{3}M-1}$	$\frac{4\pi}{9\sqrt{3}M-2\pi}$	$\frac{\pi}{3\sqrt{3}M-\pi}$
G	$\frac{M}{2M-1}$	$\frac{\pi M}{3\sqrt{3}M-\pi}$	$\frac{M}{\sqrt{3}M-1}$	$\frac{4\pi M}{9\sqrt{3}M-2\pi}$	$\frac{\pi M}{3\sqrt{3}M-\pi}$
M_{max}	$\frac{G}{2G-1}$	$\frac{\pi G}{3\sqrt{3}G-\pi}$	$\frac{G}{\sqrt{3}G-1}$	$\frac{2\pi G}{9\sqrt{3}G-4\pi}$	$\frac{\pi G}{3\sqrt{3}G-\pi}$
V_s	$(2G-1)V_{in}$	$\frac{3\sqrt{3}G-\pi}{\pi}V_{in}$	$(\sqrt{3}G-1)V_{in}$	$\frac{(9\sqrt{3}G-4\pi)}{2\pi}V_{in}$	$\frac{(3\sqrt{3}G-\pi)}{\pi}V_{in}$

Table 1. Summary of the different ST boost control methods expressions [7]

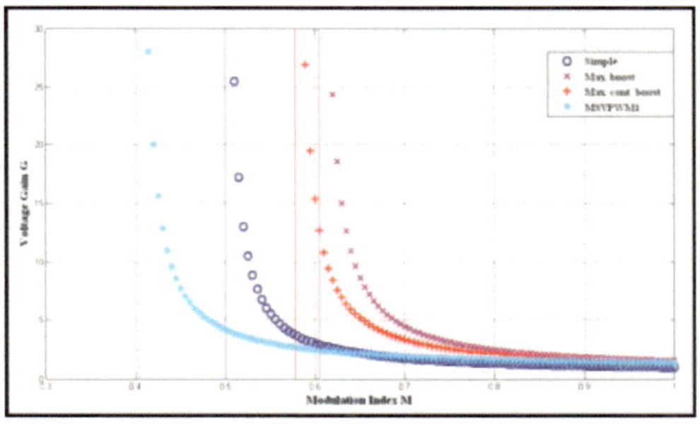

(a)

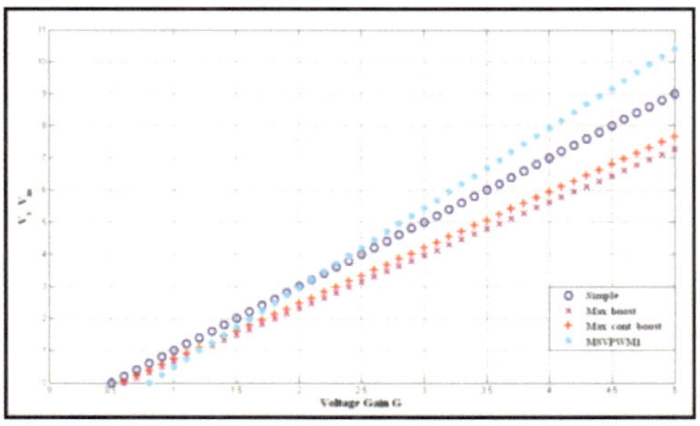

(b)

Figure 8. a) Voltage gain versus modulation index, (b) Voltage stress versus voltage gain for different PWM control methods

Efficiency evaluation is an important task during inverter design. The losses of the inverter mainly distributed on the semiconductor devices. The semiconductor device losses mainly include conduction losses and switching losses. The efficiency of the ZSI is greatly affected by the ST control methods. Figure 9 shows the losses distribution of a 10 kW ZSI at nominal input and output power, where the input diode conduction and switching losses are included, which are neglected in most publications, the extra losses of the MSVM boost method mainly come from the switching losses of IGBTs and reverse recovery losses of the input diode which are about three times of other methods [8]. Table 2 presents a comparison between the different four ST control methods. The comparison results show that the MCBC method seems to be the most suitable boost control method for the ZSI [8].

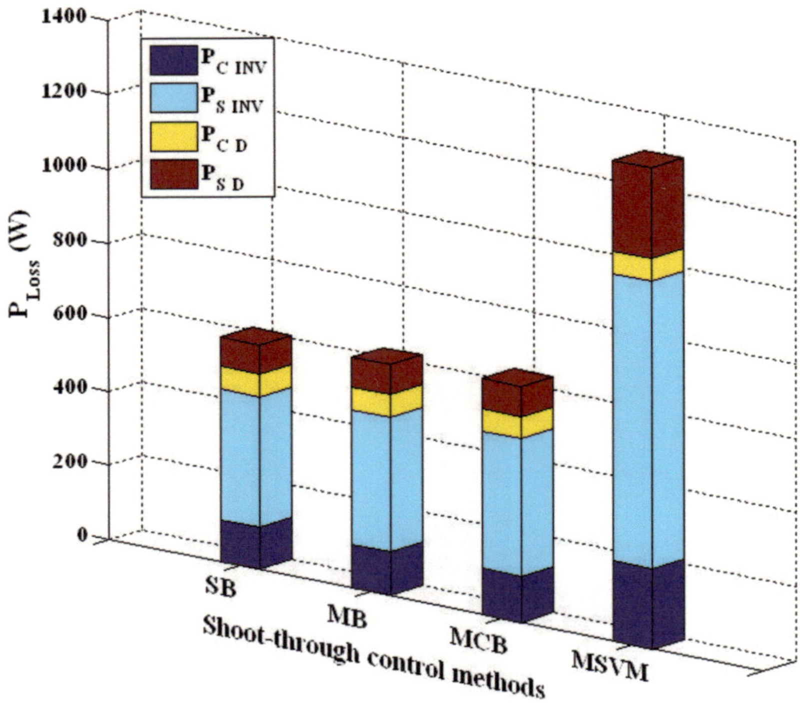

Figure 9. Losses distribution of the ZSI at 10 kW output power [8]

4. ZSI DC link voltage control

The control strategy of the ZSI is an important issue and several feedback control strategies have been investigated in recent publications [12][23]. There are four methods for controlling the ZSI dc-link voltage, which are: capacitor voltage control [12]-[19], indirect dc-link voltage control [19],[20], direct dc-link control [21],[22] and unified control [23]. Table 3, presents a review of the above mentioned control methods with their drawbacks.

ST boost control method	SBC	MBC	MCBC	MSVM
Line voltage harmonic	-	+	0	+
Phase current harmonic	0	0	+	-
DC link voltage ripples	0	-	+	0
Switch voltage stress	0	+	0	-
Inductor current ripples	0	-	+	-
Efficiency	0	+	+	-
Obtainable ac voltage	0	0	+	-
Total	-	+++--	+++++	+----

*(+, 0 and -) represents the best, the moderate and the lowest performance, respectively

Table 2. Different ST control methods comparison [8]

Reference No.	Control method	Controller type	Controller variable	Drawbacks
[12]	V_c control	Single loop control with a PID controller	D_0	The controller is designed based on a second order model
[13]	V_c control	Single loop control with a PI controller	D_0	The controller parameters is tuned by neural network (NN)
[15]	V_c control	Two control loops with a PI controller	D_0 and M	The controller is not designed or tuned by NN
[17], [18]	V_c control	Single loop nonlinear control	D_0 and M	Complex control algorithm
[19]	Indirect V_i control by controlling V_c	PID fuzzy controller	D_0	The controller is not accurate, because the relation between V_c and V_i is not linear and the controller is not designed
[20]	Indirect V_i control by controlling V_{ac}	PI controller with saturation	M	The ST is not controlled and the controller is not designed
[21]	Direct V_i control	Single loop PID controller	D_0	External complex sensing circuit and the controller is not designed
[22]	Direct V_i control	Dual loop V_i control, the outer voltage loop has a PI controller and the inner current loop has a P controller	D_0	A simplified representation of the 3-phase load of the ZSI
[23]	Unified control	Single loop with a PI controller	V_{ref} for the MSVM	Suitable for isolated operations and the controller is not designed

Table 3. Review of previous ZSI control strategies

In all the above mentioned control methods, a single-loop voltage control technique was used. However, in high power converters, a single-loop voltage control has two problems. The first problem is that, the inductor current is not regulated and can be overloaded during transient events and the limited stability limits is the second problem. Therefore, a dual-loop voltage control is preferred over a single-loop voltage control in high power converters to overcome the above mentioned problems [24]. Two new control algorithms are proposed by the authors, which are: a dual-loop capacitor voltage control [25] and a dual-loop peak DC-link voltage control [26]. These two control algorithms will be briefly presented as flows.

4.1. Capacitor voltage control

This chapter proposes a dual loop capacitor voltage control of the ZSI. The proposed control generates the ST duty ratio by controlling both the inductor current and the capacitor voltage of the ZSI as shown in Figure 10-a, where $G_M(s)$ is expressed by:

$$G_M(s) = \frac{D_0(s)}{v'_m(s)} = \frac{2}{V_{tri}} \tag{5}$$

Based on the small signal transfer functions $G_{vd}(s)$, $G_{id}(s)$ and $G_M(s)$ given by Eqs (3), (4) and (5), both controller transfer functions $G_{cv}(s)$ and $G_{ci}(s)$ can be designed. In order to design these controllers, the continuous time transfers functions are first discretized using the zero order hold (ZOH). Once the discrete transfer functions of the system are available, the digital controllers are designed directly in the Z-domain using methods similar to the continuous time frequency response methods. This has the advantage that the poles and zeros of the digital controllers are located directly in the Z-domain, resulting in a better load transient response, as well as better phase margin and bandwidth for the closed loop power converter [25]. Figure 10-b shows the entire digital closed loop control system containing the voltage loop controller, current loop controller, the zero order hold, the computational delay, the modified modulation, and the control to outputs transfer functions. In this implementation the chosen sampling scheme results in a computation delay of half the sampling period. The loop gains for inner current loop and outer voltage loop can be expressed as:

$$T_i(z) = G_{ci}(z) \cdot G_{id}(z) \tag{6}$$

$$T_v(z) = \frac{G_{cv}(z) \cdot G_{ci}(z) \cdot G_{vd}(z)}{1 + T_i(z)} \tag{7}$$

Where

$$G_{id}(z) = Z\left\{\frac{1 - e^{-T_s s}}{s} \cdot e^{-T_d s} \cdot G_M(s) \cdot G_{id}(s)\right\} \tag{8}$$

$$G_{vd}(z) = Z\left\{\frac{1-e^{-T_s s}}{s}\cdot e^{-T_d s}\cdot G_M(s)\cdot G_{vd}(s)\right\} \tag{9}$$

In this chapter, a digital PI controller with anti-windup is designed based on the required phase margin, and critical frequency, using the bode diagram of the system in the Z- domain, the transfer function of the digital PI controller in Z-domain is given by:

$$G_c(z) = K_p + \frac{K_i T_s z}{z-1} \tag{10}$$

where

$$K_p = \frac{\cos\theta}{\left|G_p(z)\right|} \tag{11}$$

$$K_i = \frac{\sin\theta\cdot f_{cz}}{\left|G_p(z)\right|} \tag{12}$$

and

$$\theta = 180^\circ + \phi_m - \angle G_p(z) \tag{13}$$

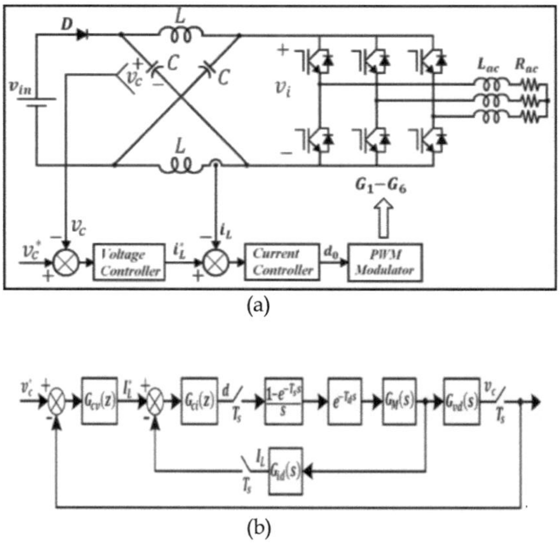

Figure 10. Dual-loop capacitor control of the ZSI (a) and its block digital block diagram (b)

Figure 11 shows the bode plots for the current loop gain and voltage loop gain, respectively, with the system parameters listed in Table 4. The plots indicate that the current loop gain has a crossover frequency as high as 1 kHz, with a phase margin of 65° and a gain margin of 10 dB. To avoid interaction between the sub-systems, low control bandwidth is used for the voltage loop. The resulting outer voltage loop has a crossover frequency of 100 Hz and a phase margin of 59° and a gain margin of 25 dB. Figs. 12-13 compare simulation and experimental results during input voltage step down by 7.5% with the same load, load increasing and decreasing by 50% and steady state operations. It is noticeable that the experimental results match the simulation results very well, which verify the performance of the proposed dual-loop capacitor voltage control for the ZSI.

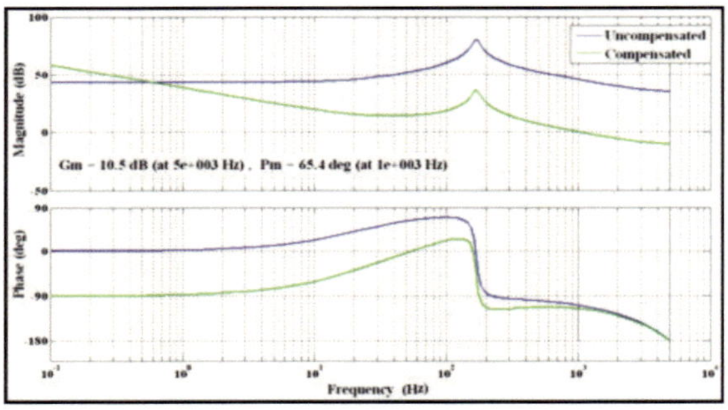

(a)

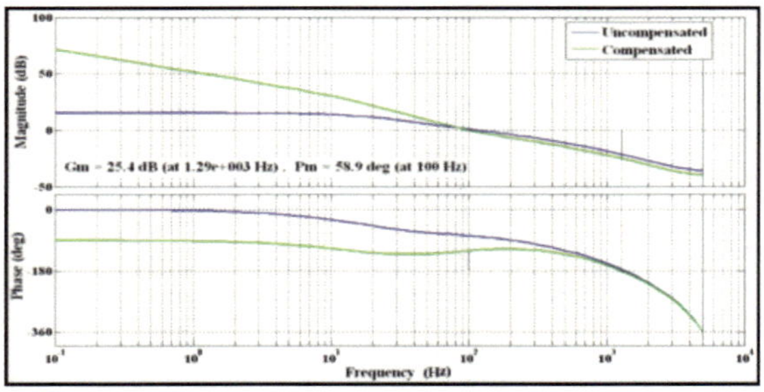

(b)

Figure 11. Bode diagrams for current (a) and voltage (b) loops for dual-loop capacitor voltage control

http://dx.doi.org/10.5772/52724

Parameter	Value
Input voltage	200 V
Capacitor reference voltage	300 V
Inductance	650 μH
Inductance internal resistance	0.22 Ω
Capacitance	320 μF
Capacitance internal resistance	0.9 mΩ
Switching frequency	10 kHz
AC load inductance	340 μH
AC load resistance	12.5 Ω

Table 4. Experimental Parameters of the ZSI

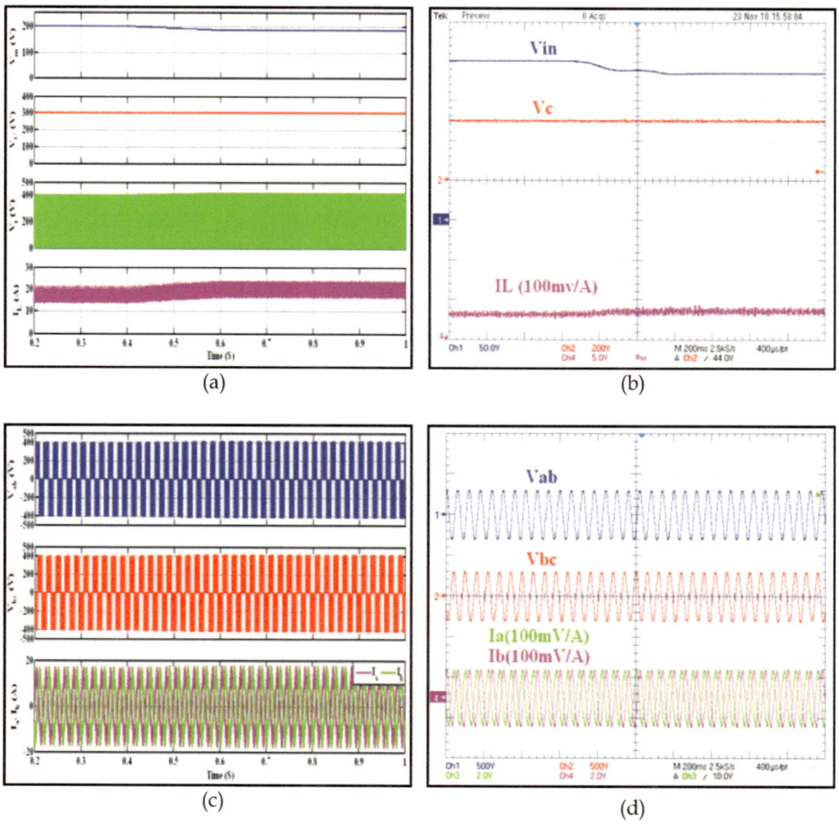

Figure 12. ZSI response during input voltage step down by 7.5% : (a and c) simulation results and (b and d) experimental results using dual-loop control technique

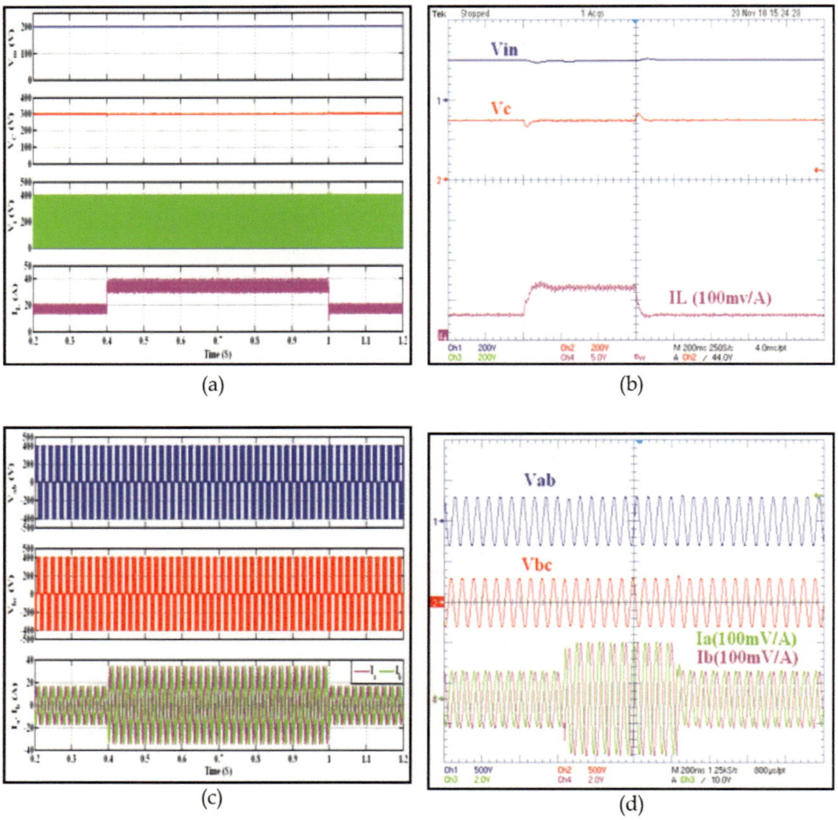

Figure 13. ZSI response during load increasing and decreasing by 50 % : (a and c) simulation results and (b and d) experimental results

4.2. Peak DC-link voltage control

The capacitor voltage, V_c, is somewhat equivalent to the peak dc-link voltage, V_{ip}, of the inverter, but the peak dc-link voltage is non-linear function of the capacitor voltage. Thus, only controlling the capacitor voltage cannot bring the high performance due to the non-linear property of the V_{ip}/V_c relation. Figure 14 shows the entire dual-loop peak dc-link voltage control technique block diagram of the ZSI. Figs. 15-16 show the simulation and experimental results of the proposed dual-loop peak dc-link voltage control technique during input voltage step and load transient. As shown in Figure 15, the input voltage stepped-down by 7.5% with the rated load, the peak dc-link voltage remains constant at 300 V, the load phase current and line voltage are not affected by input voltage decreasing and the inductor current is increased to supply the same output power. Figure 16 shows the ZSI response during load increasing and decreasing by 50%. As noticeable the inductor current is doubled during the 50% load increase and the output line voltage and the peak dc-link voltage remain unchanged.

http://dx.doi.org/10.5772/52724

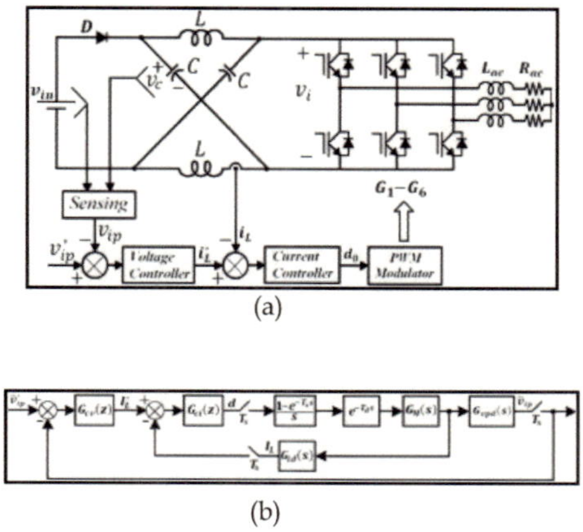

Figure 14. ZSI dual loop peak dc-link voltage control of the ZST (a) and its block diagram (b)

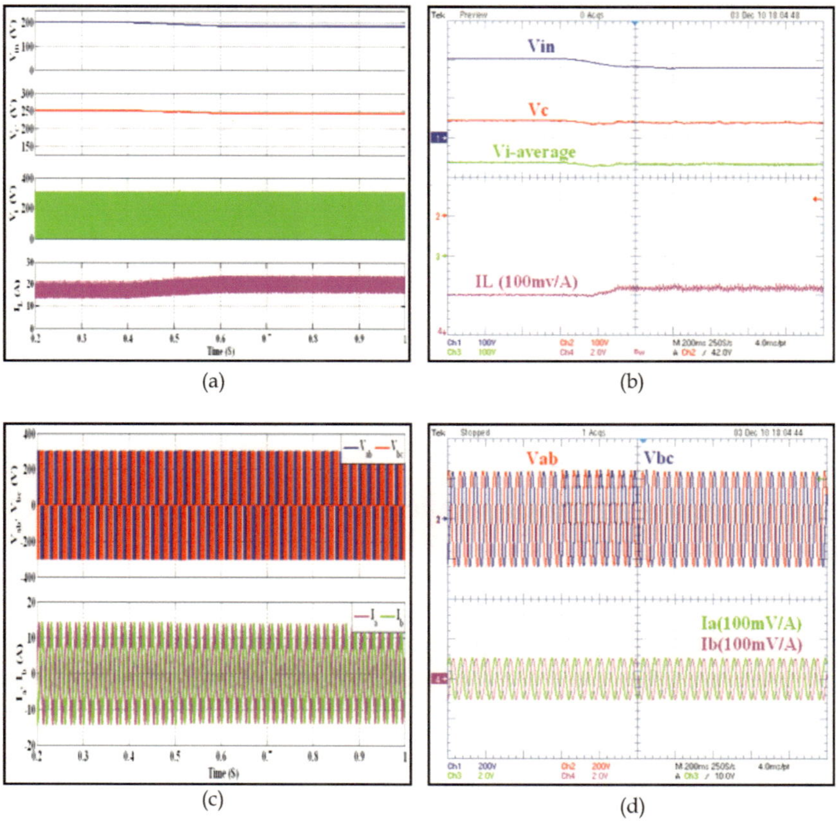

Figure 15. ZSI response during input voltage step down by 7.5%: (a, b) Z-network variables and (c, d) output variables

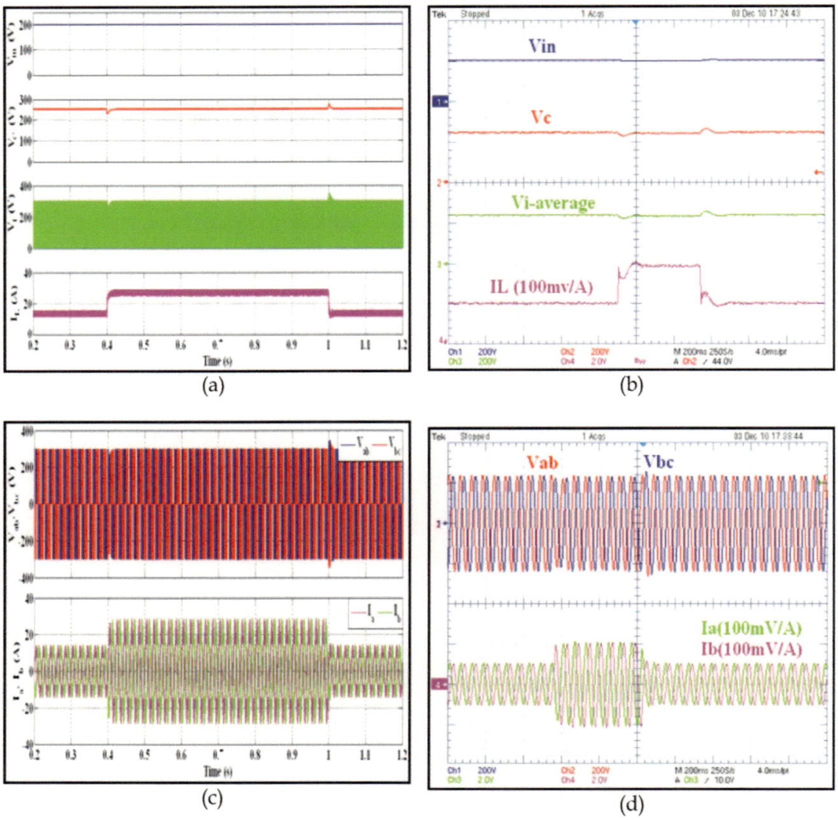

Figure 16. ZSI response during input voltage step down by 7.5% : (a, b) Z-network variables and (c, d) output variables

5. A comparative study of different control techniques for induction motor fed by a Z-source inverter

This section presents a comparative study of the most significant control methods (Scalar control (V/F), indirect field oriented control (IFOC) and direct torque control (DTC)) for an induction motor fed by a ZSI for automotive applications. The three control techniques are implemented using PWM voltage modulation. The comparison is based on various criteria including: basic control characteristics, dynamic performance, and implementation complexity. The study is done by MATLAB simulation of a 15 kW induction motor fed by a high performance ZSI (HP-ZSI). The simulation results indicates that, the IFOC seems to be the best control techniques suitable for controlling an induction motor fed by a ZSI for automotive applications.

5.1. Scalar control (V/F) technique

The closed loop speed control by slip regulation, which is an improvement of the open loop V/F control, is shown in Figure 17. The speed loop error generates the slip command ω_{sl}^{*} through a proportional integral (PI) controller with a limiter. The slip is added to the feed-back speed signal to generate the slip frequency commandω_e^{*}. Thus the frequency command generates the voltage command through a V/F generator, which incorporates the low frequency stator drop compensation. Although this control technique is simple, it provides limited speed accuracy especially in the low speed range and poor dynamic torque response.

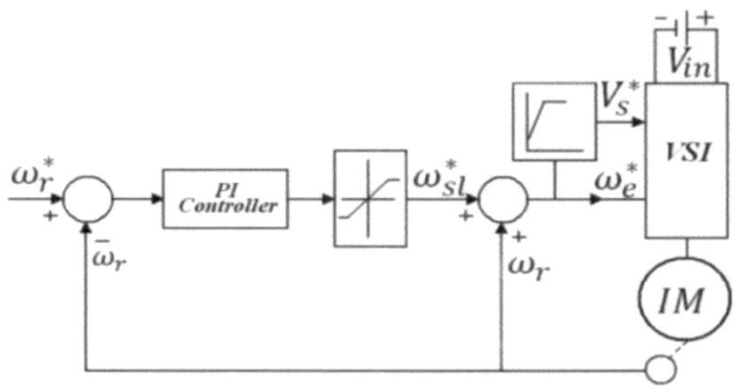

Figure 17. Block diagram of a scalar controlled induction motor

5.2. Indirect Field Oriented Control (IFOC) technique

In the indirect field oriented control method, the rotating reference frame is rotating at synchronous angular velocity, ω_e. This reference frame allows the three phase currents to be viewed as two dc quantities under steady state conditions. The q-axis component is responsible for the torque producing current, i_{qs}, and the d-axis is responsible for the field producing current, i_{ds}. These two vectors are orthogonal to each other so that the field current and the torque current can be controlled independently. Figure 18 shows the block diagram of the IFOC technique for an induction motor. The q-axis component of the stator reference current, i_{qs}^{*}, may be computed using the reference torque, T_{ref}, which is the output of a PI speed controller, as:

$$i_{qs}^{*} = \frac{2}{3}\frac{2}{p}\frac{L_r}{L_m}\frac{T_{ref}}{\psi_r} \tag{14}$$

where Ψ_r is the estimated rotor flux, which is given by:

$$\psi_r = \frac{L_m}{\tau_r s + 1} i_{ds} \tag{15}$$

where L_m, L_r and τ_r are the magnetization inductance, the rotor inductance, and the rotor time constant, respectively. The d-axis component of the stator reference current, i_{ds}^*, may also be obtained by using the reference input flux, $\psi_{r_{ref}}$, which is the output of a PI flux controller, as:

$$i_{ds}^* = \frac{\psi_{r_{ref}}}{L_m} \tag{16}$$

By using the rotor speed, ω_{rm}, and the slip frequency, ω_{sl}, which is given by:

$$\omega_{sl} = \frac{1}{\tau_r} \frac{i_{ds}^*}{i_{qs}^*} \tag{17}$$

the angle of the rotor flux, θ_e, may be evaluated as:

$$\theta_e = \int (\omega_e + \omega_{rm}) dt \tag{18}$$

Proportional integral controllers regulate the stator voltages, v_{ds}^*andv_{qs}^*, to achieve the calculated reference stator currents, i_{ds}^* and i_{qs}^*. The required voltage is then synthesized by the inverter using pulse width modulation (PWM). During motor operation the actual rotor resistance and inductance can vary. The resulting errors between the values used and the actual parameters cause an incomplete decoupling between the torque and the flux. In order to compensate for this incomplete decoupling, the values of compensation voltages are added to the output of the current controllers. This voltage compensation can improve the performance of the current control loops. The compensations terms are given by:

$$\begin{aligned} v_{dsc} &= -\omega_e \sigma L_s i_{qs}^* \\ v_{dsc} &= \omega_e \sigma L_s i_{ds}^* + \frac{L_m}{L_r} \omega_r \psi_r \end{aligned} \tag{19}$$

5.3. Direct Torque Control with Space Vector Modulation (DTC-SVM) technique

The conventional DTC scheme has many drawbacks, such as: variable switching frequency, high current and torque ripples, starting and low-speed operation problems, in addition to

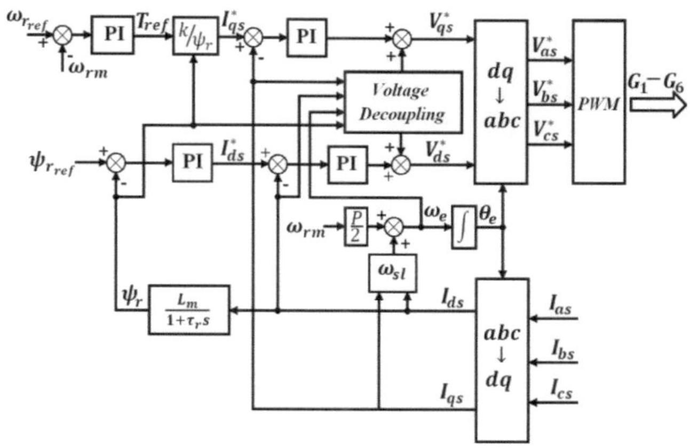

Figure 18. Block diagram of the IFOC of an induction motor

high sampling frequency needed for digital implementation of the hysteresis controllers. To overcome these drawbacks, the space vector modulation is combined with the conventional DTC scheme for induction motor drive to provide a constant inverter switching frequency. In the DTC-SVM scheme, as shown in Figure 19, the torque and flux hysteresis comparators are replaced by PI controllers to regulate the flux and torque magnitudes respectively. The motor stator flux and the motor developed torque can be estimated by:

$$
\begin{aligned}
\psi_{ds} &= \int \left(v_{ds} - R_s i_{ds} \right) dt \\
\psi_{qs} &= \int \left(v_{qs} - R_s i_{qs} \right) dt \\
\left| \psi_s \right| &= \sqrt{\psi_{ds}^2 + \psi_{ds}^2} \\
\theta_{\psi_s} &= \tan^{-1} \frac{\psi_{qs}}{\psi_{ds}}
\end{aligned}
\qquad (20)
$$

$$
T_e = \frac{3}{2} P \left(\psi_{ds} i_{qs} - \psi_{qs} i_{ds} \right) \qquad (21)
$$

The output of theses PI controllers generates the d and q components of the reference voltage command (v_{ds}^* and v_{qs}^*) in the stator flux oriented coordinates. After coordinate transformation, using the stator flux angle $\theta_{\psi s}$, we get the reference voltage vectors $v_{\alpha s}^*$ and $v_{\beta s}^*$) in the stationary frame. These two components, which can control stator flux and torque separately, are delivered to space vector modulator (SVM). The space vector modulator generates the inverter control signals, which ensures fixed inverter switching frequency. So the inverter switching frequency is significantly increased, and the associated torque ripple and current harmonics can be dramatically reduced, in comparison with the conventional switching table based DTC scheme.

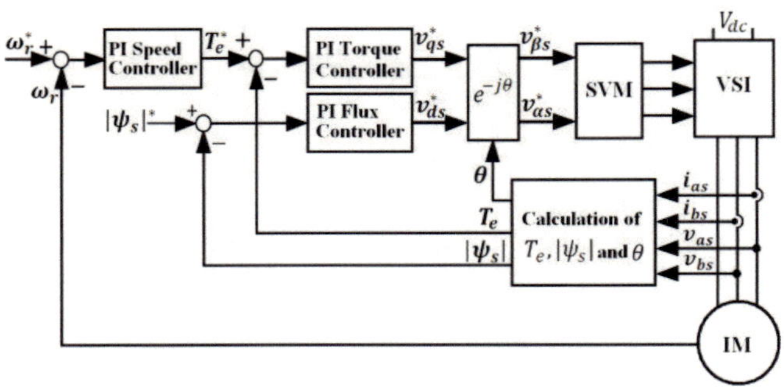

Figure 19. Block diagram of the DTC-SVM based IM drive

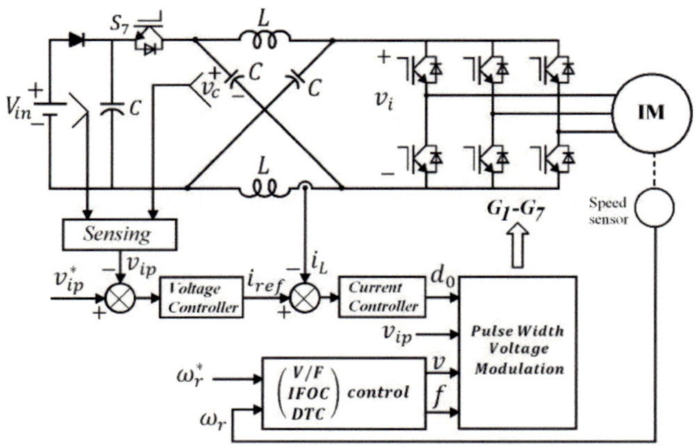

Figure 20. Closed loop speed control of three phase induction motor fed by a high performance ZSI

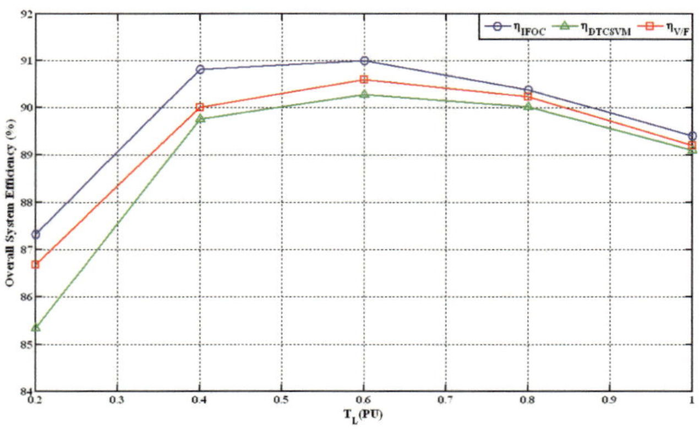

Figure 21. Overall system efficiency at different load torque values

Figure 20 shows the complete block diagram of the closed loop speed controlled IM fed by a high performance ZSI. A dual loop controller is designed to control the average value of the dc link voltage by controlling the magnitude of its peak voltage based on a small signal model of the high performance ZSI. Figure 21, shows the calculated overall system efficiency at different load torque. The three control techniques were compared on a simulated benchmark. The main results of this comparative study are summarized in Table 5.

Comparison Criterion	V/F	IFOC	DTC-SVM
Dynamic response	Poor	Good	Good
Torque ripples	Large	Small	Small
Speed error	Large	Small	Medium
ZSI performance	Good	Good	Poor
Complexity	Low	High	High
Efficiency	Medium	High	Low

Table 5. Summary of performance compression

6. Z-Source Inverter for vehicular applications

This section proposes three applications of the ZSI in the automotive field. The first application proposes the using of the bidirectional ZSI (BZSI) supplied by a battery to drive an induction motor for hybrid electric vehicle (HEV) applications, by replacing the two stages conversion. The second application proposes the using of the BZSI in plug-in hybrid electric vehicle (PHEV) applications for replacing the bidirectional battery charger, which composed of two stages conversion. By using the BSZI, the battery can be charged from the grid during night and can be discharged to the grid during peak power demand, which increase the grid stability. The third application proposes the using of the HP-ZSI for fuel cell hybrid electric vehicle (FCHEV) applications. Where the fuel cell (FC) stack and the supercapacitor (SC) module are directly connected in parallel with the HP-ZSI. The SC module is connected between the input diode and the bidirectional switch S_7 of the HP-ZSI. The SC module supplies the transient and instantaneous peak power demands and absorbs the deceleration and regenerative braking energy. The indirect field oriented control (IFOC) is used to control the speed of the IM during motoring and regenerative braking operation modes in the first and the third proposed applications. While, a proportional plus resonance (PR) controller is used to control the AC current during connecting the BZSI to the grid for battery charging/ discharging mode in the second proposed application.

6.1. ZSI applications in HEV

The ZSI is proposed to be used to replace the two stages conversion in HEV, The BZSI can replace the bidirectional DC/DC converter and the traditional VSI as a single stage converter, as shown in Figure 22. The IFOC is used for controlling the speed of the IM during motoring and regenerative braking operations and a dual loop capacitor voltage control algorithm is used to control the BZSI dc-link voltage.

6.2. ZSI applications in PHEV

The ZSI is proposed to be used to replace the two stage bidirectional battery charger in a PHEV. Figure 23 shows the entire block diagram of a grid connected BZSI containing: the battery, the BZSI, the capacitor voltage control algorithm and the AC grid current control algorithm during battery charging/discharging modes, where the capacitor voltage control generates the ST duty ratio and the AC grid current control generates the modulation index.

6.3. ZSI applications in FCHEV

In Figure 24, the FC system and the SC module are direct connected in parallel with the HP-ZSI. The SC module is connected between the input diode D and the bidirectional switch S_7. The bidirectional switch S_7 provides a path for the regenerative braking energy to be stored in the SC module during the ST state. The SC module supplies the transient and instantaneous peak power demands and absorbs the deceleration and regenerative braking energy. In addition, a dual loop control is used to control the Z-network capacitor voltage by controlling the ST duty ratio and the IFOC strategy is used to control the induction motor speed by controlling the modulation index. The proposed applications improve the vehicle efficiency and reduce its production cost due to a lower its component count, since it is a one stage converter with a reduced volume and easier control algorithm.

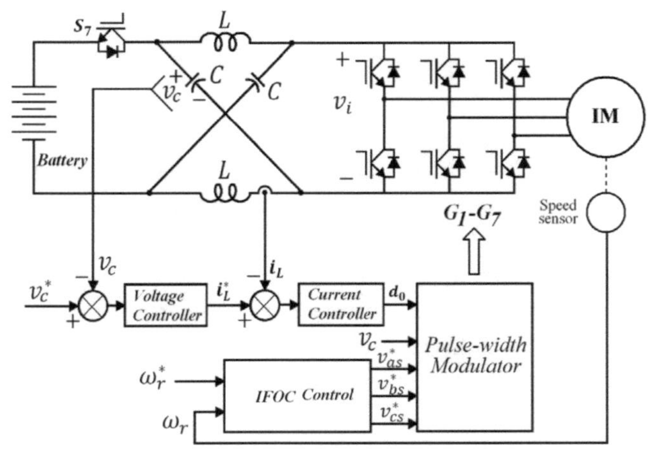

Figure 22. Using the BZSI for HEV applications

http://dx.doi.org/10.5772/52724

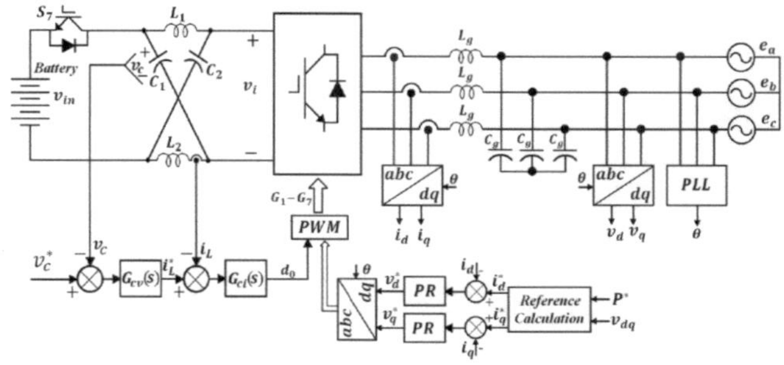

Figure 23. Using the BZSI for PHEV applications

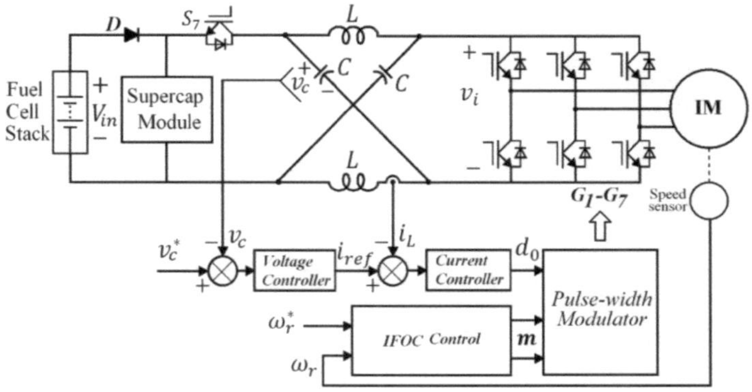

Figure 24. Using the HP-ZSI for FCHEV applications

Author details

Omar Ellabban[1*] and Joeri Van Mierlo[2*]

1 Department of Power and Electrical Machines, Faculty of Engineering, Helwan University, Cairo, Egypt

2 Department of Electric Engineering and Energy Technology, Faculty of Engineering Sciences, Vrije Universiteit Brussel, Brussels, Belgium

References

[1] Emadi, A., Young, Joo., & Lee, Rajashekara. K. (2008). Power Electronics and Motor Drives in Electric, Hybrid Electric, and Plug-In Hybrid Electric Vehicles. *IEEE Transactions on Industrial Electronics.*, 55(6), 2237-2245.

[2] Peng, FZ. (2003). Z-source Inverter. *IEEE Transactions on Industry Applications,* 39(2), 504-510.

[3] Peng, F. Z., Shen, M., & Joseph, A. (2005). Z-Source Inverters, Controls, and Motor Drive Applications. *KIEE International Transactions on Electrical Machinery and Energy Conversion Systems.,* 5, 6-12.

[4] Peng, F. Z., Shen, M., & Qian, Z. (2005). Maximum boost control of the Z-source inverter. *IEEE Transactions on Power Electronic.,* 20(4), 833-838.

[5] Shen, M., Wang, J., Joseph, A., Peng, F. Z., Tolbert, L. M., & Adams, D. J. (2006). Constant boost control of the Z-source inverter to minimize current ripple and voltage stress. *IEEE Transactions on Industry Applications,* 42(3), 770-778.

[6] Poh, Chiang., Loh, D., Mahinda, Vilathgamuwa., Yue, Sen., Lai, Geok., Tin, Chua., & Yunwei, Li. (2005). Pulse-Width Modulation of Z-Source Inverters. *IEEE Transactions on Power Electronics.,* 20(6), 1346-1355.

[7] Ellabban, O., Van Mierlo, J., & Lataire, P. (2009). Comparison between Different Shoot-Through PWM Control Methods for Different Voltage Type Z-Source Inverter Topologies. The 13th European Conference on Power Electronics and Applications.

[8] Ellabban, O., Van Mierlo, J., & Lataire, P. (2011). Experimental Study of the Shoot-Through Boost Control Methods for the Z-Source Inverter. *EPE- European Power Electronics and Drives Journal.,* 21(2), 18-29.

[9] Rajakaruna, S., & Jayawickrama, L. (2010). Steady-State Analysis and Designing Impedance Network of Z-Source Inverters. *IEEE Transactions on Industrial Electronics,* 57(7), 2483-2491.

[10] Loh PC, Vilathgamuwa DM, Gajanayake CJ, Lim YR, Teo CW. (2007). Transient Modeling and Analysis of Pulse-Width Modulated Z-source Inverter. *IEEE Transactions on Power Electronics.,* 22(2), 498-507.

[11] Liu, J., Hu, J., & Xu, L. (2007). Dynamic Modeling and Analysis of Z-Source Converter-Derivation of AC Small Signal Model and Design-Oriented Analysis. *IEEE Transactions of Power Electronics.,* 22(5), 1786-1796.

[12] Xinping, D., Zhaoming, Q., Shuitao, Y., Bin, C., & Fangzheng, P. (2007). A PID Control Strategy for DC-link Boost Voltage in Z-source Inverter. The Twenty Second Annual IEEE Applied Power Electronics Conference., 1145-1148.

[13] Rastegar, Fatemi., Mirzakuchaki, S., & Rastegar, Fatemi. S. M. J. (2008). Wide-Range Control of Output Voltage in Z-Source Inverter by Neural Network. The International Conference on Electrical Machines and Systems., 1653-1658.

[14] Tran QV, Chun TW, Kim HG, Nho EC,. (2009). Minimization of Voltage Stress across Switching Devices in the Z-Source Inverter by Capacitor Voltage Control. *Journal of Power Electronics,* 9(3), 335-342.

[15] Rostami, H., & Khaburi, D. A. (2010). Neural Networks Controlling for Both the DC Boost and AC Output Voltage of Z-Source Inverter. The 1st Power Electronic & Drive Systems & Technologies Conference. ., 135-140.

[16] Shen, M., Tang, Q., & Peng, F. Z. (2007). Modeling and Controller Design of the Z-Source Inverter with Inductive Load. The IEEE Power Electronics Specialists Conference., 1804-1809.

[17] Rajaei, A. H., Kaboli, S., & Emadi, A. (2008). Sliding-mode control of Z-source inverter. The 34th Annual Conference of IEEE Industrial Electronics., 947-952.

[18] Mo, W., Loh, P. C., & Blaabjerg, F. (2011). Model predictive control for Z-source power converter. The 8th International Conference on Power Electronics., 3022-3028.

[19] Ding, X., Qian, Z., Yang, S., Cui, B., & Peng, F. (2008). A Direct DC-link Boost Voltage PID-Like Fuzzy Control Strategy in Z-Source Inverter. The IEEE Power Electronics Specialists Conference., 405-411.

[20] Tang, Y., Xie, S., & Zhang, C. (2009). Feedforward plus feedback control of the improved Z-source inverter. *IEEE Energy Conversion Congress and Exposition.*, 783-788.

[21] Ding, X., Qian, Z., Yang, S., Cui, B., & Peng, F. (2007). A Direct Peak DC-link Boost Voltage Control Strategy in Z-Source Inverter. The Twenty Second Annual IEEE Applied Power Electronics Conference., 648-653.

[22] Sen, G., & Elbuluk, M. (2010). Voltage and Current Programmed Modes in Control of the Z-Source Converter. *IEEE Transactions on Industry Applications*, 46(2), 680-686.

[23] Yang, S., Ding, X., Zhang, F., Peng, F. Z., & Qian, Z. (2008). Unified Control Technique for Z-Source Inverter. The IEEE Power Electronics Specialists Conference., 3236-3242.

[24] Ellabban, O., Joeri Van, Mierlo. J., & Lataire, P. (2011). Capacitor Voltage Control Techniques of the Z-source Inverter: A Comparative Study. *EPE- European Power Electronics and Drives Journal.*, 21(4), 13-24.

[25] Ellabban, O., Joeri Van, Mierlo. J., & Lataire, P. (2011). Design and Implementation of a DSP Based Dual-Loop Capacitor Voltage Control of the Z-Source Inverter. *International Review of Electrical Engineering*, 6(1), 98-108.

[26] Ellabban, O., Joeri Van, Mierlo. J., & Lataire, P. (2012). A DSP Based Dual-Loop Peak DC-link Voltage Control of the Z-Source Inverter. *IEEE power Electronics Transactions.*, 27(9), 4088-4097.

[27] Peng, F. Z., Joseph, A., Wang, J., Shen, M., Chen, L., & Pan, Z. (2005). Z-Source Inverter for motor drives. *IEEE Transaction on power electronics.*, 20(4), 857-863.

[28] Ding, X., Qian, Z., Yang, S., Cui, B., & Peng, F. (2007). A New Adjustable-Speed Drives (ASD) System Based on High-Performance Z-Source Inverter. *IEEE Industry Applications Conference.*, 2327-2332.

[29] Ellabban, O., Van Mierlo, J., & Lataire, P. (2010). A new Closed Loop Speed Control of Induction Motor Fed by a High Performance Z-Source Inverter. The IEEE Electrical power and energy conference.

[30] Ellabban, O., Van Mierlo, J., & Lataire, P. (2011). Direct Torque Controlled Space Vector Modulated Induction Motor Fed by a Z-source Inverter for Electric Vehicles. *The III International Conference on Power Engineering, Energy and Electrical Drives.*

[31] Ellabban, O., Van Mierlo, J., & Lataire, P. (2011). A Comparative Study of Different Control Techniques for Induction Motor Fed by a Z-Source Inverter for Electric Vehicles. *The III International Conference on Power Engineering, Energy and Electrical Drives.*

[32] Peng, F. Z., Shen, M., & Holland, K. (2007). Z-source Inverter Control for Traction Drive of Fuel Cell- Battery Hybrid Vehicles. *IEEE Transactions on Power Electronics.*, 22(3), 1054-1061.

[33] Shen, M., Hodek, S., Fang, Z., & Peng, F. Z. (2007). Control of the Z-Source inverter for FCHEV with the battery connected to the motor neutral point. *IEEE Power Electronics Specialist Conference.*, 1485-1490.

[34] Dehghan, S. M., Mohamadian, M., & Yazdian, A. (2010). Hybrid electric vehicle based on bidirectional z-source nine-switch inverter. *IEEE Transactions on Vehicular Technology*, 59(6), 2641-2653.

[35] Ellabban, O., Van Mierlo, J., & Lataire, P. (2010). Control of a Bidirectional Z-Source Inverter for Hybrid Electric Vehicles in Motoring, Regenerative Braking and Grid Interface Operations. The IEEE Electrical power and energy conference.

[36] Ellabban, O., Van Mierlo, J., Lataire, P., & Hegazy, O. (2010). Control of a High-Performance Z-Source Inverter for a Fuel Cell/ Supercapacitor Hybrid Electric Vehicles. The 25th World Battery, Hybrid and Fuel Cell Electric Vehicle Symposium & Exhibition.

[37] Ellabban, O., Van Mierlo, J., & Lataire, P. (2011). Control of a Bidirectional Z-Source Inverter for Electric Vehicle Applications in Different Operation Modes. *Journal of Power Electronics*, 11(2), 120-131.

Chapter 11

Mathematical Analysis for Response Surface Parameter Identification of Motor Dynamics in Electric Vehicle Propulsion Control

Richard A. Guinee

Additional information is available at the end of the chapter

http://dx.doi.org/10.5772/54483

1. Introduction

This chapter addresses the topographical examination of various mean squared error (MSE) cost surface structures and selecting the most suitable MSE fitness function for accurate brushless motor drive (BLMD) dynamical parameter system identification (SI) of BLMD shaft load inertia and viscous damping for electric vehicle controlled propulsion. The parameter extraction procedure employed here is in the offline mode for optimal drive tuning purposes during the installation and commissioning phase of embedded BLMD systems in high performance electric vehicle torque, speed and position control scenarios. Two types of penalty function, based on the transient step response of the permanent magnet (PM) motor shaft velocity and its stator winding current feedback in torque control mode [1,2], are examined here for arbitration of a suitable choice of cost objective function as the response surface in the accurate extraction of the BLMD dynamics. The choice of a particular MSE cost surface as an objective function in BLMD load parameter identification is motivated by the need for reliable tuning of the proportional and integral (PI) term settings during the drive installation phase for controller robustness and optimal performance in adjustable speed drive (ASD) or torque controlled embedded PM motor applications for electric vehicle propulsion. This chapter will focus on the mathematical analysis of embedded motor drive dynamical parameter identification over an MSE multiminima response surface with the following key results obtained:

a. the development of a novel quadratic mathematical model approximation for the investigation of the (i) nature of the MSE objective function and (ii) existence of a bounded MSE global minimum stationary region, based on transient step response motor current feedback signals, for mechanical parameter identification in sensorless drive torque control of electric vehicles.

b. the examination of the phenomenon of multiminima proliferation in the MSE cost formulation due to target data choice and 'noisiness' arising from evaluation of pulse width modulated (PWM) edge transition times during BLMD simulation [1,2].

c. the measurement of cost surface selectivity based on shaft velocity and current feedback target data and the decision favouring the choice of the latter data training record as the target function for dynamical parameter identification

d. the development of a novel parameter quantization metric to overcome cost surface 'noisiness', arising from computational uncertainty in the simulated PWM edge transitions [1], for avoidance of local minima trapping in the MSE cost surface during identification of the BLMD dynamics.

e. the development of a novel parameter convergence radius measure of encirclement of the cost surface stationary region global minimum, arising from the parameter quantization metric in (d), for determination of the bounds of accuracy that can be imposed on the returned estimates of the global optimum dynamical parameter vector during BLMD identification.

1.1. Motivation

BLMD control tuning is necessary during the commissioning phase of embedded drives applications, for accurate torque and speed control in electric vehicle propulsion systems setup [3,4], accurate robotic end effector [5] or CNC tool positioning [6], where detailed apriori knowledge of expected drive load inertia and friction parameters are unknown to the electric motor drive supplier/manufacturer in the intended application beforehand. The choice of ASD [7] in high performance industrial applications, such as a small electric vehicle [4], robot manipulator [8, 9, 10] or machine tool feed drive [6,4], is usually based on consideration of a BLMD manufacturer's catalogued specifications, relating to drive performance capabilities and limitations, by the customer or embedded drive equipment designer/ manufacturer. The BLMD selection is often done independently of the motor drive manufacturer by the equipment designer for reasons of embedded systems design confidentiality and second sourcing of matching drive equivalents from different manufacturers for the purpose of cost reduction and embedded product protection from obsolescence via alternative drive substitution. The range of motor sizes available and spread of possible BLMD embedded applications has resulted in the provision of flexible drive tuning facilities with either manual or autotuning features [11] by motor drive manufacturers as a sales and marketing expedient to embedded equipment designers. This flexible approach to drive tuning policy eliminates the need for the BLMD manufacturer to participate in the detailed design of embedded drive applications except in the provision of motor drive systems with high output torque and speed ranges to cater for a range of anticipated high performance applications [12,8,13]. BLMD systems with high peak current capability and fast response times due to low PM rotor mass are designed [6, 7] to handle large inertial load torques [14] experienced in robotic applications [4, 5] and electric vehicle propulsion systems, with a no-load to full-load inertia variation [9] of 10 is to 1. It is in response to this background of applications diversity, regarding the particular design details of embedded drive products about the size of inertial loads and friction coefficients encountered [4,6,9], that the present work on cost surface analysis for parameter extraction in electric vehicle control is directed from a motor manufacturer's perspective.

Since the possible variation in the load dynamics of an intended BLMD application is unknown at the outset the initial task here for an end user is to identify the actual load inertia and friction coefficients experienced during startup of a given embedded drive in the offline mode for robust PI controller tuning [15]. In this scenario the customer has the flexibility of manually tuning the BLMD speed loop, which is provided as a PI adjustment option along with procedural details for tuning by the motor manufacturer, during the setup and commissioning phase for a particular ASD application. The challenge then posed for the motor drive designer in this instance is the provision of an automated tuning facility for the velocity or torque loop during the commissioning stage thus eliminating the need for any manual input by the customer. This feature requires the identification of a fixed embedded load configuration during setup and subsequent automated optimal configuration of the velocity controller PI terms [15]. In the absence of embedded load information the cost surfaces and identification methods investigated here focused on inertial load spreads for vehicular and robotic applications [9] of up to ten times the inherent motor shaft inertia as recommended by the BLMD manufacturer for the drive [16] modelled in [1].

The concept of a simulated cost surface is developed here [17] as an objective function to facilitate parameter extraction of the installed drive dynamics, during offline BLMD system identification, with MSE minimization. This methodology provides useful insight into the nature and formulation of the most suitable MSE objective function to be minimized, based on actual drive experimental test data available and BLMD model simulation, coupled with an effective system identification (SI) strategy for accurate motor parameter extraction [18]. This approach can also be used as an alternative means of providing the optimal set of extracted parameter estimates from inspection of the global minimum location on the simulated cost surface with embedded local minima. Furthermore it can be used as a basis for comparison of the effectiveness of the actual identification search strategy deployed in terms of the accuracy of returned parameter estimates. The problem of inertia (J) and friction (B) parameter extraction of an actual BLMD system over a sinc function shaped multiminima cost surface [19], based on step response feedback current (FC) target data which has a constant amplitude swept frequency characteristic, is investigated as a test case using response surface simulation.

The global minimum estimation, from response surface simulation discussed in section 2.0 below, is targeted towards offline identification of the fixed dynamical load possibly encountered by an embedded BLMD system during the setup and commissioning phase. This is necessary for optimal tuning of the installed BLMD velocity and position loops in any high performance electric vehicle and industrial application. The present work on optimal parameter estimation is mainly concerned with the offline identification of the worst case inertial load that could possibly be experienced by an installed embedded BLMD. This is articulated here through BLMD simulation in torque control mode, using the full reference model developed in [1, 2], and drive experimental step response measurements with three known test cases of shaft load inertia, for validation of the accuracy of the parameter identification strategy, corresponding to:

a. the no-load rotor inertial value $J_T = J_m$,

b. medium shaft load inertia $J_T \sim 4J_m$ and

c. large shaft load inertia $J_T \sim 7J_m$.

where J_T is the total inertia consisting of rotor J_m and additional shaft load J_l with

$$J_T = J_l + J_m. \tag{1}$$

The problem of a numerically 'noisy' multiminima cost function resulting in non optimal parameter convergence because of local minimum trapping, associated with the adoption of the BLMD reference model in [1, 2] during simulation, in motor parameter identification is examined [20, 21]. An explanation is provided as to the existence of 'false' local minima plurality with inaccurate resolution of PWM edge transition times, associated with the choice of fixed step sizes Δt in BLMD model simulation, in both the current feedback I_{fj} and shaft velocity $V_{\omega r}$ MSE objective functions. An explanation is also furnished as to the existence and proliferation of genuine local minima with the observed feedback current (FC) target data used in penalty cost surface generation, which will be shown to posses an inverted sinc function-like shape. Details are presented, through MSE response surface simulation with coarse step sizes chosen initially for the inertia J and friction B parameters employing shaft velocity (parabolic cost surface) and feedback current (sinc-like surface) experimental target data respectively, to shed light on the numerical noise problem for SI purposes. Both simulated MSE response surfaces reveal on a macro-scale the presence of a 'line minimum' of possible feasible solutions in a stationary region, enveloping a global extremum within the central surface fold, principally in the B-parameter direction. A novel mathematical approximation [17], which provides verification of the cost surface shape in both cases, is given and is used to provide information on the existence of a unique global minimum with an accompanying optimal parameter set $\overline{X}_{opt} = \{\overline{J}_{opt}, \overline{B}_{opt}\}^T$ instead of a multiplicity of candidate options, $\overline{X}^j_{opt} = \{\overline{J}_{opt}, \overline{B}^j_{opt}\}^T$, along a '$B$ - line minimum', for $j = 1,2$ .... Details of BLMD model simulation at a finer parameter step size δX, which illuminates the problem of a noisy cost surface, are also provided for both objective functions. An independent statistical analysis appraisal of the computation 'noise' voltage engendered in the search for accurate PWM transitions, based on a novel theoretical estimation [18] for the random error pulse energy expectation associated with PWM replication with chosen simulation step size Δt, is also provided. This probability analysis in itself provides a useful insight into the induced noise mechanism with chosen time step size and highlights the magnitude contribution of the random error 'noise' voltage with PWM resolution to the overall accuracy in the BLMD model simulation exercise. The effect of inherent 'noisy' evaluation of the PWM edge transition times during BLMD simulation is transferred as a lack of smoothness in the simulated construction of the MSE cost surface at the micro-scale for very low step changes δX in the BLMD dynamical parameters J and B.

A novel mathematical analysis [21] is presented, via embedded quadratic curve fitting in the MSE cost surface, to establish the worst case parameter quantization step size δX^{L} necessary to overcome cost function 'noisiness'. This analysis also provides a radius of convergence r_X in parameter space about the global minimum for any parameter identification search strategy and establishes a bound on the limits of accuracy for the returned optimal parameter estimate $\hat{X}_{opt} = \{\hat{J}_{opt}, \hat{B}_{opt}\}^T$. Furthermore this methodology provides a sensitivity measure of the MSE cost surface selectivity for both the step response shaft velocity and current feedback response surfaces in the neighbourhood of the global extremum $\overline{X}_{opt}$. This surface variability metric dependency on elemental parameter variation δX can then be used to decide on the best objective function for parameter extraction purposes based on the accuracy of the returned estimate. The choice of the FC target data is explained for its excellent coherence properties, based on frequency and phase attributes from step response tests, in checking BLMD model fidelity and accuracy and also for its high selectivity in penalty cost function formulation for accurate parameter identification. Furthermore it will be shown that there is an improvement in FC cost surface selectivity with longer data training records while the converse effect is manifested for shaft velocity target data with measurement data length in the reduction of cost surface curvature in the vicinity of the global minimum. These current feedback step response attributes arbitrate in its favour as the most suitable choice of target data in MSE cost function formulation.

In the absence of embedded drive application details from the BLMD manufacturer [17] no precise limits on the desired accuracy of the returned J and B parameter estimates could be affixed to the parameter identification strategy for velocity controller tuning purposes in the commissioning phase. However the use of a quantized metric δX^{L}, as mentioned previously, in parameter space puts a limit on the parameter resolution accuracy possible during identification of the BLMD dynamics in electric vehicles. It should be noted that without the imposition of this parameter quantization strategy there is a risk of false minimum trapping of the identification search algorithm in a 'noisy crevice' [18] in a side-wall of a cost surface, besides local minimum capture, well away from the global minimum estimate. This novel quantization procedure in parameter space, which eliminates the effect of simulation step size related computation induced noise, results in the availability of a smooth cost surface over which a parameter identification search algorithm will work and converge to an optimal estimate [17,18,21]. One other benefit of the parameter quantization process is that it divides up parameter space and restricts the identification strategy to a countable number of parameter lattice points [18] and thus minimizes the search time to global optimality.

A further aspect of concern besides false minimum trapping is that all optimization algorithms for BLMD parameter identification proceed in a continuous search of parameter space to a convergence estimate of the parameter vector sought with an end stopping criterion [22,23]. The norm of cessation of the optimal parameter search strategy is generally based on the smallness of cost reduction over successive iterations within a specified error bound ε at termination. The termination criteria are generally not focused on the smallest percentage variation of the parameter estimates acceptable. However with the quantization δX^L of parameter space for response surface smoothness, limits for parameter resolvability can be imposed by restricting the identification search process to an integral number k of quantum steps $k\delta X$ commensurate with the percentage accuracy $\%X$ required in absolute terms such that $\%X = k\delta X$. This restricted step approach, in terms of the specified parameter accuracy sought for BLMD tuning purposes during the setup and commissioning phase, can reduce the SI computation time to optimality [18].

2. Response surface simulation and analysis [17]

The concept of a simulated response surface (RS) is presented as an aid to motor dynamical parameter optimization in high performance Brushless Motor Drive (BLMD) identification with a multiminima objective function. This methodology provides useful information concerning the formulation and nature of the most suitable objective function to be minimized, based on actual drive experimental test data available and BLMD model simulation, coupled with an effective system identification (SI) strategy for accurate motor parameter extraction. This simple approach, although computationally intensive, can also be used as an alternative means of providing the optimal set of parameter estimates from inspection of the global minimum location on the simulated cost surface with embedded local minima. Furthermore it can be used as a basis for comparison of the effectiveness of other identification search strategies deployed, such as the Powell Conjugate Direction search method [18] and Fast Simulated Diffusion algorithm [20,21], in terms of the accuracy of returned parameter estimates. The problem of inertia J and viscous friction B parameter extraction of an actual BLMD system over a sinc-function (sinx/x) shaped multiminima cost surface, based on step response feedback current (FC) target data which has a constant amplitude swept frequency characteristic, is investigated using response surface simulation. The choice of the FC target data is based on its excellent coherence properties [24] from step response testing, for checking BLMD model fidelity and accuracy and for the penalty cost function formulation in SI. This difficulty with a multiminima objective function converging to a non optimal parameter estimate, associated with the adoption of the FC target data for motor parameter identification, is examined in the FSD method [20, 21]. An explanation is provided as to the existence of local minima plurality with the observed FC target data used in the sinc-like penalty cost surface generation. All classical optimization techniques [22], with the exception of modern statistical methods [21], are known to have difficulty with this type of cost surface in identifying the optimal parameter vector. The problem arises with initialization of the search strategy far from the global minimum resulting in possible local minimum trapping and non optimal convergence of the cost minimization algorithm during the parameter extraction process. This response surface [RS] methodology, however, provides a simple and effective alternative to classical methods in acquiring an accurate estimate of the global minimum. Results are presented, which demonstrate the efficacy and reliability of the RS method in returning accurate estimates for 'known' values of the BLMD shaft dynamics. The application of this FC step response related multiminima cost function in parameter ex-

http://dx.doi.org/10.5772/54483

traction is compared with the alternative parabolic shaped shaft velocity objective function for cost surface selectivity in the vicinity of the global minimum and for accuracy of the returned identified parameter estimates. A mathematical approximation analysis is provided for verification of the cost surface shapes resulting from the deployment of step response FC and shaft velocity as target data in objective function formulation.

2.1. Cost function formulation [18]

Response surface simulation is a useful graphical tool [25] in system identification and can easily be applied to motor parameter extraction and BLMD model validation. This visual concept, which has been used in process control optimization [25], provides an intuitive insight into the topographical structure of the cost function to be minimized and the rapid location of the global minimum. It also provides information on the most suitable identification search strategy that should be adopted in parameter space to obtain an accurate estimate $\hat{\boldsymbol{X}}_{opt}=\{\hat{x}_{1\,opt},\ \hat{x}_{2\,opt}\}^T$ of the motor dynamics where the inertia $J \equiv x_1$ and viscous friction $B \equiv x_2$ are the coded variables. The location of the global minimum stationary point can be obtained by inspection from the simulated cost surface. This approach, although computationally expensive, can be used to secure an independent alternative optimal estimate $\overline{X}_{opt}=\{\bar{J}_{opt},\ \overline{B}_{opt}\}^T$ as a reference against which the accuracy of other parameter identification search schemes such as the Fast Simulated Diffusion [26] can be judged.

BLMD parameter extraction is generally based on the minimization of the errors of fit e_k between the observed motor drive target data and BLMD model responses in terms of the controlled parameter vector X. This identification process results in the adjustment of the J and B parameters towards global optimality. The search strategy is performed in the neighbourhood of the global extremum using the least squares error criterion in the cost function formulation between each value of a time series

$$t=\left\{t_k \middle| t_k=kT,\ T=\Delta t,\ 1\le k\le n,\ k\in N\right\} \tag{2}$$

of n experimental sample points of the actual motor drive response $g(t_k)$ as the target data reference and the corresponding simulated model response $f(X,\ t_k)$. The objective function $E(X)$ is defined as the mean square error (MSE) from the residual vector as

$$e^T=\left[e_1,e_2,\ldots,e_n\right] \tag{3}$$

as

$$E(\mathbf{X})=\tfrac{1}{n}\mathbf{e}^T\mathbf{e}=\tfrac{1}{n}\sum\nolimits_{k=1}^{n}e_k^2 \tag{4}$$

where

$$e_k = g(t_k) - f(\mathbf{X}, t_k) \tag{5}$$

The *MSE* generates an error response cost surface in parameter X space based on target data from one of the internal test points in [1]. The Powell Conjugate Direction (PCD) [23] and Fast Simulated Diffusion optimization techniques [27] can be applied in conjunction with the BLMD model in [1] to the response surfaces corresponding to motor shaft velocity $V_{\omega r}$ and winding FC I_{fa} target data respectively, obtained in torque control mode for different shaft load inertia listed in [1], for optimal parameter X_{opt} extraction.

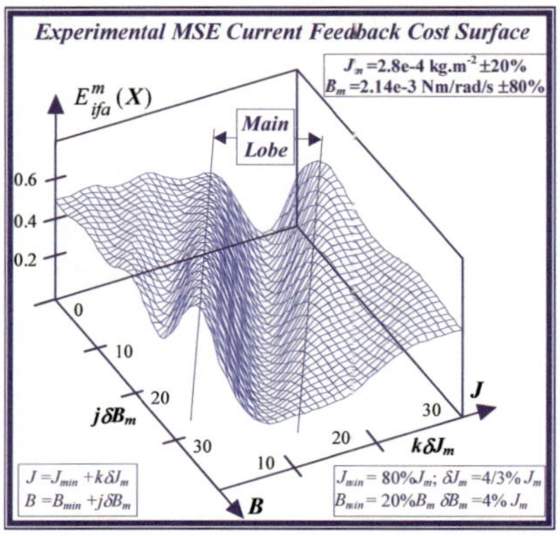

Figure 1. Experimental FC Cost Surface

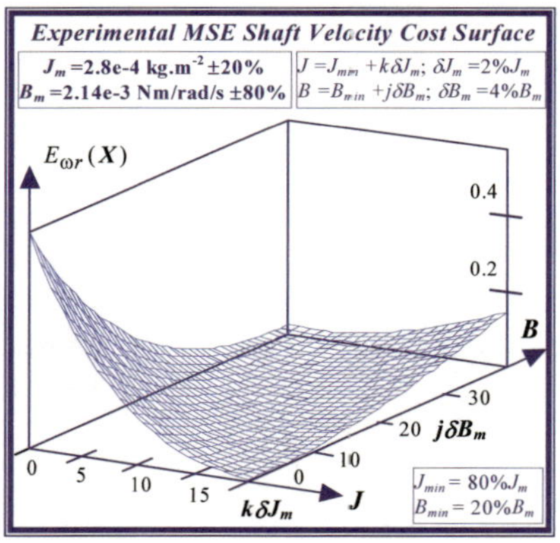

Figure 2. Shaft Velocity Cost Surface

The simulated response surfaces $E(X)$ are derived from BLMD simulation, with a fixed time step of 1μs and appropriate decimation factor, using the model test point o/p $f(X,t_k)$ in conjunction with the sampled experimental target data

$$g(t_k) \in \{V_{\omega_r}(kT), I_{fa}(kT)\} \tag{6}$$

as the target reference. These penalty cost functions are depicted in Figs.1 and 2 for zero shaft load conditions over parameter space $\mathbf{X}=[J,B]^{T}$ with a crude mesh size δX chosen as per Table 1 to initially determine surface shape, according to the rotor inertia and friction tolerances likely to be encountered in practice. The experimental test data training records used in the MSE formulation for each objective function are displayed in Figs.1 and 2.

MSE Cost Surface Type $E(X)$	Current Feedback: $E_{ifa}(J, B)$		Shaft Velocity: $E_{\omega r}(J, B)$	
Data Training Record $g(t_k)$	Current Feedback: $I_{fa}(t_k)$		Shaft Velocity: $V_{\omega r}(t_k)$	
No. of Data Points N_d @ 20μs	4095		4095	
BLMD Parameter varied x	J_m (kg.m^2)	B_m (Nm/rad/sec)	J_m (kg.m^2)	B_m (Nm/rad/sec)
Nominal Parameter Value x_m	2.8×10^{-4}	2.14×10^{-3}	2.8×10^{-4}	2.14×10^{-3}
Parameter Tolerance Band $\mathbf{\Delta} x$	±20%	±80%	±20%	±80%
Crude Parameter Step size $\boldsymbol{\delta} x$	1.33%	4%	2%	4%
No. of Parameter Steps N_x	30	40	20	40
Parameter Value Returned	2.99×10^{-4}	~1.54×10^{-3}	3.024×10^{-4}	~1.626×10^{-3}
Assumed Optimal Parameter Vector X_o for Response Surface Analysis				
X_o	3.0×10^{-4}	2.14×10^{-3}	3.0×10^{-4}	2.14×10^{-3}

Table 1. Experimental Cost Surface Formulation for Zero Shaft Load (NSL) Conditions

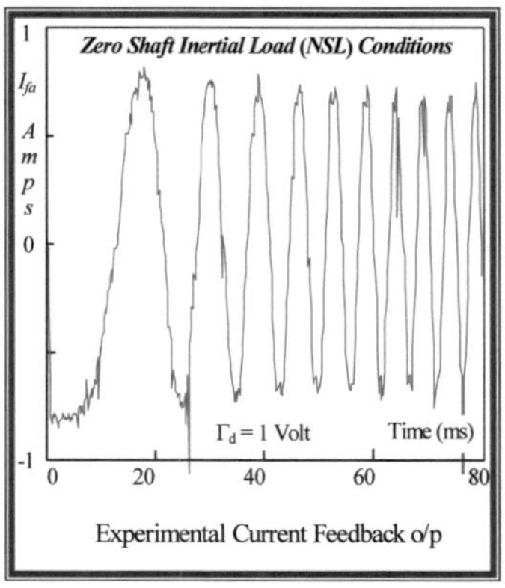

Figure 3. BLMD Current Feedback

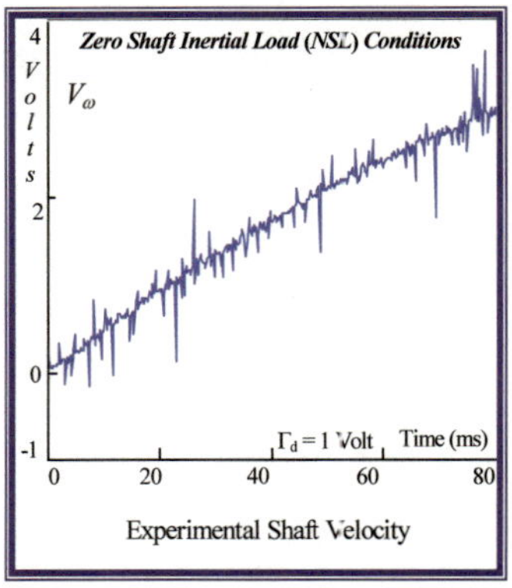

Figure 4. Rotor Shaft Velocity V_{ω}

The anticipated variation in the search cost, likely to be encountered during BLMD system identification (SI) over the parameter tolerance band of interest, can be gauged from cross sections through the chosen response surface at nominal values of the rotor parameters $[J_m, B_m]^T$. The cost variations associated with specific dynamic parameters are illustrated in Figs.5 and 6 for motor current feedback and in Figs.7 and 8 for shaft velocity target data. These cross sections provide important information regarding the surface shape and curvature and consequently about the nature of the stationary points found and type of SI search algorithm that should be deployed over such hitherto surface 'terra incognita'.

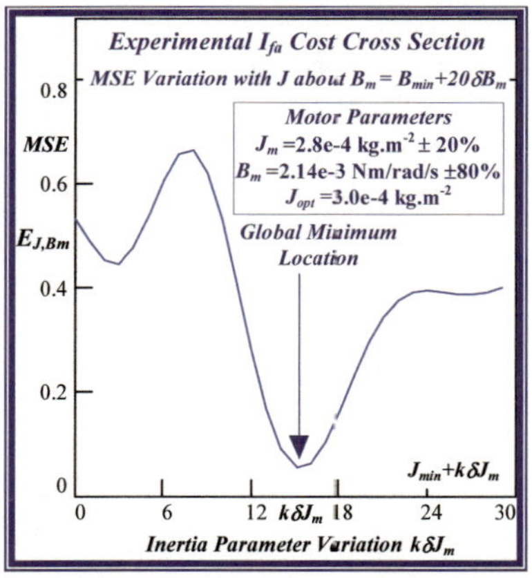

Figure 5. MSE-I_{fa} Cross section at B_m

http://dx.doi.org/10.5772/54483

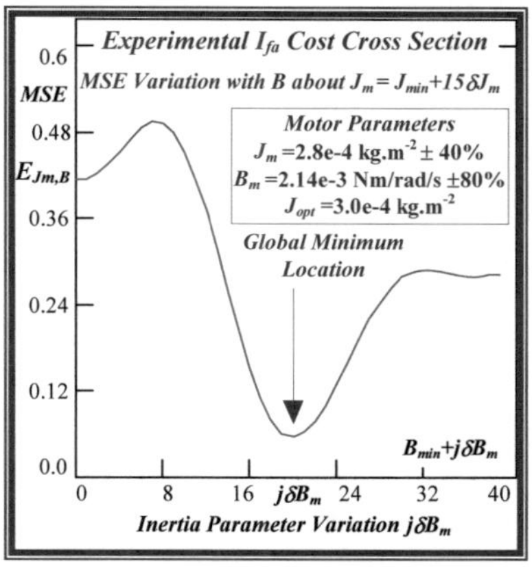

Figure 6. MSE-I_{fa} Cross section at J_m

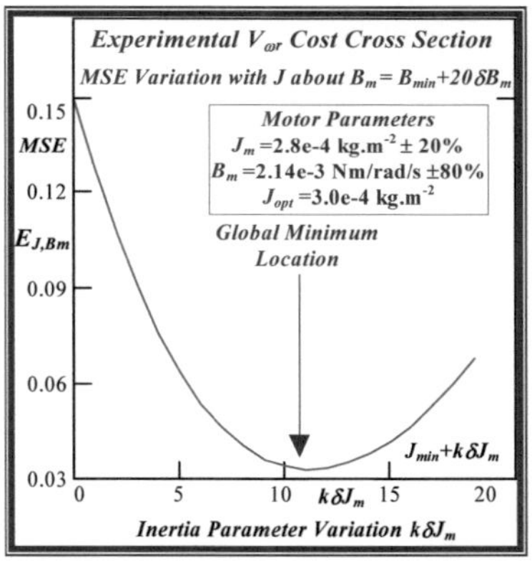

Figure 7. MSE-$V_{\omega r}$ Cross section at B_m

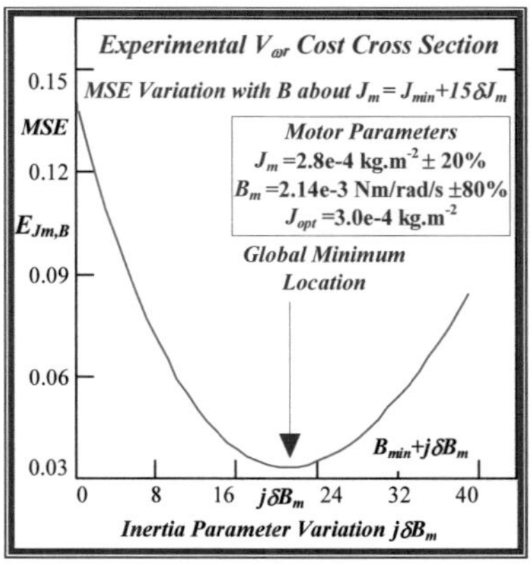

Figure 8. MSE-$V_{\omega r}$ Cross section at J_m

The FC cost 'landscape' highlights the existence of several parabolic shaped ridges, interspersed with embedded synclines within its sinc-like folded topography, with a consequent plurality of local minima. The cost terrain also shows the presence of a stationary elliptical shaped ridge system centrally located in the contour map of Fig. 9 with the possible existence of a 'line minimum' [25] along the principal/major axis. These multiminima folds are manifested in the constructive and destructive interference patterns encountered in the frequency ramp up of the FC sinusoid, when compared with the optimal parameter reference or test data waveform, during the transient phase of motor acceleration. The shaft velocity cost surface is parabolic shaped as seen from the contour map in Fig. 10 but is less selective than its FC equivalent in the vicinity of the global minimum when the respective cost surface cross sections with equivalent parameter grid sizes are compared.

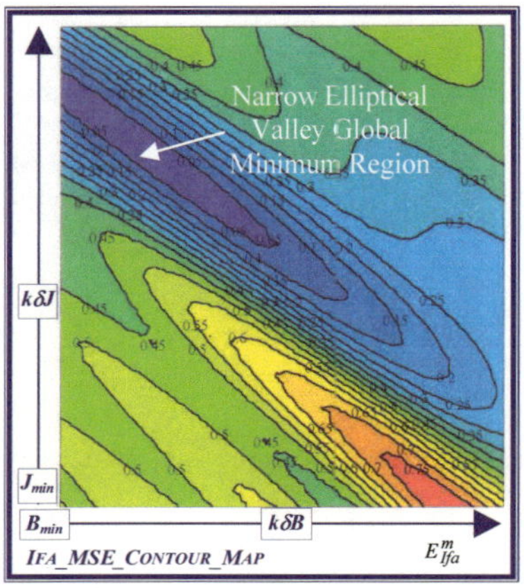

Figure 9. Experimental MSE-I_{fa} Contour Map

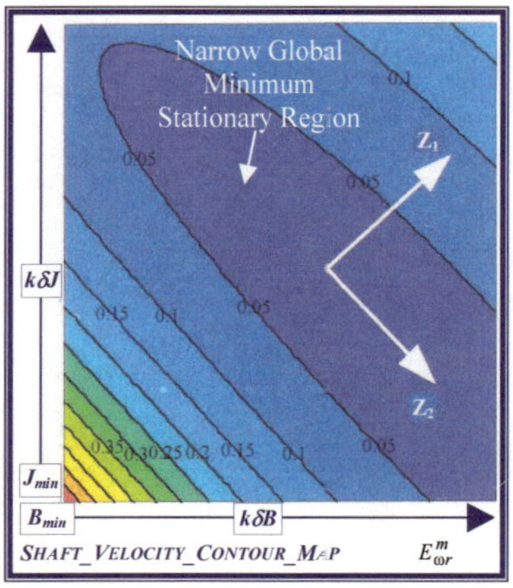

Figure 10. Experimental MSE-$V_{\omega r}$ Contour Map

It is evident from Figs. 9 and 10 that both objective functions possess long wedge shaped stationary valleys in the response surfaces with no 'apparent' clearly defined global minimizer. The observed near linear dependence of the surface shape on the parameters in a 'line minimum' along the valley floor indicates that *B* is commensurate with *J* in the ratio *J*/*B* which is the dynamical time constant τ_m of the motor. The *B* parameter, which is the least likely of the two to vary in the electromechanical drive applications [8,9] can to be acquired from dynamic testing as per [1] to free the other parameter for identification purposes. This reduces the identification problem to single parameter extraction in *J* or alternatively in τ_m, where parameter decoupling is non essential, for controller design purpose.

Response surface simulation provides an alternative route of accurately estimating the optimal parameter vector X_{opt} by means of inspection of the surface minimum cost. This method, although computationally expensive, can be used as a yardstick by which the overall convergence performance of other identifications schemes [21] can be contrasted, such as FSD, over a range of motor shaft inertial loads. The response surfaces can be simulated initially using a coarse parameter mesh size, for a range of supposedly 'unknown' motor inertial load test cases for shaft velocity and current feedback MSE objective functions, for rapid location of the global minimum. Further refinement in mesh size can be made down to the parameter step sizes necessary in the vicinity of the global minimum for accurate resolution of the optimal parameter set. Results, which demonstrate the accuracy and effectiveness of RS simulation, are presented for global minimum estimates of motor shaft inertia which are in close agreement with known test inertial load values.

2.2. Novel mathematical analysis of response surface [18] – Modelling and simulation

Response surfaces can be generated for the BLMD shaft velocity and current feedback step responses, as the mean squared error cost function between an actual drive experimental target data record and simulated model responses, by varying *J* and *B* over the two dimensional dynamical parameter space of interest. This graphical procedure is then used to shed light on the shape of the respective cost surfaces and to make a decision as to the most efficient parameter identification strategy to be deployed in each case. Inspection of each of the 2-D MSE response surfaces reveal the existence of 'open' wedge shaped stationary regions principally in the *B*-parameter direction containing what appears to be a global 'line' minimum in both cases. From a parameter identification perspective such open stationary regions would mean an infinite number of admissible solutions and thus uncertainty in the parameters extracted. The presence of such a difficulty would require careful measurement of one the parameters, in this case the friction as this is the principal direction that the line minimum appears to exists, in order to free the other (*J*) for identification. A novel mathematical analysis is presented in this chapter to determine whether or not these embedded stationary regions are open. This approach is articulated by formulating a simple quadratic model approximation of the cost surface stationary regions over a small neighbourhood of parameter space, with interacting *J* and *B* terms, for proposed model accuracy. The BLMD model step responses are also approximated by simple analytical expressions over response time spans that are very short by comparison with the dynamical time constant τ_m for validation and accuracy of the response surface quadratic model approximation. These simple step response representations, in which the parameters *J* and *B* can be adjusted over the space of interest for local cost surface generation and analysis of the stationary region, are included along with the relevant experimental target data in the cost surface quadratic model approximation. This mathematical analysis, employing the simplified quadratic model for both cost surfaces, can be used to show:

- that the stationary regions for the current feedback and shaft velocity objective functions are closed and bounded indicating the presence of a trapped global minimum,
- how closely the dynamical *J* and *B* parameters are coupled by making a comparison of the extracted quadratic model eigenvalues,
- that a line minimum exists principally in the *B* parameter direction and quantifies the extent of this *B*-line minimum by the eigenvalue ratio
- establishes the degree of ill conditioning for the global minimum solution parameter vector estimate X_S extracted from the minimized quadratic model.

Furthermore this analysis also demonstrates that the current feedback response surface has better selectivity in the global stationary region than the shaft velocity equivalent with increasing data record lengths. This outcome helps in the decision analysis that favours the use of current feedback target data in cost function formulation for dynamical parameter identification.

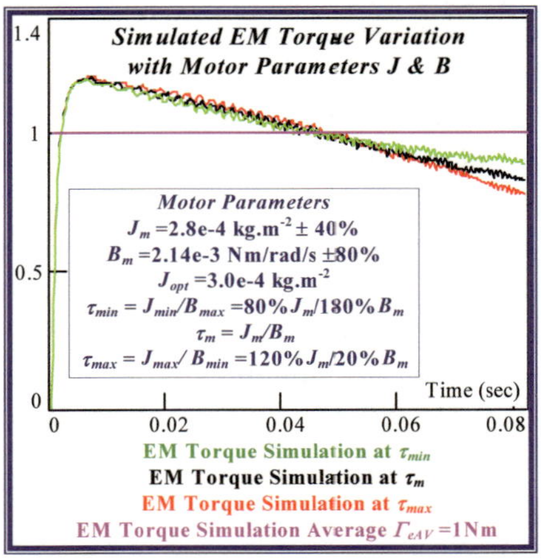

Figure 11. EM Torque Variation with B_m & J_m

The observed topographical features in the above penalty response surfaces can be anticipated from the following approximation analysis. Initially the developed electromagnetic torque Γ_e is at a maximum for unit torque demand step input Γ_d and remains so for a very short time as per the BLMD model simulation in Fig.11 until the shaft speed starts to build up exponentially as in Fig. 12 with time constant τ_m. The back-emf term v_{ej} in [1] becomes substantial causing a

http://dx.doi.org/10.5772/54483

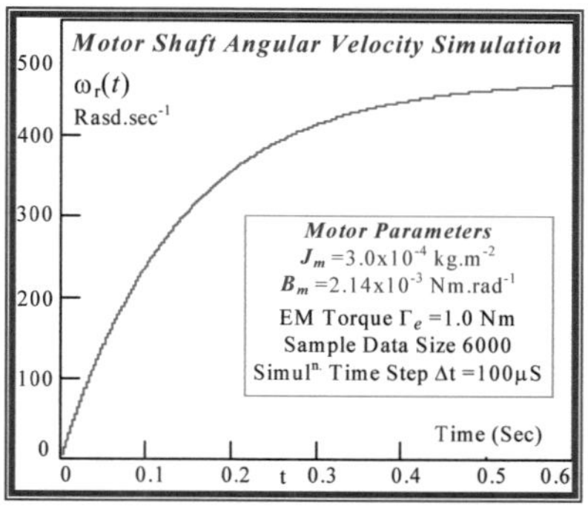

Figure 12. Simulated Shaft Velocity

decrease in winding current i_{js} which reduces the applied torque. Furthermore the increased rotor angular velocity ω_r causes the machine impedance angle ϕ_z in [1] to approach $\pi/2$ and forces the winding currents into quadrature with the current command signals i_{dj} with subsequent torque reduction as in [1]. The variation in applied motor torque with the worst case spread of dynamical time constant τ_m values, observed for the parameter tolerance ranges in Table 1 with zero shaft load conditions, is small over the motor acceleration period ($\hat{t}$ =0.08sec≈60%τ_m) shown in Fig.11. The average value of applied mechanical torque Γ_{em} is 1Nm and is assumed constant over the period $\hat{t}$ for tractability reasons in the following analysis of the cost surfaces used in the PCD and FSD methods of parameter extraction. However this value deteriorates over longer time spans as the winding current moves out of phase alignment with current demand as motor speed increases and thus with the back *EMF*. The simulated shaft speed variation with time, based on the nominal parameter vector X_m in Table 1 and displayed in Fig.12 for a step i/p torque demand Γ_d (~1v) is given by (7)-(a)

$$\omega_r^m(t) = K_m(1-e^{-t/\tau_m}) \quad \text{with} \quad K_m = \Gamma_{em}/B_m \quad \text{and} \quad 0 \le t \le \hat{t} \qquad (a)$$

$$\omega_r^0(t) = K_0(1-e^{-t/\tau_0}) \quad \text{with} \quad K_0 = \frac{\Gamma_{em}}{B_0}; \; \tau_0 = \frac{J_0}{B_0}. \qquad (b) \qquad (7)$$

Similarly the corresponding shaft speed variation with time at the *assumed* optimum parameters $\{J_o, B_o\}$ in Table 1, which are be identified from cost surface trial analysis, is given by (7)-(b)

The sampled motor speed 'test' data $\omega_r^o(t_k)$ generated via (7)-(b) can now be used as target reference 'test' data in the simulated trial cost function $E_{\omega r}^{O}$ for analytical purposes. The optimal parameter set X_o is supposedly unknown and the task here is to obtain a good estimate $[\bar{J}_o, \bar{B}_o]^T$ of this vector in the following cost surface analysis for verification of the RS strategy. The variation in the time constant τ_m over the permitted parameter tolerance ranges employed in the response surface generation, such as those in Figs 1 and 2 relying on experimental test data, is insufficient to cause departure from nominal applied torque Γ_{em} for the short time span shown in Fig.11. The shaft speed variation in this instance is approximated by

$$\omega_r(t) = K(1 - e^{-t/\tau}) \quad \text{with} \qquad K = \frac{\Gamma_{em}}{B}; \;\; \tau = \frac{J}{B}. \tag{8}$$

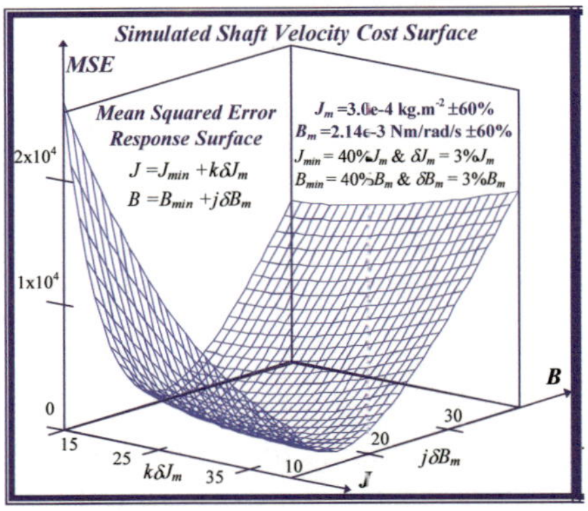

Figure 13. Simulated Velocity Cost Surface

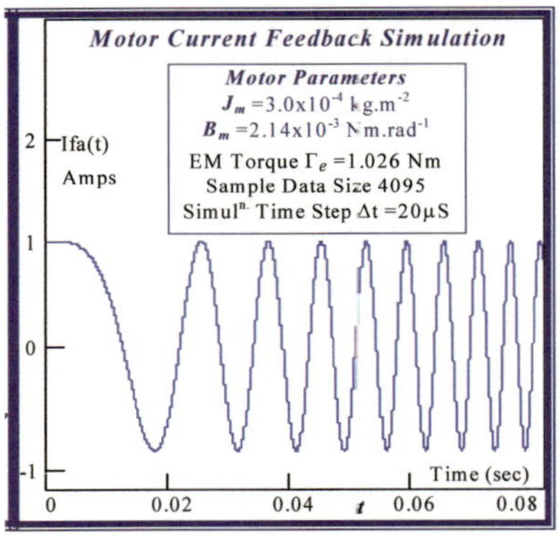

Figure 14. BLMD Current Feedback

The resulting *MSE* cost function construct, illustrated in Fig.13 with details in Table 2, is for simulation purposes given by

$$E^m_{\omega r}(\mathbf{X}) = \frac{1}{N_d}\sum(\omega_r - \omega_r^m)^2 \tag{9}$$

with target data ω_r^m. The parabolic cost variations associated with specific dynamic parameters for shaft velocity target data are illustrated in Figs.15 and 16. The corresponding winding current feedback $i_{fa}(t)$ has the characteristics of a frequency modulated sinusoid during the exponential buildup of motor shaft speed in that it exhibits the features of a constant amplitude swept frequency waveform as shown in Fig.14. The effect of shaft speed increase on the phase angle ϕ of the FC response is determined from (8) as

$$\varphi = \int_0^t \omega_r(x)dx = K\left\{t + \tau(e^{-t/\tau} - 1)\right\} = Kt - \tau\omega_r(t) \tag{10}$$

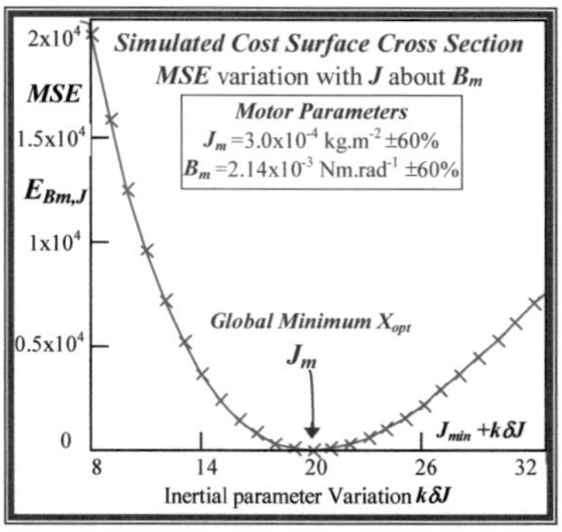

Figure 15. MSE-$V_{\omega r}$ Cross section at B_m

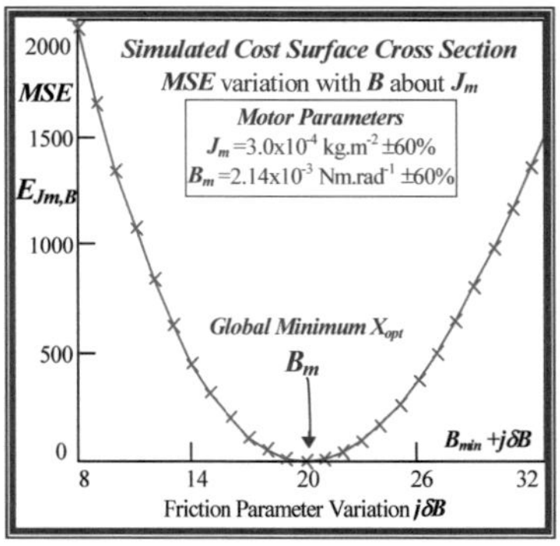

Figure 16. MSE-$V_{\omega r}$ Cross section at J_m

The frequency modulated FC, which is current regulated by G_I in [1], is given by

$$I_{fa}(t) = I_f \cos p\varphi = I_f \cos p\left(Kt - \tau\omega_r(t)\right) \tag{11}$$

with $I_f \approx 1$ amp for a unit step torque demand i/p. The resultant FC cost surface generated from simulation in Fig.17, with parameter grid sizes in Table 1, is based on the target shaft velocity $\omega_r^m(t)$ in (7)-(a) for nominal values of the dynamical parameters X_m with

$$E_{Ifa}^m(\mathbf{X}) = \frac{1}{N_d}\sum\left(I_{fa} - I_{fa}^m\right)^2 \tag{12}$$

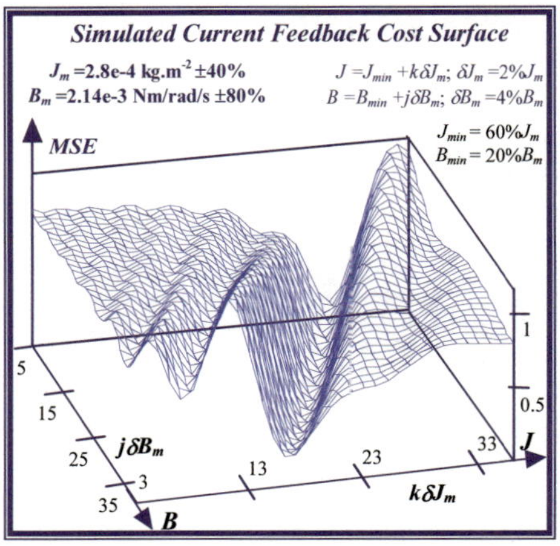

Figure 17. Simulated FC Cost Surface

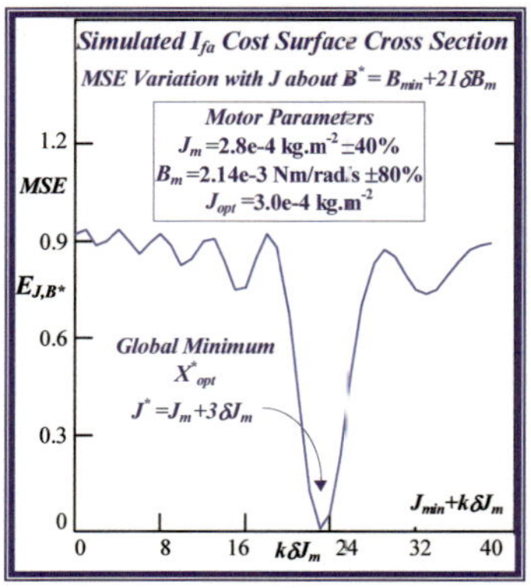

Figure 18. MSE-I_{fa} Cross section at B^*

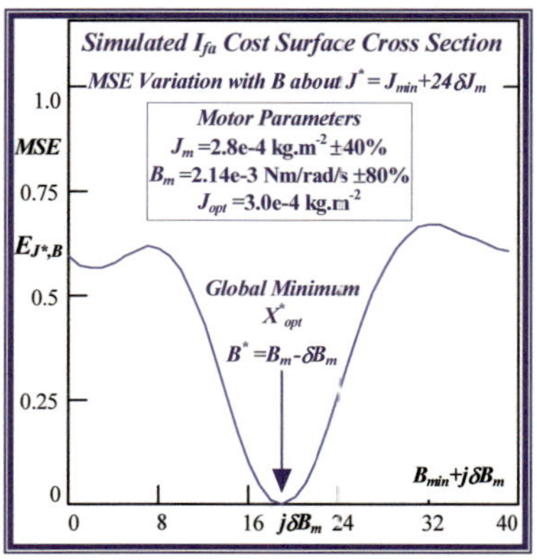

Figure 19. MSE-I_{fa} Cross section at J^*

The sinc-profile cost variations associated with specific dynamic parameters for motor current feedback target data at nominal values of the BLMD parameters $[J_m, B_m]^T$ are illustrated in Figs.18 and 19. The *MSE* penalty cost function can described in a more general form about X_m as

$$E_f^m(\mathbf{X}) = \tfrac{1}{N_d}\sum_k \left(f(t_k) - f^m(t_k)\right)^2 \tag{13}$$

with either target data training record deployed using the representation

$$f(\mathbf{X}, t_k) \in \left\{\omega_r(t_k), I_{fa}(t_k)\right\} \tag{14}$$

The nature of the global stationary region embedded in either cost surface, described by (9) or (12), can be explored in canonical form [25] by fitting a quadratic model using a Taylor series. This two dimensional truncated series expansion, with quadratic terms measuring the surface curvature, is anchored at the nominal value X_m to establish the principal axes/directions in parameter space for global minimum search. It is assumed that the expansion pivot X_m is in proximity to the supposed global optimum X_O in the case of the FC objective function as this consists of parallel ridges interlaced with folds enveloping local minima regions which obscure the global extremum position. The response surface model $\Re^f$ can be expressed, in either case with (14), in terms of the variables $J \equiv x_1$ & $B \equiv x_2$ and low order interactive terms β_{ij} about X_m as

$$\begin{aligned}\Re^f &= \beta_0 + \beta_1(x_1 - x_{1m}) + \beta_2(x_2 - x_{2m}) + \tfrac{\beta_{11}}{2!}(x_1 - x_{1m})^2 \\ &+ \tfrac{\beta_{22}}{2!}(x_2 - x_{2m})^2 + \beta_{12}(x_1 - x_{1m})(x_2 - x_{2m}) + \varepsilon\end{aligned} \tag{15}$$

with random modelling error ε. The surface model can alternatively be approximated in compact matrix form as

$$\hat{\Re}^f = \beta_0 + \mathbf{B}^{\mathbf{T}}(\mathbf{X} - \mathbf{X}_m) + \tfrac{1}{2}(\mathbf{X} - \mathbf{X}_m)^{\mathbf{T}}\hat{\mathbf{G}}(\mathbf{X} - \mathbf{X}_m) \tag{16}$$

with constant coefficient matrices determined from the cost at X_m, based on target data

$$f^0(t_k) = f(\mathbf{X}_0, t_k) \in \left\{\omega_r^0(t_k), I_{fa}^0(t_k)\right\} \tag{17}$$

by the gradient vector B given by

$$\begin{bmatrix} \beta_1 \\ \beta_2 \end{bmatrix} = \nabla E_f^0(\mathbf{X}_m) = \begin{bmatrix} \frac{\partial E_f^0}{\partial x_1} \\ \frac{\partial E_f^0}{\partial x_2} \end{bmatrix}_{\mathbf{X}_m} \tag{18}$$

and the symmetric Hessian matrix $\hat{\boldsymbol{G}}$

$$\begin{bmatrix} \beta_{11} & \beta_{12} \\ \beta_{12} & \beta_{22} \end{bmatrix} = \nabla^2 E_f^0(\mathbf{X}_m) = \begin{bmatrix} \frac{\partial^2 E_f^0}{\partial x_1^2} & \frac{\partial^2 E_f^0}{\partial x_1 \partial x_2} \\ \frac{\partial^2 E_f^0}{\partial x_1 \partial x_2} & \frac{\partial^2 E_f^0}{\partial x_2^2} \end{bmatrix}_{\mathbf{X}_m} \tag{19}$$

which determines the curvature in the vicinity of a local minimum via

$$E_f^0(\mathbf{X}_m) = \frac{1}{N_d} \sum_k \left(f^m(t_k) - f^0(t_k) \right)^2 \tag{20}$$

The set of constant coefficient differential equations pertaining to (15) are obtained via (13), using either target data record (7)-(b) or (11) with $I_{fa}(t)\,|_{\omega_r(t)=\omega_r^0(t)}$, as

$$\beta_0 = E_f^0(\mathbf{X}_m) = \frac{1}{N_d} \sum (f^m - f^0)^2 \tag{21}$$

$$\beta_1 = \frac{\partial E_f^0}{\partial x_1}(\mathbf{X}_m) = \frac{2}{N_d} \sum (f^m - f^0) \frac{\partial f}{\partial J}\Big|_{\mathbf{X}_m} \tag{22}$$

$$\beta_2 = \frac{\partial E_f^0}{\partial x_2}(\mathbf{X}_m) = \frac{2}{N_d} \sum (f^m - f^0) \frac{\partial f}{\partial B}\Big|_{\mathbf{X}_m} \tag{23}$$

$$\beta_{11} = \frac{\partial^2 E_f^0}{\partial x_1^2}(\mathbf{X}_m) = \frac{2}{N_d} \sum \left[\left(\frac{\partial f}{\partial J} \right)^2 + (f^m - f^0) \frac{\partial^2 f}{\partial J^2} \right]_{\mathbf{X}_m} \tag{24}$$

$$\beta_{22} = \frac{\partial^2 E_f^0}{\partial x_2^2}(\mathbf{X}_m) = \frac{2}{N_d} \sum \left[\left(\frac{\partial f}{\partial B} \right)^2 + (f^m - f^0) \frac{\partial^2 f}{\partial B^2} \right]_{\mathbf{X}_m} \tag{25}$$

$$\beta_{12} = \beta_{21} = \frac{\partial^2 E_f^0}{\partial x_2^2}(\mathbf{X}_m) = \frac{2}{N_d} \sum \left[\frac{\partial f}{\partial J} \cdot \frac{\partial f}{\partial B} + (f^m - f^0) \frac{\partial^2 f}{\partial J \partial B} \right]_{\mathbf{X}_m} \tag{26}$$

The required first and second order partial differential equations, based on the shaft velocity ω_r, to substantiate expressions (22) to (26) are given by

$$\frac{\partial \omega_r}{\partial J} = \left(\frac{t}{\tau}\right)\left(\frac{\omega_r - K}{J}\right) \tag{27}$$

$$\frac{\partial \omega_r}{\partial B} = -\left(\frac{\omega_r}{B} + \tau \frac{\partial \omega_r}{\partial J}\right) \tag{28}$$

$$\frac{\partial^2 \omega_r}{\partial J^2} = -\left(\frac{1}{J}\right)\left(2 - \frac{t}{\tau}\right)\frac{\partial \omega_r}{\partial J} \tag{29}$$

$$\frac{\partial^2 \omega_r}{\partial B^2} = \left(\frac{2}{B^2}\right)\omega_r + \left(\frac{\tau}{B}\right)\left(2 + \frac{t}{\tau}\right)\frac{\partial \omega_r}{\partial J} \tag{30}$$

$$\frac{\partial^2 \omega_r}{\partial B \partial J} = -\left(\frac{t}{J}\right)\frac{\partial \omega_r}{\partial J} \tag{31}$$

Target Data Record Length N_d with Time Step 20μs	2000 Points - $\omega_r^0(t_k)$ $\hat{t}$=0.04sec ~31%τ_m	2000 Points - $I_{fa}^0(t_k)$ $\hat{t}$=0.04sec ~31%τ_m
Target Data Parameters $X_0=[J_0,B_0]^T$ *"To be identified"*	*Shaft Velocity Reference Data* $[3.0\text{x}10^{-4}, 2.14\text{x}10^{-3}]^T$	*Current Feedback Reference Data* $[3.0\text{x}10^{-4}, 2.14\text{x}10^{-3}]^T$
Quadratic Model Fulcrum $X_m=[J_m,B_m]^T$	Model Surface $\hat{R}^{\omega r}$: $[2.8\text{x}10^{-4}, 2.14\text{x}10^{-3}]^T$	Model Surface $\hat{R}^{Ifa}$: [2.8x10-4, 2.14x10-3]T
Model Cost $\hat{R}_m^f$ at X_m	19.553	0.098
Constant β_0 via (21)	19.553	0.098
Gradient Vector B $[\beta_1, \beta_2]^T$ via (22/3)	$[-2.079\text{x}10^{6}, -3.279\text{x}10^{4}]^T$	$[-9.827\text{x}10^{3}, -113.345]^T$
Hessian Matrix $\hat{\boldsymbol{G}}\begin{bmatrix}\beta_{11} & \beta_{12}\\ \beta_{12} & \beta_{22}\end{bmatrix}$ via (24/5/6)	$\begin{bmatrix}1.238x10^{11} & 1.958x10^{9}\\ 1.958x10^{9} & 8.873x10^{7}\end{bmatrix}$	$\begin{bmatrix}4.592x10^{8} & 5.114x10^{6}\\ 5.114x10^{6} & 1.08x10^{5}\end{bmatrix}$
Stationary Point $X_s=[J_s,B_s]^T$ via (38)	$[2.968\text{x}10^{-4}, 2.138\text{x}10^{-3}]^T$	$[3.005\text{x}10^{-4}, 2.216\text{x}10^{-3}]^T$
Slope at X_s via (37)	$[9.313\text{x}10^{-10}, 1.455\text{x}10^{-11}]^T$	$[0, 1.421\text{x}10^{-14}]^T$
Model Cost $\hat{R}_s^f$ at X_s	2.086	$-7.654\text{x}10^{-3}$
Quadratic Form $Q(X_s-X_m)$ via (39)	34.934	0.211
Normal Form of $\hat{\boldsymbol{G}}$ $\Lambda=\begin{bmatrix}\lambda_1 & 0\\ 0 & \lambda_2\end{bmatrix}$	Eigenvalues $\begin{bmatrix}1.238x10^{11} & 0\\ 0 & 5.774x10^{7}\end{bmatrix}$	Eigenvalues $\begin{bmatrix}4.593x10^{8} & 0\\ 0 & 5.109x10^{4}\end{bmatrix}$
Transformation/Modal Matrix T with $\boldsymbol{T}^{-1}\hat{\boldsymbol{G}}\boldsymbol{T}=\Lambda$	Normalized Eigenvectors $\begin{bmatrix}999.875 & -15.825\\ 15.825 & 999.875\end{bmatrix}\cdot 10^{-3}$	Normalized Eigenvectors $\begin{bmatrix}999.938 & -11.137\\ 11.137 & 999.938\end{bmatrix}\cdot 10^{-3}$
Co-ordinate Rotation θ	-1.813°	-1.276°

Table 2. Summary of Cost Surface Quadratic Modelling Details at X_m

Target Data Record Length N_d with Time Step 20μs	4095 Points $\hat{t}$=0.082sec ~62.6%τ_m	4095 Points $\hat{t}$=0.082sec ~56.5%τ_m
Target Data Parameters $\bar{X}_{opt}=[\bar{J}_{opt}, \bar{B}_{opt}]^T$ *"To be identified"*	Shaft Velocity Reference Data for zero shaft Inertial load (NSL) Fig. 32; Ref [1] below	*Current Feedback Reference Data for zero shaft Inertial load (NSL)* Fig. 29; Ref [1] below
Quadratic Model Fulcrum $X_m=[J_m,B_m]^T$	Model Surface $\hat{R}^{\omega r}$: $[2.8\text{x}10^{-4}, 2.14\text{x}10^{-3}]^T$	Model Surface $\hat{R}^{Ifa}$: [3.1x10-4, 2.14x10-3]T
Model Cost $\hat{R}^f_m$ at X_m	66.543	0.081
Constant β_0 via (21)	66.543	0.081
Gradient Vector B $[\beta_1, \beta_2]^T$ via (22/3)	$[-6.842\text{x}10^6, -2.422\text{x}10^5]^T$	$[1.865\text{x}10^4, 449]^T$
Hessian Matrix $\hat{\boldsymbol{G}}$	$\begin{bmatrix} 4.02x10^{11} & 1.376x10^{10} \\ 1.376x10^{10} & 8.801x10^8 \end{bmatrix}$	$\begin{bmatrix} 1.809x10^9 & 4.061x10^7 \\ 4.061x10^7 & 2.843x10^6 \end{bmatrix}$
Stationary Point $X_s=[J_s,B_s]^T$ via (38)	$[2.964\text{x}10^{-4}, 2.159\text{x}10^{-3}]^T$	$[3.00\text{x}10^{-4}, 2.124\text{x}10^{-3}]^T$
Slope at X_s via (37)	$[4.657\text{x}10^{-9}, 1.746\text{x}10^{-10}]^T$	$[4.002\text{x}10^{-11}, 5.116\text{x}10^{-13}]^T$
Model Cost $\hat{R}^f_s$ at X_s	8.232	-0.015
Quadratic Form $Q(X_s-X_m)$ via (39)	116.624	0.193
Normal Form of $\hat{\boldsymbol{G}}$ $\Lambda=\begin{bmatrix} \lambda_1 & 0 \\ 0 & \lambda_2 \end{bmatrix}$	Eigenvalues $\begin{bmatrix} 4.025x10^{11} & 0 \\ 0 & 4.086x10^8 \end{bmatrix}$	Eigenvalues $\begin{bmatrix} 1.81x10^9 & 0 \\ 0 & 1.931x10^6 \end{bmatrix}$
Spectral Condition No. η	$0.985\text{x}10^3$	$0.937\text{x}10^3$
Contour sign check (51/2)	$-1.645\text{x}10^{20}$	$-3.495\text{x}10^{15}$
Contour Eccentricity e	$999.999\text{ x}10^{-3}$	$999.999\text{ x}10^{-3}$
Modal Matrix T	$\begin{bmatrix} 999.413 & -34.246 \\ 34.246 & 999.413 \end{bmatrix} \cdot 10^{-3}$	$\begin{bmatrix} 999.748 & -22.461 \\ 22.461 & 999.748 \end{bmatrix} \cdot 10^{-3}$
Co-ordinate Rotation θ	-1.9624°	-1.2874°

Table 3. Details of Cost Surface Quadratic Model Fit at X_m based on actual BLMD Experimental Test Data shown in [1] for Zero Shaft Inertial Load Conditions

The corresponding set of partial derivatives with FC I_{fa} are obtained via (11) as

$$\frac{\partial I_{fa}}{\partial J} = \sin p\left(Kt - \tau\omega_r\right) \cdot p\left(\frac{\omega_r}{B} + \tau\frac{\partial\omega_r}{\partial J}\right) \tag{32}$$

$$\frac{\partial I_{fa}}{\partial B} = \sin p\left(Kt - \tau\omega_r\right) \cdot p\left(\tfrac{t}{B}\left[2K - \omega_r\right] - 2\left(\tfrac{\tau}{B}\right)\omega_r\right) \tag{33}$$

http://dx.doi.org/10.5772/54483

$$\frac{\partial^2 I_{fa}}{\partial J^2} = -I_{fa} \cdot p^2 \left(\frac{\omega_r}{B} + \tau \frac{\partial \omega_r}{\partial J}\right)^2 + \sin p\left(Kt - \tau\omega_r\right) \cdot p\left(\frac{t}{B\tau}\right)\frac{\partial \omega_r}{\partial J} \tag{34}$$

$$\begin{aligned} \frac{\partial^2 I_{fa}}{\partial B^2} &= -I_{fa} \cdot p^2 \left(\frac{t}{B}\left[2K - \omega_r\right] - 2\left(\frac{\tau}{B}\right)\omega_r\right)^2 \\ &\quad - \sin p\left(Kt - \tau\omega_r\right) \cdot p\left(2\frac{t}{B}\left[\frac{3K-2\omega_r}{B}\right] - 6\left(\frac{\tau}{B^2}\right)\omega_r - \tau\left(\frac{t}{B}\right)\frac{\partial \omega_r}{\partial J}\right) \end{aligned} \tag{35}$$

$$\begin{aligned} \frac{\partial^2 I_{fa}}{\partial J \partial B} &= I_{fa} \cdot p^2 \left(\frac{t}{B}\left[2K - \omega_r\right] - 2\left(\frac{\tau}{B}\right)\omega_r\right) \cdot \left(-\frac{\omega_r}{B} - \tau \frac{\partial \omega_r}{\partial J}\right) \\ &\quad - \sin p\left(Kt - \tau\omega_r\right) \cdot p\left(2\frac{\omega_r}{B^2} + \left(\frac{t}{B}\right)\frac{\partial \omega_r}{\partial J} + 2\left(\frac{\tau}{B}\right)\frac{\partial \omega_r}{\partial J}\right) \end{aligned} \tag{36}$$

The variation of the directed contour gradient over the fitted cost surface model, given by

$$\nabla \hat{\Re}^f = \mathbf{B} + \hat{\mathbf{G}}(\mathbf{X} - \mathbf{X}_m) \tag{37}$$

is used to locate the global optimum X_0 in the parameter hyperspace region of interest. The condition necessary [22] for the presence of a stationary point X_s is the existence of a vanishing gradient in the neighborhood of X_m located within the parameter tolerance band ΔX with

$$\mathbf{X}_s = \mathbf{X}_m - \hat{\mathbf{G}}^{-1}\mathbf{B} \tag{38}$$

from (37) and the nature of which is determined by the local curvature from the sign of the quadratic form [28]

$$Q(\mathbf{X} - \mathbf{X}_m) = (\mathbf{X} - \mathbf{X}_m)^{\mathbf{T}} \hat{\mathbf{G}}(\mathbf{X} - \mathbf{X}_m) \tag{39}$$

The parametric details, which include estimates of the gradient vectors and Hessian matrices at X_m for the indicated data record lengths, of the fitted models to the cost surfaces illustrated in Figs.13 and 17 are summarized in Table 2. Similar parametric quantities, employing BLMD experimental test data, are given in Table 3 for cost surface models shown in Figs.1 and 2.

2.2.1. *Novel analysis of global minimum estimation and response surface selectivity [18]*

An estimate of the cost surface global minimum $\hat{X}_{opt}$ is provided in each case by inference from the vanishing gradient in (37) with location of the fitted model stationary point X_s in

(38). A sufficient condition for the existence of a global minimizer at X_s is that $Q(X_s - X_m)$ must be positive-definite [28] in which $Q(X_s - X_m) > 0$ for $X_s \neq X_m$. This is verified by the sign of the eigenvalues λ_i of $\hat{G}$ in Table 2 which are determined from the characteristic equation

$$det\left[\hat{\mathbf{G}} - \lambda \mathbf{I}\right] = 0 \tag{40}$$

The accuracy of global estimates returned in each case for the inertial parameter J in Table 2 admit to the quality and goodness of fit of the models employed for cost surface approximation in the vicinity of the global extremum. The contributory effect of parameter interaction in model approximation in both cases is not insignificant with coefficients β_{ij} comparable in magnitude to the geometric mean of the eigenvalues of Hessian $\hat{G}$ in Table 4 defined by

$$\hat{\lambda} = \sqrt[n]{\lambda_1 \cdot \lambda_2 \ldots \lambda_n} \tag{41}$$

	Uniqueness of Global Minimum Estimate	
	Shaft Velocity Reference Data	*Current Feedback Reference Data*
Cond $_{\infty \hat{G}(X_m)}$	2.2126×10^3	9.1878×10^3
Spectral Condition No. η	2.144×10^3	8.99×10^3
Geometric Mean $\hat{\lambda}$	2.674×10^9	4.844×10^6
	Cost Surface Selectivity and Fitted Model Re-evaluation at Global Minimum Estimate X_s	
Fitted Model Fulcrum X_s	$[2.968 \times 10^{-4}, 2.138 \times 10^{-3}]^T$	$[3.005 \times 10^{-4}, 2.216 \times 10^{-3}]^T$
Model Constant β_0 at X_s	0.376	4.275×10^{-4}
Gradient Vector at X_s	$[3.562 \times 10^4, 579.78]^T$	$[11.535, 0.081]^T$
Re-evaluation of $\hat{G}$ at X_s	$\begin{bmatrix} 9.142x10^{10} & 1.437x10^9 \\ 1.437x10^9 & 3.145x10^7 \end{bmatrix}$	$\begin{bmatrix} 4.295x10^8 & 4.99x10^6 \\ 4.99x10^6 & 6.08x10^4 \end{bmatrix}$
Global Estimate Update X_{s1}	$[2.996 \times 10^{-4}, 2.138 \times 10^{-3}]^T$	$[2.998 \times 10^{-4}, 2.159 \times 10^{-3}]^T$
Slope at X_{s1} via (37)	$[1.717 \times 10^{-9}, 2.547 \times 10^{-11}]^T$	$[8.413 \times 10^{-12}, 9.948 \times 10^{-14}]^T$
Eigenvalues of $\hat{G}$ at X_s	$\begin{bmatrix} 9.144x10^{10} & 0 \\ 0 & 8.846x10^6 \end{bmatrix}$	$\begin{bmatrix} 4.295x10^8 & 0 \\ 0 & 2.819x10^3 \end{bmatrix}$
Modal Matrix T	$\begin{bmatrix} 999.876 & -15.723 \\ 15.723 & 999.876 \end{bmatrix} \cdot 10^{-3}$	$\begin{bmatrix} 999.933 & -11.618 \\ 11.618 & 999.933 \end{bmatrix} \cdot 10^{-3}$
Residual Cost $\hat{R}_s^f$ at X_{s1}	6.439×10^{-3}	-3.996×10^{-6}
Quadratic Form $Q(X_{s1} - X_s)$	0.738	8.629×10^{-4}

Table 4. Results Derived From Cost Surface Quadratic Fit in Table 2

The relative magnitudes of the Hessian curvature components provide information about the uniqueness of the solution X_s in (38), via the matrix condition number in Table 3, based on the infinity norm defined as

$$cond_{\infty}(\hat{\mathbf{G}}) = \left\|\hat{\mathbf{G}}\right\|_{\infty} \left\|\hat{\mathbf{G}}^{-1}\right\|_{\infty} \text{ where } \left\|\hat{\mathbf{G}}\right\|_{\infty} = \max_{1 \le i \le n}\left(\sum_{j=1}^{n}\left|\beta_{ij}\right|\right) \tag{42}$$

The matrix condition number is much greater than unity in both cases, with the highest value associated with the FC response surface, which indicates a sizeable measure of ill conditioning in the extraction of the global estimate in (38). The curvature component β_{11} associated with the *J*-parameter is much greater than that associated with damping *B* by about three orders of magnitude which indicates greater selectivity of the solution J_s along the *J* axis. This suggests the presence of many potential solutions to (38) along the *B*-parameter co-ordinate direction due to poorer selectivity or smaller curvature component β_{22}. A more complete interpretation of the nature of the cost surface syncline containing the stationary point region is obtained from the spectral condition number η of $\hat{G}$ [22] as

$$\eta = \lambda_{max} / \lambda_{min} \tag{43}$$

The relative magnitude η of the eigenvalues indicate that a 'line minimum' of potential solutions, which explains the degree of ill conditioning in the global solution estimate, is feasible due to the 'long' elliptical shape of the contour map associated with the stationary point zone of convergence in both cases as shown in Figs 9 and 10. The elliptical character of the response surface model in the vicinity of the global minimum estimate can be visualized by a coordinate translation of the parameter axes to X_s as pivot with

$$\mathbf{V} = X - X_S \tag{44}$$

resulting in the modified representation from (16) as

$$\begin{aligned} \hat{\mathsf{R}}^f &= \beta_0 + \mathbf{B}^{\mathbf{T}}(\mathbf{V} + \mathbf{X}_s - \mathbf{X}_m) + \tfrac{1}{2}(\mathbf{V} + \mathbf{X}_s - \mathbf{X}_m)^{\mathbf{T}}\hat{\mathbf{G}}(\mathbf{V} + \mathbf{X}_s - \mathbf{X}_m) \\ &= \hat{\mathsf{R}}_s^f + \tfrac{1}{2}\mathbf{V}^T\hat{\mathbf{G}}\mathbf{V} \qquad \text{(a)} \\ \text{where } \hat{\mathsf{R}}_s^f &= \beta_0 + \mathbf{B}^{\mathbf{T}}(\mathbf{X}_s - \mathbf{X}_m) + \tfrac{1}{2}(\mathbf{X}_s - \mathbf{X}_m)^{\mathbf{T}}\hat{\mathbf{G}}(\mathbf{X}_s - \mathbf{X}_m) \qquad \text{(b)} \end{aligned} \tag{45}$$

The normalized eigenvectors T_i, associated with the distinct eigenvalues of the symmetric Hessian $\hat{G}$ as

$$\mathbf{T}^{-1}\hat{\mathbf{G}}\mathbf{T} = \Lambda \tag{46}$$

in Table 2, can be used as an orthonormal basis to transform the parameter axes along the principal directions of the elliptical shaped contour system.

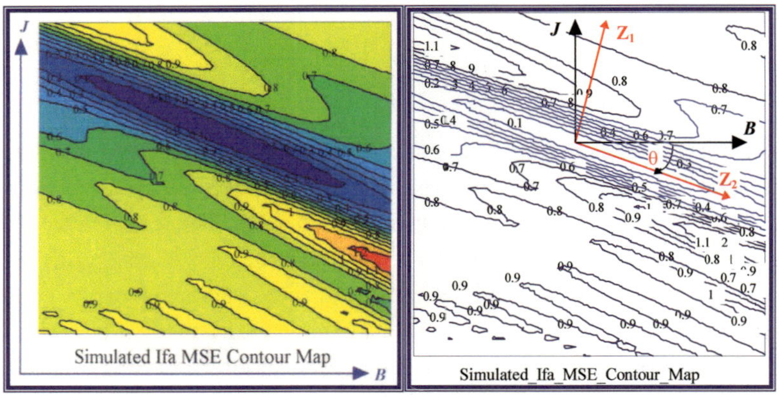

Figure 20. A: Simulated MSE-I_{fa} Contour Map B: I_{fa} Contour Map with Canonical Variables

This rotation of co-ordinates, with origin anchored to X_s, is displayed in Figs 20 and 10 for both simulated cost surfaces to eliminate the interactive terms β_{12} in $\hat{G}$. The X co-ordinate angular displacement θ in Figure 20B can be evaluated from the conical expression [29] as

$$\cot 2\theta = \frac{\beta_{22}-\beta_{11}}{2} \Big/ \beta_{12} \tag{47}$$

using (15) with values listed in Tables 2 and 3 which are very small. The rotation can also be deduced from the directional cosines of the unit column vectors $[\hat{a}_{z1} \;\; \hat{a}_{z2}]$ constituting the modal matrix T via

$$\mathbf{T} = \begin{bmatrix} \hat{\mathbf{a}}_{z1} & \hat{\mathbf{a}}_{z2} \end{bmatrix} = \begin{bmatrix} \hat{\mathbf{a}}_{x1} & \hat{\mathbf{a}}_{x2} \end{bmatrix} \begin{bmatrix} \cos\theta & -\sin\theta \\ \sin\theta & \cos\theta \end{bmatrix} \tag{48}$$

The normal form of the response surface model can be expressed with substitution of the canonical variable

$$Z = T^T V \tag{49}$$

$$\begin{aligned} \text{As} \qquad \hat{\mathsf{R}}^f &= \hat{\mathsf{R}}_s^f + \tfrac{1}{2}\mathbf{Z}^{\mathbf{T}}\Lambda\mathbf{Z} \\ &= \hat{\mathsf{R}}_s^f + \tfrac{1}{2}\sum_{k=1}^{2}\lambda_k z_k^2 = \hat{\mathsf{R}}_s^f + \tfrac{\lambda_1}{2} z_1^2 + \tfrac{\lambda_2}{2} z_2^2 \end{aligned} \tag{50}$$

http://dx.doi.org/10.5772/54483

with model cost $\hat{R}_s^f$ at the global estimate given in Table 4. The nature of the fitted quadratic model can be deduced from the shape of the embedded contours, in the vicinity of the global minimum estimate X_s, by inference from the sign of the scalar discriminant invariant which pertains to elliptical conic sections [29] as

$$\beta_{12}^2 - \beta_{11}\beta_{22} < 0 \tag{51}$$

in the x_1x_2 frame or

$$-\lambda_1\lambda_2 < 0 \qquad \text{with} \qquad \lambda_{12} = 0 \tag{52}$$

in z_1z_2 normal co-ordinates. These contours are elliptical for both choices of observed BLMD target data with a negative discriminant in Table 3 based on the model cost at nominal parameter value X_m. Consequently the stationary region enveloping the global minimum is encircled and thus bounded by elliptical contours rather than contained within an open wedge shaped response surfaces with no define convergence zone. The degree of elliptical eccentricity e of the trapped stationary zone quantifies the extent of the 'line minimum' of global minimum convergence, congruent with the major axis, as notionally illustrated in Fig. 21.

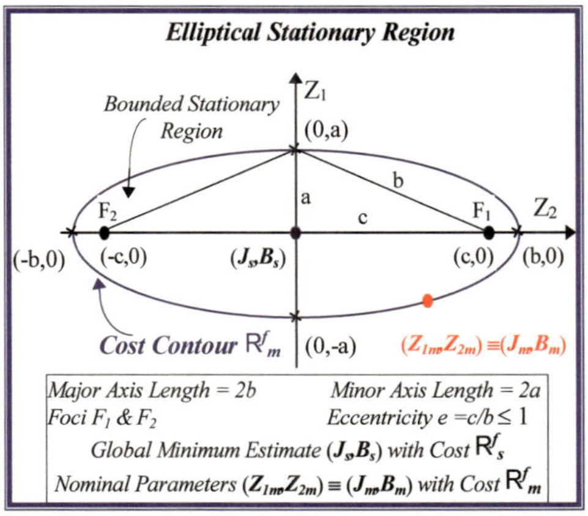

Figure 21. Elliptical Contour Bounded Stationary Zone

This can be determined from consideration of Fig. 21 by recasting the expression for the normal form of the cost surface model in (50) into that for an elliptical contour, evaluated at the nominal parameter value X_m with origin at X_s, as

$$\frac{z_1^2}{a^2} + \frac{z_2^2}{b^2} = 1 \tag{53}$$

for model cost differential

$$\Delta\hat{\mathrm{R}}^{f}_{m,s} = \left(\hat{\mathrm{R}}^{f}_{m} - \hat{\mathrm{R}}^{f}_{s}\right) \tag{54}$$

with intrinsic parameters

$$a^2 = 2\Delta\hat{\mathrm{R}}^{f}_{m,s}/\lambda_1 \text{ and } b^2 = 2\Delta\hat{\mathrm{R}}^{f}_{m,s}/\lambda_2 \tag{55}$$

The eccentricity e, which measures the degree of 'flatness' of the oblate model contour $\hat{R}^{f}_{m}$ specified at X_m and thus the 'linear extension' of the global minimum X_0, is given in terms of the lateral displacement c of the elliptical foci from the global estimate X_s relative to the length $2b$ of the major axis as

$$e = \frac{c}{b} = \frac{\sqrt{b^2 - a^2}}{b} = \sqrt{1 - \left(\frac{a}{b}\right)^2} = \sqrt{1 - \left(\frac{\lambda_2}{\lambda_1}\right)^2} \tag{56}$$

The eccentricity of the contours for the model target data in Table 3 is almost unity, as a consequence of the large spectral condition number η in each case, indicating a very flat distended ellipse. The elliptical contours approximates an extended pencil-like global minimum predominantly in the B parameter co-ordinate direction because the inclination angle θ in (47) is less than 2º. This is qualified by the magnitude of the axial ratio (AR) which defines the extent $2b$ of the 'line minimum valley' along the principal direction of the ellipse in relation to its girth $2a$, given by the minor axis length, as

$$AR = \frac{b}{a} = \frac{\lambda_1}{\lambda_2} = \eta \tag{57}$$

This contrast of stationary region 'feature sizes' in parameter X-space is readily identified as the spectral condition number η in Table 3 with a 'line minimum' extension ratio of about three orders of magnitude for each target data training record used in the MSE cost surface description.

The slope and curvature matrix $\hat{G}$ of the fitted cost model including its associated eigenvalues are re-evaluated at the acquired global estimate X_s as summarized in Table 4 to gauge the response surface selectivity either along the parameter co-ordinate directions or the principal axes of the normal form. The second iterative estimate X_{s1}, along with the residual costs given in Table 4, is very close to the global minimum target X_0 listed in Table 2 for both cost surface models despite the large condition number in each case. A quantifiable measure of the fitted model selectivity in an ill conditioned stationary region, tagged by a large spectral number η, at discerning the global minimum can be obtained from the surface curvature κ_j along a particular parameter co-ordinate direction x_j as

$$\kappa_j = \left|\beta_{jj}\right| \Big/ \sqrt{1 + \left.\frac{\partial \hat{R}^f}{\partial x_j}\right|_{\mathbf{X}_s}} \approx \left|\beta_{jj}\right| \tag{58}$$

where

$$\frac{\partial \hat{R}^f}{\partial x_j} << 1 \ \text{ at } \boldsymbol{X} = \boldsymbol{X}_s \tag{59}$$

or alternatively in normal form as

$$\kappa_j = \lambda_j \tag{60}$$

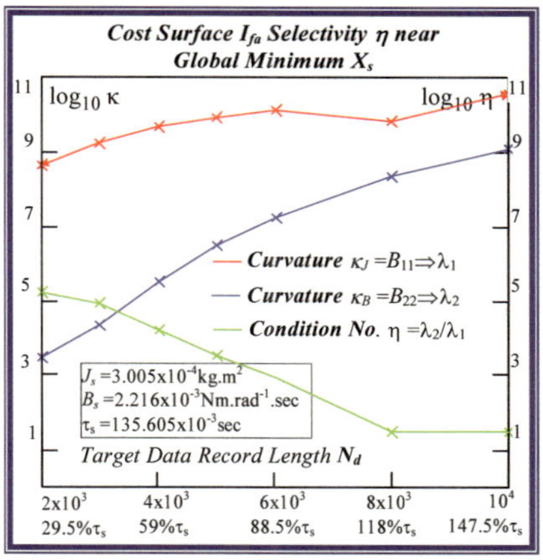

Figure 22. I_{fa} Cost Surface Selectivity

The degree of model selectivity is three orders of magnitude greater in the case of the motor inertia for actual measured target data employed in both response surface approximations as evidenced from the spectral condition number in Table 3 and in Table 4 for simulated target data trials. Consequently this selectivity margin renders a more accurate estimate in the extracted *J*-parameter which is mirrored by the arguments leading to the feature size ratio in (57). The cost surface selectivity improves along the principal axes of the normal form when the target data length N_d is extended as indicated by the increased magnitudes of the eigenvalues in Tables 4 and 3. This trend in enhanced *J* parameter selectivity, which is a measure of the accompanying increase in curvature at the global extremum, is displayed in Figs. 22 and 23 for increasing data record lengths and is a manifestation of the narrowing of the cost surface fold containing the directed 'line minimum' principally in the *B*-parameter direction.

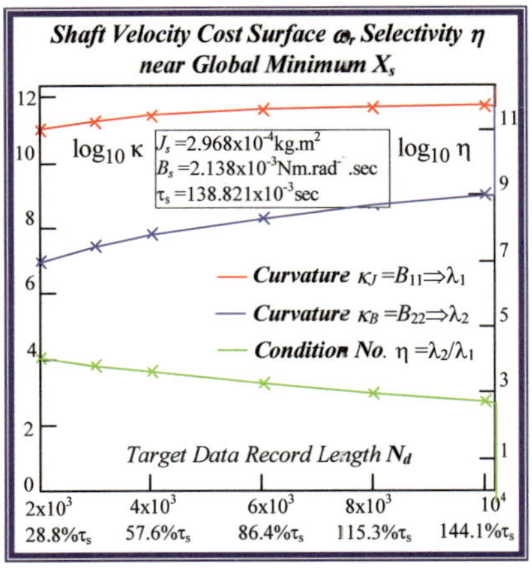

Figure 23. ω_r Cost Surface Selectivity

The selectivity improvements are greater for increased FC step response data record lengths in Fig. 22 than those for shaft velocity target data in Fig.23. This due to the appearance of more FC cycles with reduced periodicity as motor speed increases demanding a greater degree of fitted model accuracy, with smaller margins of error in terms of frequency and phase coherence at the global minimum value, in the extraction of the optimum parameter vector X_O during system identification. The shaft velocity step response by contrast losses its excitation persistence with transient speed decay as it evolves towards steady state conditions with increased data capture time. After a sufficient time elapse the target data transient information, responsible for velocity cost surface folding, is submerged by the steady state onset of maximum motor speed conditions. This irretrievable loss of target velocity signal amplitude variation with time results in a reduction of surface selectivity with parameter variation near the global minimum. These considerations admit to a better choice in the current feedback as a suitable candidate for MSE objective function formulation where accurate parameter extraction is essential during the identification phase of optimal controller design in high performance adaptive BMLD systems for electric vehicle mobility. Furthermore the increasing trend towards motor sensorless control [30] obviates the need for separate rotor position sensors with essential information obtained from the motor signature current via FC sensing at the inverter controller o/p. This adoption of sensorless operation in motor drive systems lends added importance to observed FC data as a suitable target function during parameter identification.

3. Response surface noise and parameter quantization

The computation 'noise' inherent in the *MSE* penalty function construct, based on simulated target data at nominal machine parameter values, is manifested as response surface roughness in parameter space. This is due to model nonlinearities and coarseness of evaluation of the PWM switching instants and results in 'false' local minima proliferation in the neighborhood of the global minimizer.

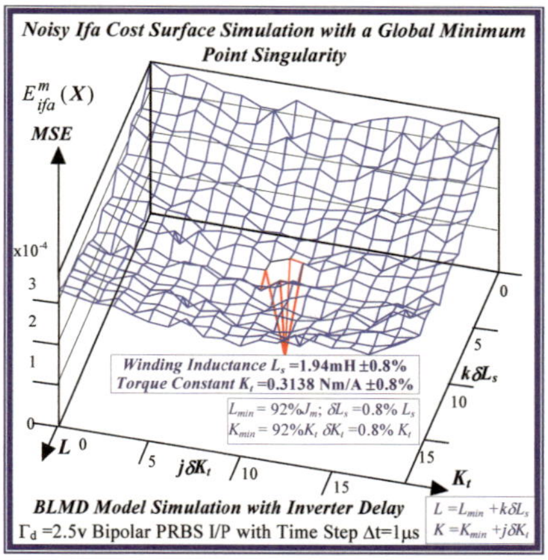

Figure 24. Noisy I_{fa} Cost Surface with PRBS I/P

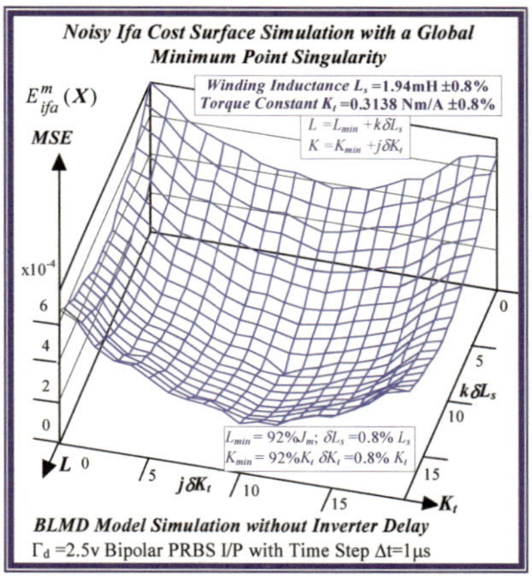

Figure 25. Noisy I_{fa} Cost Surface with PRBS I/P

A typical example of this is illustrated in Figs 24 and 25 for BLMD model simulations, with and without inverter turn-on delay δ considered, for small step change variations in the stator winding inductance L_S and torque constant K_t parameters. These response surfaces were obtained from BLMD simulation, using FC target data for nominal parameter values as in [1], with a 4095 bit maximal length 2.5 volt bipolar pseudorandom binary sequence (PRBS) input stimulus. The response surface in Fig.24 has a very shallow paraboloidal shape for the small parameter tolerance ranges chosen with a rough noisy texture peppered with local minima in the vicinity of the point-like global minimum. The response surface for simulated

FC target data is relatively smooth in the absence of inverter delay turn-on with a point-like singularity at the global minimum as shown in Fig.25. The cost functions pertaining to simulated step response FC I_{fa} and shaft velocity ω_r target data, displayed in Figs.26 and 27 for the dynamic parameters $\{J,B\}$, are also noisy with point-like multiminima scattered around the 'pinhole' stationary point as in the former case. These surfaces are parabolic for very small tolerance ranges selected near the global minimum as in the main lobe of Fig.1 for the FC corrugated surface.

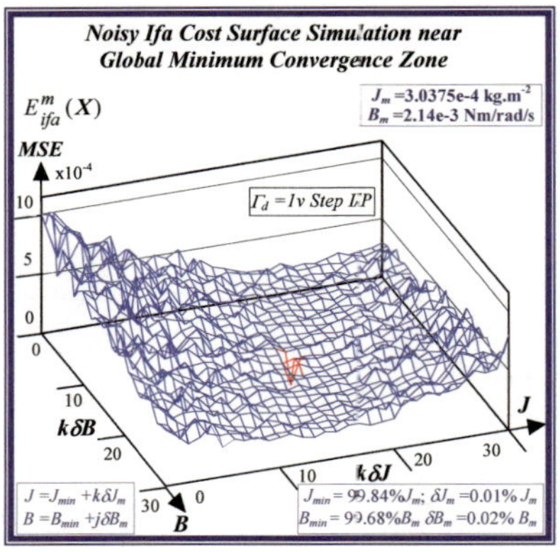

Figure 26. Noisy I_{fa} Cost Surface with Step I/P

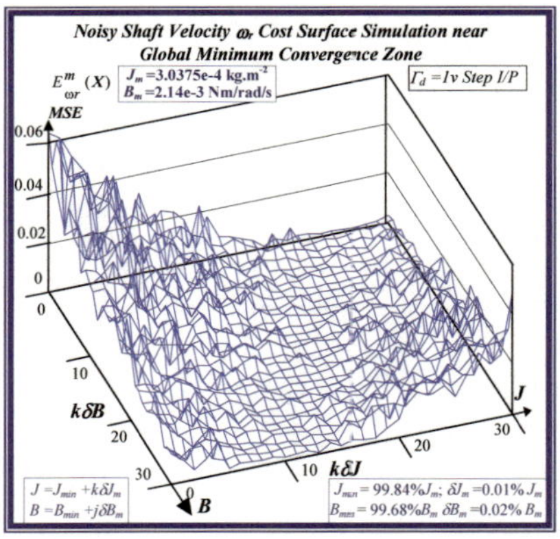

Figure 27. Noisy Velocity Cost Surface with Step I/P

The side elevations of the MSE cost functions in Figs.28 and 29 demonstrate very effectively the fractal landscape with multiminima plurality disposed about the global extremum in the FC case.

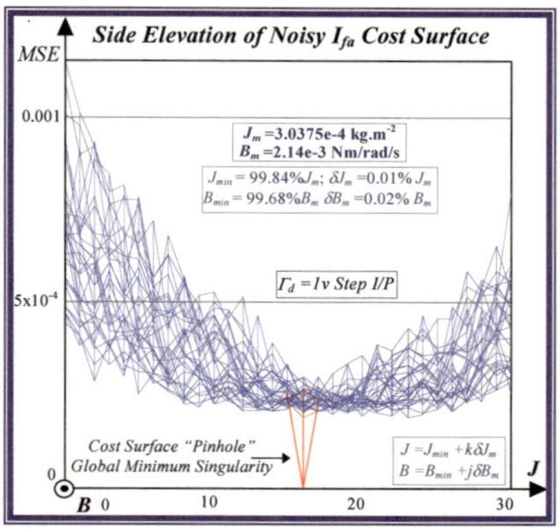

Figure 28. Noisy I_{fa} Cost Surface Side Elevation

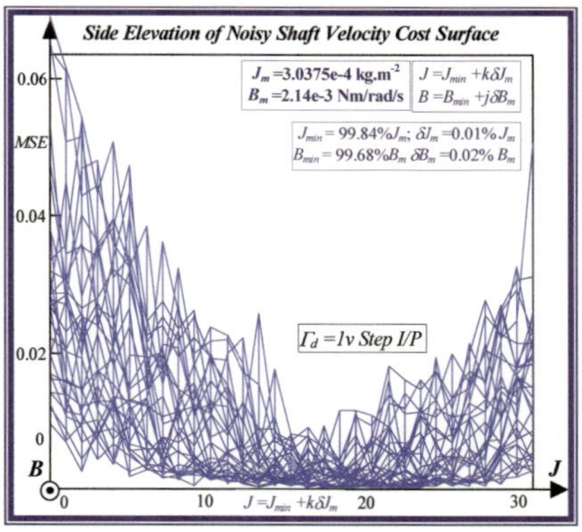

Figure 29. Noisy Velocity Cost Surface Side Elevation

In the simulated velocity response surface shown in Fig.29 a stationary region exists at zero floor cost with no definite observable global minimum point. An alternative perspective of the minimum stationary regions is provided by the contour maps shown in Figs.30 and 31 for FC and shaft velocity target data respectively. The existence of the point-like global minimum singularity with surface noisiness is clearly evident from the level contours in the FC surface relief map. In the case of the shaft velocity response surface the presence of 'noisy' local minima strewn over the 'river bed' syncline of the global minimum stationary region is clearly defined by the contour map in Fig.31. The occurrence of 'noisy' local minima in the above error surfaces presents a difficulty to any classical optimization method in acquiring the global minimizer where fine parameter resolution is concerned.

A more detailed examination of the effect of inverter delay, achieved through BLMD model simulation without current controller o/p saturation using a 1 volt torque demand step i/p, on the one dimensional *MSE* response surface in Fig. 32 for very small inductance variation reveal a granulated profile which is less pronounced than that in Fig. 33 with the absence of delay.

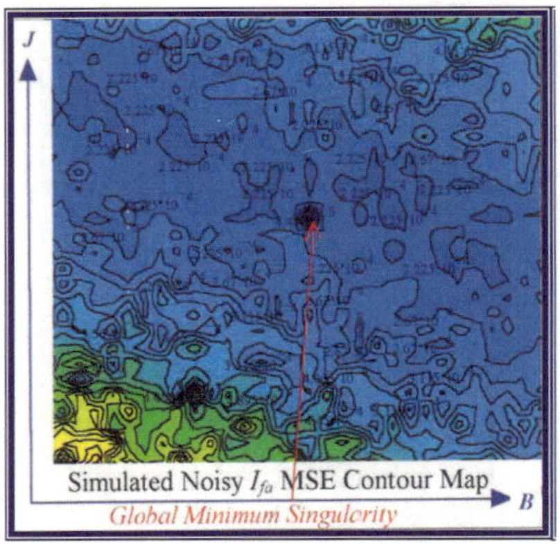

Figure 30. Noisy Local Minima I_{fa} Contour Map

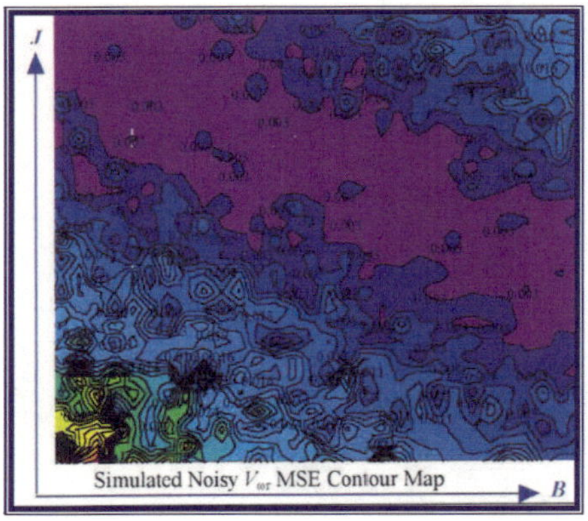

Figure 31. Noisy Local Minima ω_r Contour Map

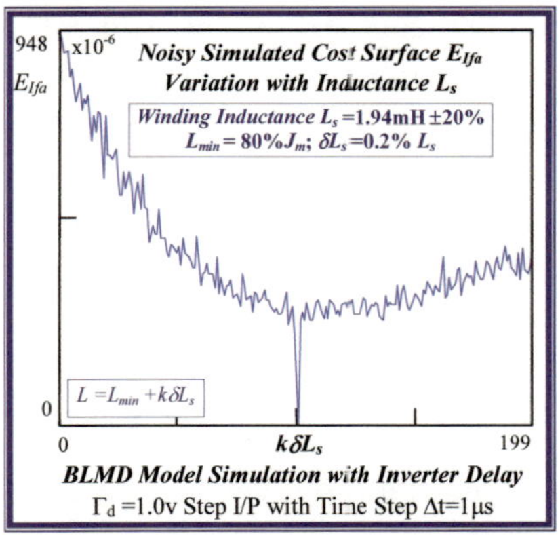

Figure 32. Cost Simulation with Delay δ

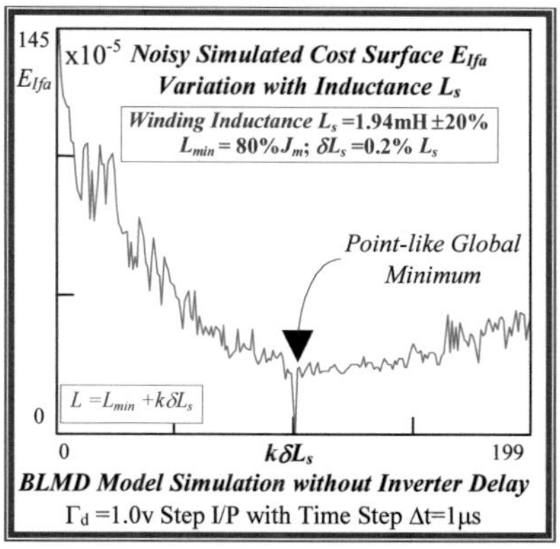

Figure 33. Cost Simulation without Delay

The use of a PWM switch transition time search, based on a single iteration of the regula-falsi method to keep simulation time overhead low, marginally reduces the response error as in Figs.34 and 35. The sensitivity [31] of the error response E with inductance L_s

$$S_{L_s}^{E} = \left(\frac{L_s}{E}\right)\left(\frac{\partial E}{\partial L_s}\right) \tag{61}$$

is very low in all cases and for a ±12% inductance variation gives a change of $\Delta E=1.5\times10^{-4}$ in 1.75×10^{-4} for E. Poor cost surface selectivity will ensue in such cases of low sensitivity over a large parameter tolerance range with possible local minimum convergence if the search process is initiated far from the global minimizer with a noisy cost function.

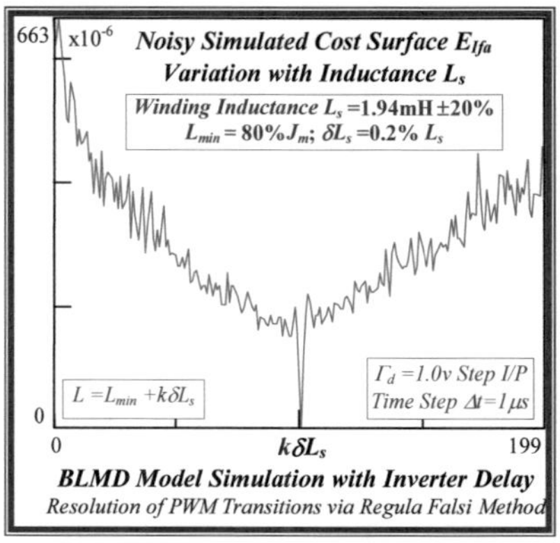

Figure 34. BLMD Delay & Reg.-Fal. Method

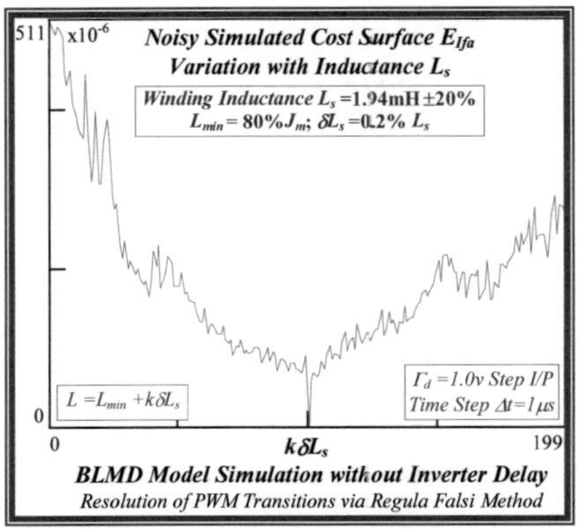

Figure 35. Zero Delay & Reg.-Fal. Method

3.1. Novel theoretical estimation of PWM edge transition computation noise [18]

A measure of the computation 'noise', induced through inaccurate resolution of the PWM edge transition within a simulation time step Δt, can be ascertained from the associated error in random pulsed energy delivery by the inverter to the stator winding within Δt. Since there is one PWM edge transition every half-switching interval $T_S/2$ of the inverter the expectation in the power delivery error to the stator can be obtained [18], from the error in pulse energy dispatch during the time step interval Δt, as

$$E(\boldsymbol{P}) = E(\boldsymbol{E})\Big/\frac{\boldsymbol{T_s}}{2} = \frac{\boldsymbol{E}_{max}}{2}\Big/\frac{\boldsymbol{T_s}}{2} = U_d^2 \Delta t / T_S \tag{62}$$

The expected random voltage v_n error associated with inaccurate resolution in PWM inverter switching during BLMD simulation is thus given by

$$\begin{aligned} v_n &= \sqrt{E(\boldsymbol{P})} = U_d\sqrt{\Delta t/T_S} && (a) \\ v_n &= 310\sqrt{1/200} = 21.92 \text{ volts} && (b) \end{aligned} \tag{63}$$

If the chosen simulation time step Δt is 1μs and the inverter switching parameters in [1] are substituted into (63)-(a) the expected uncertainty v_n in the inverter output voltage V_{jg} per phase j can be obtained as (63)-(b)

This value of voltage uncertainty in the inverter output is not insignificant as its magnitude is 7.1% of the inverter HT voltage U_d for a simulation time step size of 1μs. The error can be reduced by decreasing the simulation step size Δt for a more accurate resolution of the pulse edge transition time, once its occurrence has been flagged, or alternatively by means of an accurate search using the regula-falsi method as described in [1].

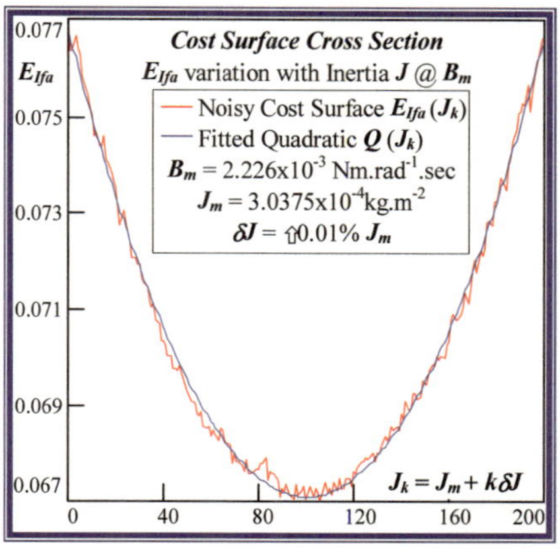

Figure 36. I_{fa} Surface Noise with Inertia Variation

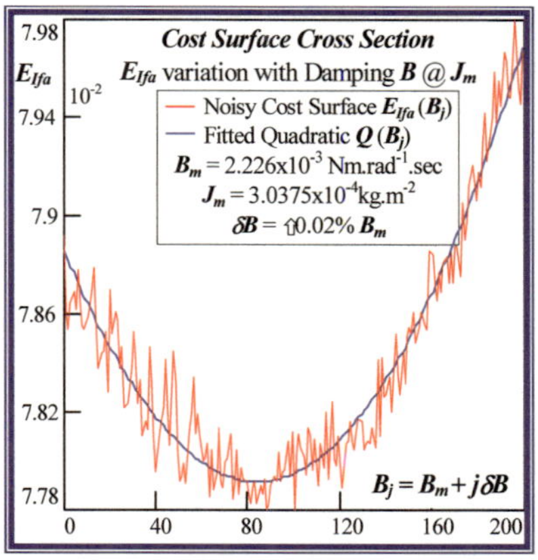

Figure 37. Ifa Surface Noise Variation with *B*

The statistical considerations of pulsed energy delivery by the PWM inverter in [18], arising from BLMD simulation with a fixed time step size, illuminates the origin of computation 'noisiness' and its subsequent manifestation as cost surface roughness as shown in Figs.36 and 37. For fixed shaft load inertia changes, encountered in motive power applications and electric vehicles, the use of a coarse quantization step size δJ in the inertial parameter variable about the nominal value J_m results in smooth generated and noise-free response surfaces as shown in Figs.1 and 2 for actual FC and shaft velocity target data. However for a sufficiently small step size variation in the inertia J_m and damping B_m a 'noisy' cost surface with a proliferation of local minima results in both cases as shown in Figs 36 and 37 for corresponding target test data. The degree of resolution of the parameter step size, that can be obtained and then used in an identification search strategy, depends upon the onset of cost surface irregularity.

3.2. Novel mathematical analysis of quadratic curve fitting to noisy mse cost surface [18,21]

The accuracy of any classical identification scheme used in terms of parameter resolving capabilities can be gauged by fitting a quadratic [26] to the response surfaces in each test case and determining rms deviation of the PWM computation noise related residuals. The quadratic fit employed

$$Q(x_k) = b_0 + b_1 x_k + b_2 {x_k}^2; \qquad 1 \le k \le N \tag{64}$$

for N steps in the indexed parameter x_k as shown in Fig.38 with

$$x_k = x_m + (k - m)\delta x; \qquad x_m \in \{J_m, B_m\} \tag{65}$$

is based on an infinitesimal step size δx, in model simulation to reflect response surface roughness, and centred in a tolerance band $\pm\Delta x_m$ within indexed range m of the nominal value x_m as

$$m = \Delta x_m / \delta x \tag{66}$$

for

$$\Delta x_m = \left(x_m - x_{min}\right) = \left(x_{max} - x_m\right) \tag{67}$$

The nature of the residual error

$$W_f(x_k) = E_f(x_k) - Q_f(x_k) \tag{68}$$

associated with the least squares quadratic fit to the various cost functions

$$f \in \{\omega_r, I_{fa}\} \tag{69}$$

for example in Figs.39 and 40 for the FC cost surface, is demonstrated by the autocorrelation (ACR) functions

$$\Re_W^f(j) = \tfrac{1}{N}\sum_{k=1}^{N} W_f(x_k) \cdot W_f(x_{k+j}) \tag{70}$$

shown in Figs.41 and 42, which are mainly of the impulse type at zero offset, indicating a white noise-like characteristic.

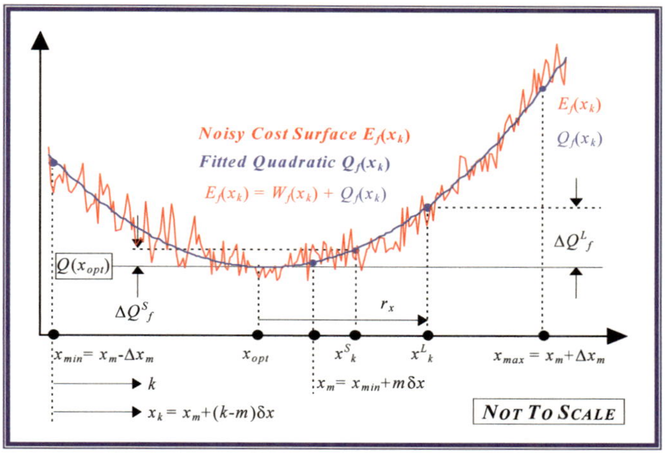

Figure 38. Parameter Step Size Resolution

FC Target Data with reference to details in Figs.36 and 37		
Parameter varied x	*Jm*	*Bm*
Nominal value x_m	3.0375×10^{-4}kg.m^2	2.226×10^{-3}Nm.rad^{-1}.sec
Parameter Step Size δx	0.01% *Jm*	0.02% *Bm*
No. of Steps N	200	200
Tolerance Index m	100	100
b_0	96.864	3.286
b_1	-6.373×10^5	-2.891×10^3
b_2	1.049×10^9	6.516×10^5
Residual Error $W_{lfa}(x) = E_{lfa}(x) - Q_{lfa}(x)$ illustrated in Figs.39 and 40		
Standard Deviation $\hat{\sigma}$	1.916×10^{-4}	1.127×10^{-4}
Mean	-3.309×10^{-8}	-3.306×10^{-9}
Peak Absolute Deviation	5.203×10^{-4}	3.355×10^{-4}

Table 5. FC Cost Function $E_{lfa}(x)$ and Quadratic Fit $Q_{lfa}(x)$

The various cost function details with accompanying quadratic fits and corresponding residuals are summarized in Tables 5 and 6, based on FC and shaft velocity test data respectively, for independent parameter variation in the BLMD shaft inertia and damping factor. The quadratic polynomials fitted to the noisy shaft velocity cost surface sections are displayed in Figs. 43 and 44 with coefficients given in Table 6. The corresponding cost residuals associated with the fitted velocity profiles, which appear to be random, are shown for each of the dynamical parameter variables in Figs. 45 and 46.

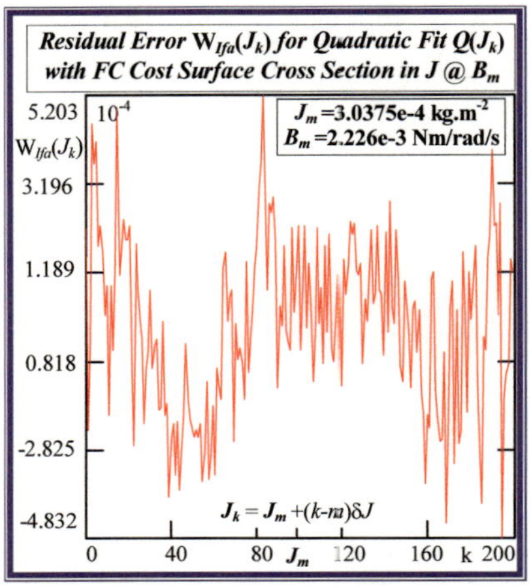

Figure 39. Quadratic Error in J @ B_m

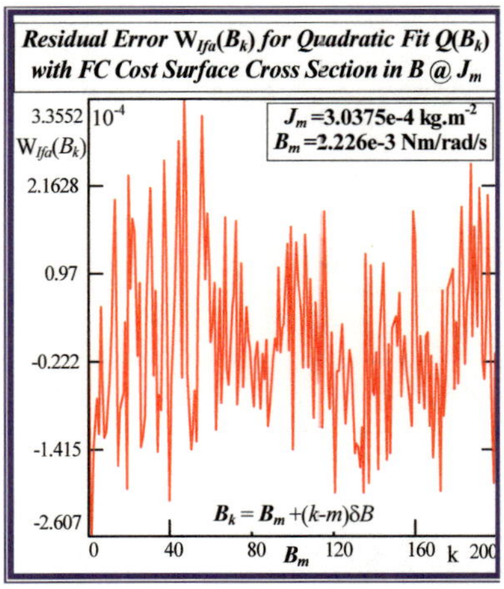

Figure 40. Quadratic Error in *B* @ J_m

http://dx.doi.org/10.5772/54483

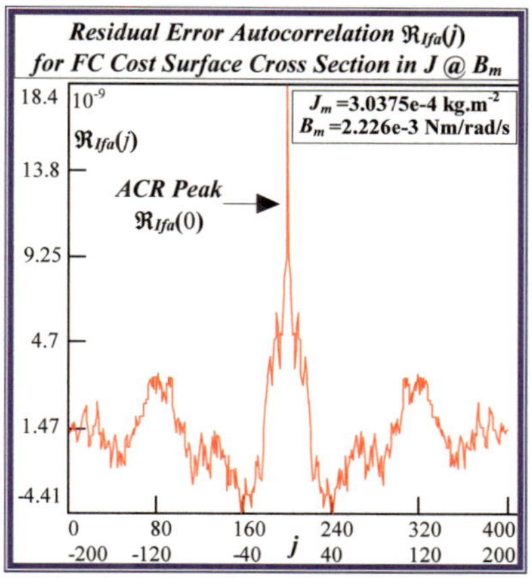

Figure 41. Error Autocorrelation in J @ B_m

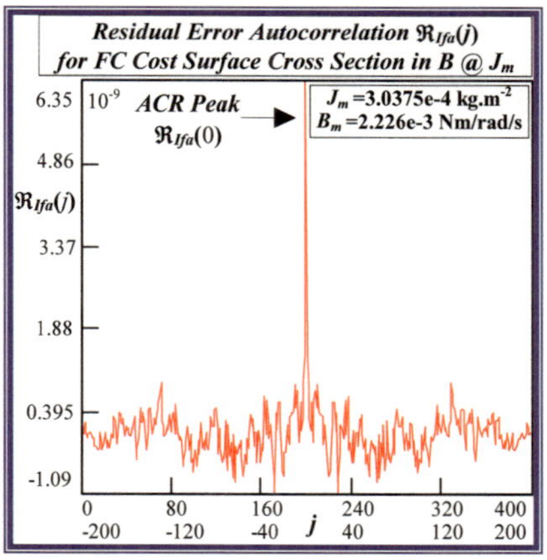

Figure 42. Error Autocorrelation in B @ J_m

The white noise-like nature of the error-of-fit in the case of the shaft velocity cost surface sections is demonstrated by the impulse characteristic of the ACR spike functions in Figs.47 and 48. The errors-of-fit can thus be considered as a random entity, with an ACR related noise signature, associated with the BLMD simulation model at very high parameter resolution for each of the observed target data records used in the *MSE* cost formulation. This manifestation is attributed to some residual uncertainty in the BLMD model simulation of the PWM edge transitions at the comparator o/p with dead time, despite the single iteration cycle of the regula-falsi search, which are magnified in the three phase inverter o/p before being fed to the stator winding.

Shaft Velocity Target Data with reference to details in Figs.43 and 44		
Parameter varied x	Jm	Bm
Nominal value x_m	3.0375×10^{-4}kg.m^2	2.14×10^{-3}Nm.rad^{-1}.sec
Parmeter Step Size δx	0.01% Jm	0.02% B_m
No. of Steps N	370	600
Tolerance Index m	100	100
b_0	1.698	0.157
b_1	-1.079×10^4	-1.098×10^2
b_2	1.747×10^7	2.465×10^4
Residual Error $W_\omega(x)=E_\omega(x)-Q_\omega(x)$ illustrated in Figs 45 and 46		
Standard Deviation $\hat{\sigma}$	1.028×10^{-5}	8.974×10^{-4}
Mean	6.982×10^{-10}	-1.415×10^{-11}
Peak Absolute Deviation	3.476×10^{-5}	2.871×10^{-5}

Table 6. Shaft Velocity Cost Function $E_\omega(x)$ and Quadratic Fit $Q_\omega(x)$

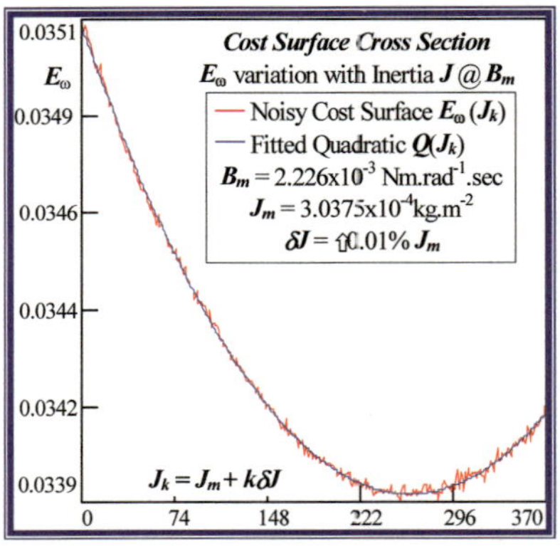

Figure 43. Velocity Cost Noise Variation with J

The effect of lowering the drive model simulation time step Δt, as shown in Figs. 49 and 50 for very small parameter variation in the vicinity of the global singularity, translates into a reduction of the MSE as well as gradual removal of response surface roughness. This tangible decrease in surface roughness with time step size, evident form Fig.51, is measured in terms of the standard deviation of the residual errors associated with various quadratic polynomials fitted to each of the FC cost sections. However the computational effort in terms of CPU time increases in proportion with the decrease in time step size for a given simulation trace length. The requirement for surface noise reduction with the elimination of false local minima plurality has to be balanced with a tradeoff in simulation run time in an attempt to reduce computation costs where BLMD model tractability is an issue in parameter identification and as a simulator in practical applications for performance related prediction of proposed embedded drive systems. A Taylor series expansion of the quadratic fit about the parabolic vertex x_{opt} as

http://dx.doi.org/10.5772/54483

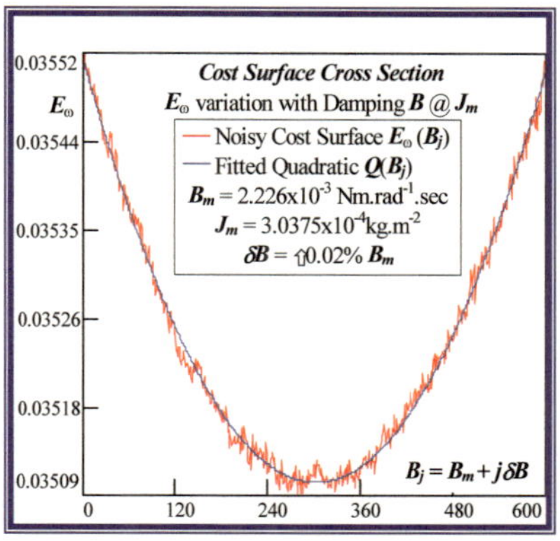

Figure 44. Shaft Vel. Surface Noise with B Variation

$$Q_f(x) = Q_f(x_{opt}) + b_2(x - x_{opt})^2 \tag{71}$$

with gradient

$$\left.\frac{\partial Q_f}{\partial x}\right|_{x_{opt}} = b_1 + 2b_2 x_{opt} = 0 \tag{72}$$

can now be used to check the limit of parameter resolution and the "radius" of convergence for worst case conditions [19].

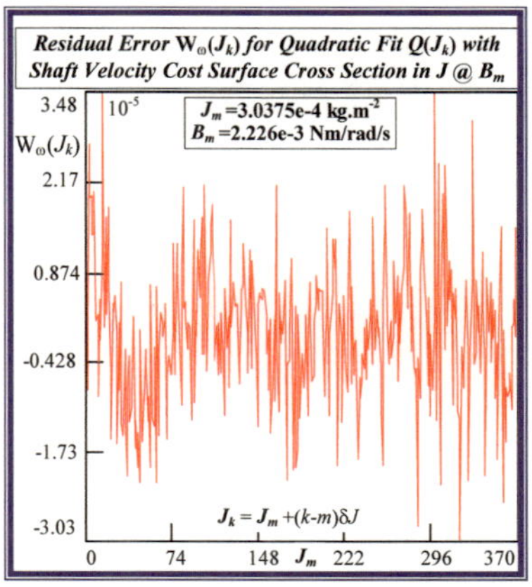

Figure 45. Quadratic Error in J @ B_m

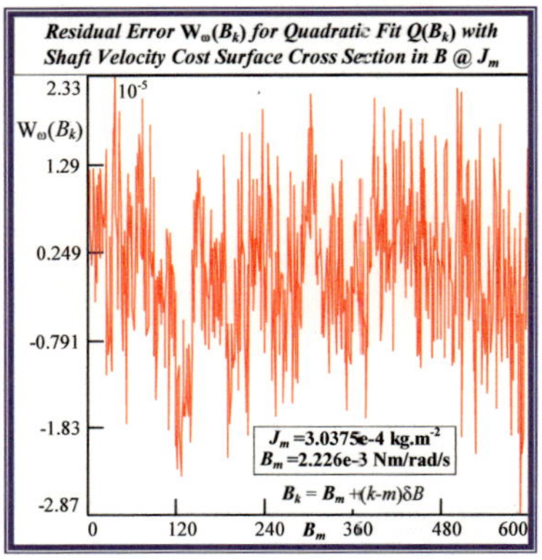

Figure 46. Quadratic Error in *B* @ J_m

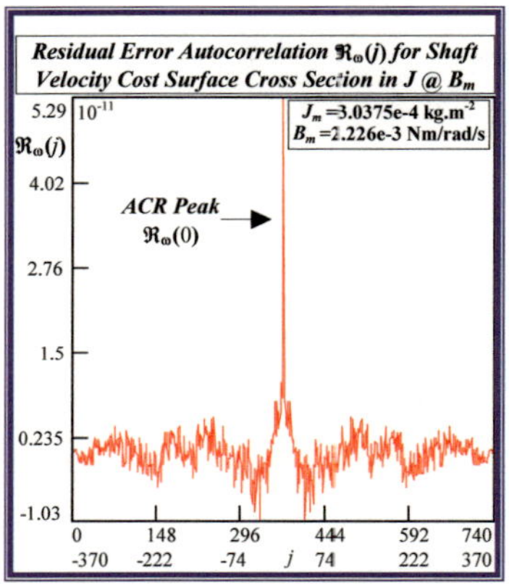

Figure 47. Error Autocorrelation in *J* @ B_m

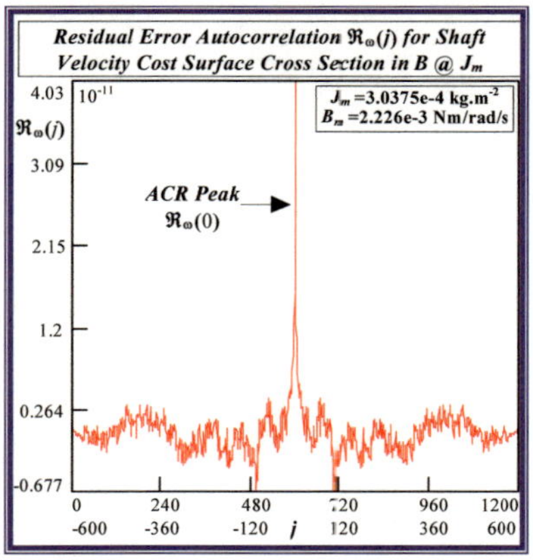

Figure 48. Error Autocorrelation in *B* @ J_m

At best the smallest parameter threshold step size required in simulation to overcome response surface noise, with rms sample estimate $\hat{\sigma}$, is determined from that value x_k^s near the global minimum as in Fig.38 such that

$$\Delta Q_f^S = \Delta Q_f(x_k^S) = Q(x_k^s) - Q(x_{opt}) = b_2(x_k^s - x_{opt})^2 \geq \hat{\sigma} \tag{73}$$

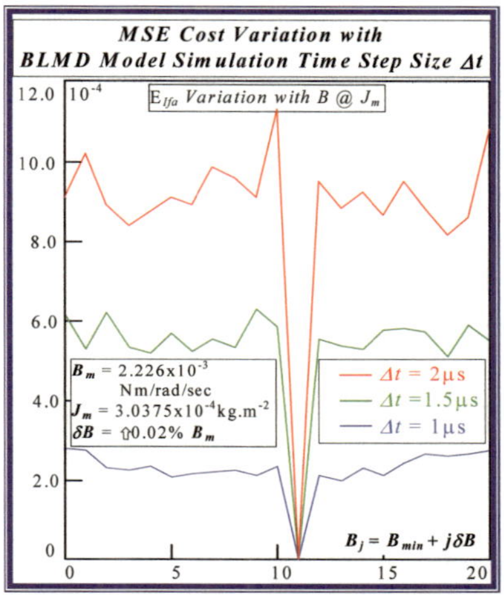

Figure 49. Error of Fit Vs Time Step

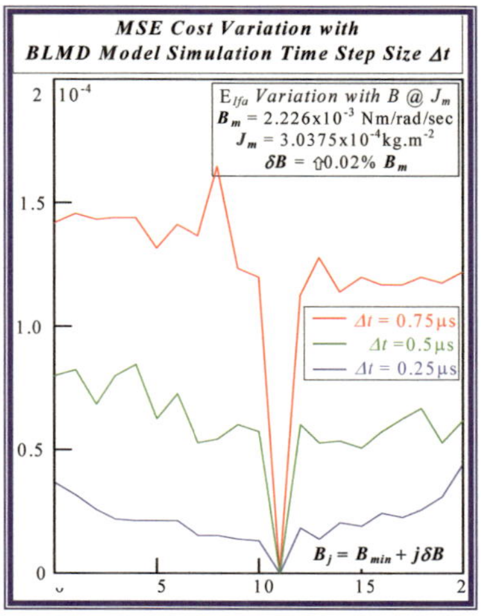

Figure 50. Error of Fit Vs Time Step

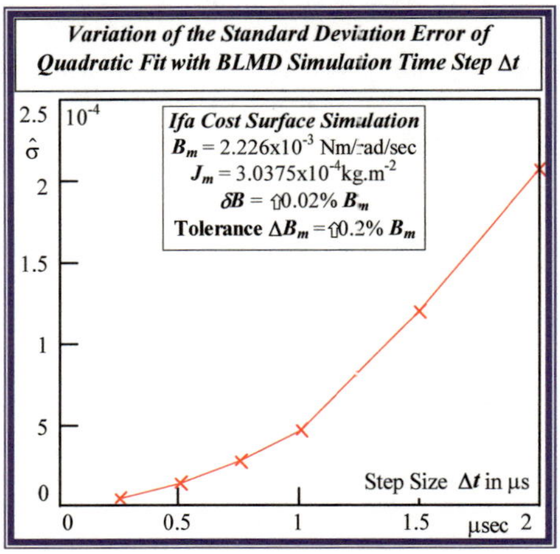

Figure 51. Error Variation with Time Step

3.2.1. Parameter quantization and radius of convergence estimation for system identification

The largest threshold step size estimate can be determined by applying Chebyshev's theorem for statistical measurements [32] to the response surface noise sample [19, 26]. This theorem indicates that at least the fraction $1-(1/h^2)$ of all the residuals W_{f_k} in any sample lie within h standard deviations of the mean $\hat{\mu}$ with probability

$$Prob\{(\hat{\mu}-h\hat{\sigma}) \le W_f(x_k) \le (\hat{\mu}+h\hat{\sigma})\} = 1-(1/h^2) \tag{74}$$

and for h = 4, which exceeds the tabulated peak absolute deviation in all cases in Tables 5 and 6, is 94%. Thus a measure of the worst case parameter resolution is provided by the inequality

$$\Delta Q_f^L = Q_f(x_k^L) - Q_f(x_{opt}) = b_2(x_k^L - x_{opt})^2 \ge 4\hat{\sigma} \tag{75}$$

for some large x_k^L via the quadratic minimiser

$$x_{opt} = -b_1/2b_2 \tag{76}$$

in (72) as

$$x_k^L \ge x_{opt} \pm \sqrt{\frac{4\hat{\sigma}}{b_2}} \tag{77}$$

The parameter resolution limit in terms of quantization step size δx necessary to overcome cost surface noisiness and local minimum trapping in BLMD parameter identification, which is also a measure of the convergence radius r_x about the global minimizer in Fig.38, is given by

$$\delta x^L = r_x = \sqrt{\frac{4\hat{\sigma}}{b_2}} \tag{78}$$

Parameter varied x	Jm	Bm
Minimizer x_{opt}	3.038×10^{-4}kg.m^2	2.219×10^{-3}Nm.rad^{-1}.sec
Minimizer Offset $(m - k_{opt})$	-0.243	16.199
Threshold Locations k^L	72 & 129	24 & 143
Worst Relative Step Size $\frac{\delta x^L}{x_m}$	0.281%	1.182%

Table 7. Quantized Step Sizes for FC Cost Function in Figs. 36 & 37

If measurements are referenced to the nominal value x_m at the centre of the parameter tolerance range the relative step sizes

$$\delta x_m^L = (k^L - m)\delta x \tag{79}$$

of which there are two pending the sign of the quadratic surd in (77), must be corrected by allowance for the global minimum offset

$$x_m - x_{opt} = (m - k_{opt})\delta x \tag{80}$$

Parameter varied x	Jm	Bm
Minimizer x_{opt}	3.007×10^{-4}kg.m^2	2.097×10^{-3}Nm.rad^{-1}.sec
Minimizer Offset $(m - k_{opt})$	-161.146	-201.448
Threshold Locations k^L	210 & 312	212 & 391
Worst Relative Step Size $\frac{\delta x^L}{x_m}$	0.505%	1.783%

Table 8. Quantized Step Sizes for Shaft Velocity Cost Function in Figs.43 & 44

The dynamical parameter threshold step sizes δx^{L}, which are by default the convergence radii measures for reliable global parameter estimation, are tabulated for the response surface cross sections in Tables 7 and 8. The resolution of the motor shaft inertia from the tabulated step sizes, which is the most likely to vary and more essential to identify in high performance applications, is higher when the FC cost function is used instead of the shaft velocity equivalent. Convergence of the inertia parameter estimates to the global minimum is enhanced in the former case with a lower uncertainty due to the smaller step size. The degree of selectivity of the fitted response surfaces with respect to the parameter variability [31] given by

$$V_x = \frac{\Delta x}{x} \tag{81}$$

can be determined through the sensitivity coefficient

$$S_x^{Q_f} = \left(\frac{x_{opt}}{Q_{opt}}\right)\left(\frac{\partial Q_f}{\partial x}\right) \tag{82}$$

in the vicinity of the global minimum x_{opt}. This measure can then be usefully employed as a performance index to decide on the best target test data available to use in a motor parameter identification strategy. The sensitivities for 2% parameter variability, greater than the largest threshold step size encountered, of the various fitted surfaces are summarized in Table 9. These sensitivity considerations indicate the suitability of FC test data in the objective function formulation for accuracy in parameter identification.

Parameter varied *x*	*Jm*	*Bm*
	FC Response Surface Sensitivities Figs.36 and 37	
Sensitivity	28.88	0.82
	Shaft Velocity Response Surface Sensitivities Figs.43 and 44	
Sensitivity	0.98	0.07

Table 9. Response Surface Sensitivity

The above method of parameter quantization, employed to surmount cost surface noise and resultant avoidance of local minimum capture during system identification, reduces the search time in parameter space to global optimality. This is due to the reduction of N-Dimensional parameter space into a finite sized hypercube of countable lattice points N_C to be searched, within the imposed parameter tolerance bounds $\pm\Delta x_m$, using an interstitial 'distance' equivalent to the step size variability in Tables 7 and 8 as

http://dx.doi.org/10.5772/54483

$$N_C = 2^N \prod_{j=1}^{N} \frac{\Delta x_m^j}{\delta x_j^L} = 2^2 \left(\frac{\Delta J_m}{\delta J^L}\right)\left(\frac{\Delta B_m}{\delta B^L}\right) \tag{83}$$

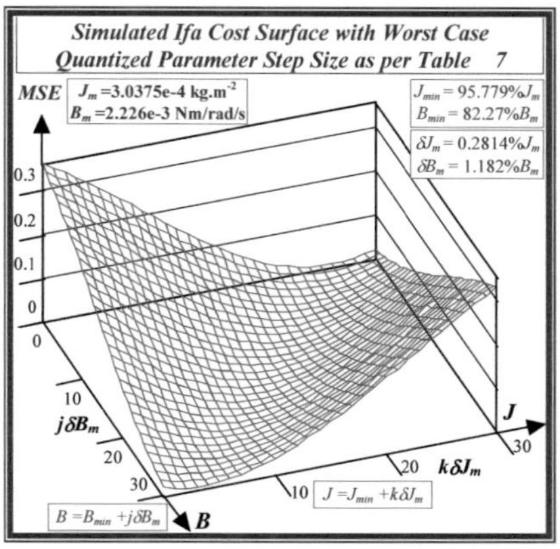

Figure 52. Simulated *Ifa* Cost Surface

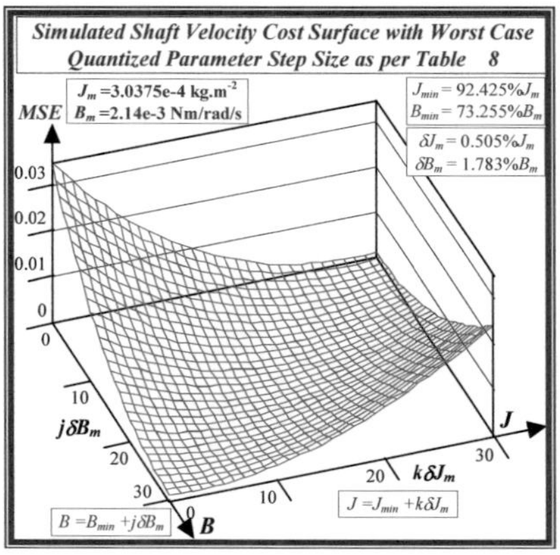

Figure 53. Simulated Shaft Velocity Cost Surface

The application of the tabulated parameter threshold sizes δx^L in the objective function simulation results in smooth noise-free response surfaces in the stationary region enclosing the global minimum as displayed in Figs.52 and 53. The degree of accuracy achieved by parameter quantization, with restricted step size during dynamical system identification, in acquiring the global extremum X_{opt} is determined from the critical values in Tables 7 and 8 as the estimate

$$\bar{\mathbf{X}}_{opt} = \mathbf{X}_{opt} \pm \delta \mathbf{X}^L \tag{84}$$

The accuracy of the estimate in (84) can be improved with the selection of FC target data because of its greater cost surface sensitivity and smaller relative step size. If the length of the test data record is extended with more data values collected, accompanied by a corresponding transient response decay in the observed variables, the selectivity of the FC response surface improves with genuine local minima proliferation while the parabolic $V_{\omega r}$ surface concavity decreases. Thus a more accurate global estimate $\overline{X}_{opt}$ is obtained for reference purposes with increased data record length and improved FC surface selectivity. A suitable identification method can then be applied in conjunction with the BLMD model to the response surfaces corresponding to either motor shaft velocity $V_{\omega r}$ or winding FC I_{fa} target data, obtained in torque mode control for different shaft load inertia, in the parameter search process of the optimal estimate $\hat{X}_{opt}=[\hat{J}_{opt}, \hat{B}_{opt}]^T$. The Powell conjugate direction method [22,23] and FSD [19] parameter extraction techniques can be applied, for example, to the respective $V_{\omega r}$ and I_{fa} cost surfaces to obtain $\hat{X}_{opt}$ [18].

4. Conclusions

Response surface simulation has been theoretically investigated and shown to be a useful graphical tool in motor parameter identification with a multiminima objective function and BLMD model validation for electric vehicle systems. This visual concept provides an intuitive insight into the topographical structure of the cost function to be minimized, the location of the global minimum, and the relevant identification search strategy to be adopted in parameter space to obtain an accurate estimate. It also provides an alternative parameter measurement strategy against which the accuracy of other parameter identification search techniques can be judged. A novel mathematical analysis of the competing shaft velocity and current feedback response surfaces, for identification purposes, has revealed the existence of a 'line' minimum of possible solutions principally in the *B*-parameter direction via a comparison of the eigenvalues derived from the quadratic model fit of the global stationary region. This analysis also shows that the global stationary region is closed and bounded by elliptical shaped MSE contours, which guarantees the existence of an optimal parameter vector solution. Furthermore a comparison of the quadratic model eigenvalues, for the competing cost surfaces, illustrates the dominance of the current feedback response selectivity and its acceptance as the most suitable objective function during SI for accurate parameter extraction.

The quantization of parameter space to remove 'false' local minima proliferation has been examined and demonstrated to be effective in surmounting cost surface 'noisiness' engendered during BLMD simulation, with a finite step size, of the PWM natural sampling process. A probability analysis has shown that the error incurred in resolution of PWM edge transition times during BLMD simulation, which is responsible for cost surface granularity, is dependent on the step size and is manifested as a random error voltage at the PWM inverter output. The effect of cost surface selectivity with choice of target data in MSE penalty

http://dx.doi.org/10.5772/54483

cost function formulation, for usage in BLMD parameter identification, has been examined with motor current feedback being the preferred option.

Acknowledgements

The author wishes to acknowledge

i. Eolas – The Irish Science and Technology Agency – for research funding.

ii. Moog Ireland Ltd for brushless motor drive equipment for research purposes.

Author details

Richard A. Guinee

Department of Electrical and Electronic Engineering, Cork Institute of Technology, Cork, Ireland

References

[1] R.A Guinee, Extended Simulation of an Embedded Brushless Motor Drive (BLMD) System for Adjustable Speed Control Inclusive of a Novel Impedance Angle Compensation Technique for Improved Torque Control in Electric Vehicle Propulsion Systems in Electric Vehicles - Modelling and Simulations, ISBN: 978-953-307-477-1, InTech ; 2011.

[2] R.A Guinee, Mathematical Modelling and Simulation of a PWM Inverter Controlled Brushless Motor Drive System from Physical Principles for Electric Vehicle Propulsion Applications in Electric Vehicles - Modelling and Simulations, ISBN: 978-953-307-477-1, InTech ; 2011.

[3] Miller, J.; (2010). *Propulsion Systems for Hybrid Vehicles*, IET, Renewable Energy, 2nd Edition.

[4] R.M. Crowder, *Electric Drives and their Controls*, 1995, Clarendon Press, Oxford.

[5] A.J. Critchlow, *Introduction to Robotics*, Macmillan Pub. Co. NY, 1985.

[6] H. Gross, *Electrical Feed Drives for Machine Tools*, 1983 by Siemens, J. Wiley & Sons.

[7] *Moog Brushless Technology User Manual:D31X-XXX Motors,T158-01X Controllers,T157-001 Power Supply*, Moog GmbH, D-7030 Böblingen, Feb 1989.

[8] H. Asada and K. Youcef-Toumi, *Direct-Drive Robots* Theory and Practice, 1987, MIT Press.

[9] R.P. Paul, *Robot Manipulators: Mathematics, Programming and Control,* The MIT Press, Camb, Mass, USA, 1986.

[10] W.E. Snyder, *Industrial Robots: Computer Interfacing and Control,* PHI, 1985

[11] M.A. El-Sharkawi and S. Weerasooriya, "Development and Implementation on Self-Tuning Tracking Controllers for DC Motors", *IEEE Trans. on Energy Conv.*, Vol. 5, No. 1, Mar 1990.

[12] N.A. Demerdash, T.W. Nehl and E. Maslowski, "Dynamic modelling of brushless dc motors in electric propulsion and electromechanical actuation by digital techniques", *IEEE/IAS Conf. Rec.* CH1575-0/80/0000-0570, pp. 570-579, 1980.

[13] H. Dohmeki and M. Nasu, " Development of a Brushless DC Motor for Incremental Motion Systems", Proc. 14th IMCSD annual Symp., pp.63-71, 1985

[14] J.Y.S. Luh, "Conventional Controller Design for Industrial Robots – A Tutorial", IEEE Trans. On Systems, Man, and Cybernetics, Vol. SMC-13, No. 3, May/June 1983.

[15] K.J. Astrom and T. Hagglund, *Automatic Tuning of PID Controllers,* Instr. Soc. Amer, 1988, ISBN 1-55617-081-5.

[16] *Moog Brushless Technology:Brushless ServodrivesUser Manual* D310.01.03 En/De/It 01.88, Moog GmbH, D-7030 Böblingen, Germany, 1988.

[17] R.A. Guinee and C. Lyden, "Motor Parameter Identification using Response Surface Simulation and Analysis", Proc. of American Control Conference, ACC-2001,June 25-27, 2001, Arlington, VA, USA.

[18] Guinee, R.A., "Response Surface Methodology", *Modelling, Simulation, and Parameter Identification of a Permanent Magnet Brushless Motor Drive System,* Chapter 3, pages 125 – 206, Ph. D. Thesis, 2003, National University of Ireland – University College Cork.

[19] R.A. Guinee and C. Lyden, "A Novel Application of the Fast Simulated Diffusion Algorithm for Dynamical Parameter Identification of Brushless Motor Drive Systems". IEEE-ISCAS 2000, The 2000 IEEE International Symposium on Circuits and Systems, May 28-31, Geneva, Switzerland.

[20] R.A. Guinee and C. Lyden, "A Novel Application of the Fast Simulated Diffusion Optimization Technique for Brushless Motor Parameter Extraction" UKACC International Conference on Control, Cambridge Univ., Sep 2000.

[21] R.A. Guinee and C. Lyden, "Parameter Identification of a Brushless Motor Drive System using a Modified Version of the Fast Simulated Diffusion Algorithm", Proc. of American Control Conference – IEEE ACC-1999, San Diego, June 1999, pp.3467-3471

[22] R. Fletcher, *Practical Methods of Optimization,* 2nd edition,1993, J.Wiley & Sons.

http://dx.doi.org/10.5772/54483

[23] W.H. Press, B.F. Flannery, S.A. Teukolsky and W.T. Vetterling, *Numerical Recipes in C*, 1990, CUP.

[24] Guinee and C. Lyden, "Accurate Modelling And Simulation Of A DC Brushless Motor Drive System For High Performance Industrial Applications", IEEE ISCAS'99 - *IEEE International Symposium on Circuits and Systems*, May/June 1999, Orlando, Florida

[25] R.H. Myers and D.C. Montgomery, *Response Surface Methodology - Process and Product Optimization Using Designed Experiments*, 1995, J.Wiley & Sons, NY.

[26] R.A. Guinee and C. Lyden, "A Novel Application of the Fast Simulated Diffusion Algorithm in Brushless Motor Parameter Identification", The 3rd IEEE European Workshop on *Computer-Intensive Methods in Control and Data Processing*, Sep 7-9, Prague, Czech Republic.

[27] R.A. Guinee and C. Lyden, "Parameter Identification of a Motor Drive using a Modified Fast Simulated Diffusion Algorithm", *Proc. of the IASTED Intern. Conf. on Modelling and Simulation*, pp 224-228, May. 1998, Pittsburgh, Pa., USA.

[28] H.A. Taha, *Operations Research*, Macmillan Publishing Co., NY., 1971.

[29] Protter and Murray, *Calculus and Analytic Geometry*

[30] J. Holtz, "Sensorless Position Control of Induction Motors - an Emerging Technology", invited paper, IEEE-IECON'98, Proc. of the 24th Annual Conf. of the IEEE Indus. Electronics Society, Aug 31 - Sep 4, 1998, Aachen, Germany.

[31] G. Daryanani, *Principles of Active Networks Synthesis and Design*, 1976, J. Wiley & Sons.

[32] F. Mosteller, R.E.K. Rourke and G.B. Thomas, *Probability with Statistical Applications*, 1961, Addison-Wesley Publ. Co.